Microorganisms for Sustainability

Volume 11

Series Editor

Naveen Kumar Arora, Environmental Microbiology, School for Environmental Science, Babasaheb Bhimrao Ambedkar University, Lucknow, Uttar Pradesh, India

More information about this series at http://www.springer.com/series/14379

Naveen Kumar Arora • Jitendra Mishra • Vaibhav Mishra
Editors

Microbial Enzymes: Roles and Applications in Industries

Springer

Editors
Naveen Kumar Arora
Babasaheb Bhimrao Ambedkar University
Department of Environmental Science
Lucknow, Uttar Pradesh, India

Jitendra Mishra
DST-CPR, Babasaheb Bhimrao Ambedkar
(Central) University
Lucknow, Uttar Pradesh, India

Vaibhav Mishra
Babasaheb Bhimrao Ambedkar University
Department of Environmental Microbiology
Lucknow, Uttar Pradesh, India

ISSN 2512-1901 ISSN 2512-1898 (electronic)
Microorganisms for Sustainability
ISBN 978-981-15-1709-9 ISBN 978-981-15-1710-5 (eBook)
https://doi.org/10.1007/978-981-15-1710-5

This Springer imprint is published by the registered company Springer Nature Singapore Pte Ltd.
The registered company address is: 152 Beach Road, #21-01/04 Gateway East, Singapore 189721, Singapore

Preface

Practical applications of microbial enzymes are of immense value to the industries. The current understanding of fundamentals and advancements in microbial system physiology have provided better process control, higher yield, and enhanced recovery of microbial enzymes. Nevertheless, a systematic approach for interpreting the current usage of microbial enzymes in food, agriculture, and medical fields is needed.

This book is meant to provide comprehensive information on microbial enzymes having wide industrial applicability. The volume reflects contributions from researchers in the domain of microbial enzymology. The tome contains 13 chapters covering various aspects and applications of microbial enzymes. The volume covers commercial importance of some important enzymes such as fructosyltransferase, laccase, fungal peroxidases, chitinases, proteases, cholesterol oxidases, and oxalate decarboxylase in the food processing, nutraceuticals, textiles, agriculture, bioremediation, and biomedical sectors. Integration of modern tools in enhancing the yield and optimization of bioprocess technology are the essential components of the enzyme biotechnology. The volume highlights major advancements happening in the field of industrial biotechnology in recent times.

The major limitations of microbial enzymes production and use in industry are the yield, desired characteristics, workability under diverse conditions, and high capital investment. Hence, despite great catalytic activity, and eco-friendly nature, above-mentioned constraints are responsible for the lower acceptance of these enzymes in various formulations and fields. Advantages and progressions made in the fermentation process have now resulted in the production of necessary enzymes with low capital investment. However, in some instances where they are exclusively used as key components, such as in clinical and biomedical sectors, more advanced and sophisticated technology is the need of the hour. The volume highlights essential knowledge on the cutting-edge technologies that are being employed for the low-cost production of the microbial enzymes for their mass production at industrial level and utilization of these biomolecules in novel fields. Integration of these technologies may overcome many other constraints which industries are currently facing. The volume also enlightens the multifaceted application of microbial enzymes in different streams and provides a new direction to researchers to conduct novel research in the domain of microbial biotechnology.

Overall, the whole idea of the volume is to seek the potential of the microbial enzymes in solving issues related to sustainable environment, agriculture, environment, and human health.

The volume provides thoroughly updated content on some industrially relevant microbial enzymes. The quest for improving the existing enzyme technology related to procedure development and improvement, raw material utilization, and broad applicability in different industries is also explained.

The editors are indebted to all the contributors for providing their valuable inputs on different aspects which made the book more informative and full of updated information in the focus area. The editors also convey gratitude to Dr. Mamta Kapila and Ms. Akanksha Tyagi from Springer Nature (India) for their continuous support.

We are also indebted to everyone whosoever has supported the project and helped in the compilation of the book. We hope that the book will be very useful to researchers, students, and industrialists particularly from the field of microbial technology, biotechnology, and biochemistry and provide an opportunity for them to become familiar with remarkable developments happening in the era of industrial biotechnology.

Lucknow, Uttar Pradesh, India — Naveen Kumar Arora
Lucknow, Uttar Pradesh, India — Jitendra Mishra
Lucknow, Uttar Pradesh, India — Vaibhav Mishra

Contents

About the Editors

Naveen Kumar Arora, Ph.D. (Microbiology), FISEB Professor and Head in the Department of Environmental Science, Babasaheb Bhimrao Ambedkar University (a Central University), Lucknow, Uttar Pradesh, India, is a renowned researcher in the field of Environmental Microbiology and Biotechnology. His specific area of research is rhizosphere biology and plant growth-promoting rhizobacteria (PGPR). He has more than 60 research papers published in premium international journals. He is the editor of 21 books published by Springer Nature.

He is member of several national and international societies, in the editorial board of 4 journals and reviewer of several international journals. He is Editor-in-Chief of the journal *Environmental Sustainability* published by Springer Nature and Series Editor of book series *Microorganisms for Sustainability*. He has delivered lectures in conferences and seminars around the globe. He has been an advisor to 11 doctoral students. He has also received awards for excellence in research by the Honourable Governor of Uttar Pradesh, Asian PGPR Society, and Samagra Vikas Welfare Society. He is Fellow of International Society for Environmental Botanists and delivered talks as Key Note and Plenary Speaker in conferences around the world. Although an academician and researcher by profession, he has a huge obsession for the wildlife and its conservation and has authored a book, *Splendid Wilds*. He is President of Society for Conservation of Wildlife and is also Secretary of Society for Environmental Sustainability (website: www.ses-india.org).

Jitendra Mishra has attained Ph.D. in the field of Agricultural Microbiology. Dr. Mishra has also qualified ASRB-NET. He completed his master's degree in the Environmental Microbiology (with Gold Medal) from Babasaheb Bhimrao Ambedkar University, Lucknow. He gathered a comprehensive experience in terms of research as well as teaching. He has worked as Assistant Professor (Guest) in the Department of Environmental Microbiology, Babasaheb Bhimrao Ambedkar University, Lucknow. He is also holding the post of Assistant Editor in the journal *Environmental Sustainability* published by Springer Nature. He has published his research finding in the form of research articles, reviews, book chapters, and opinion in several journals of repute. Currently, he is Research Associate in DST-Centre for Policy Research, BBAU, Lucknow.

Vaibhav Mishra has completed his Ph.D., in Applied Microbiology from Banaras Hindu University, Varanasi, India, and has expertise in Microbiology and Gastroenterology. While working in the Department of Pharmacology, CSIR-Central Drug Research Institute, Dr. Mishra gathered a comprehensive research experience in herbal drug formulation against gastric ulceration. Currently, he is working as Senior Post-Doctoral Researcher at Louisiana State University, USA. After attaining the prestigious SERB-National Post-Doctoral Fellowship (N-PDF), he did his post-doctoral research from the Department of Biochemistry, King George Medical University, Lucknow. Previously, he was employed as an Assistant Professor (Guest), in the Department of Environmental Microbiology, Babasaheb Bhimrao Ambedkar University, Lucknow. He has published many research papers in acclaimed journals and has authored two books, many book chapters, and review articles.

1 Fructosyltransferase Enzymes for Microbial Fructan Production

Nazlıcan Tezgel, Onur Kırtel, Wim Van den Ende, and Ebru Toksoy Oner

Abstract

Fructans are fructose-based oligo- and polysaccharides synthesized using sucrose as substrate. Depending on the glycosidic bonds in their structure, they are classified as inulin and levan types or a mixture of these, namely graminans and agavins. Fructans constitute one of the most widespread functional biomolecules in nature and they occur in microbes and plants and to a lesser extent in some fungi and certain algal species. The escalating number of evidence on their health-promoting effects made fructans an important class of platform chemicals. In fact, they have the largest market share among the natural functional additives in the food sector. Plants are the main resources of inulin-, graminan-, and agavin-type fructans but levan type of fructans are commercially produced by microorganisms. In microbes, levan and inulin are synthesized by extracellular fructosyltransferase (FT) enzymes named levansucrase (EC 2.4.1.10) and inulosucrase (EC 2.4.1.9), respectively. Although microbial levan producers are widespread in nature, microbial inulin production is only limited in few

Nazlıcan Tezgel and Onur Kırtel contributed equally with all other contributors.

N. Tezgel
AbdiBio QC Department, Abdi Ibrahim Pharmaceutical Company, Istanbul, Turkey

Bioengineering Department, Industrial Biotechnology and Systems Biology Research Group, Marmara University, Istanbul, Turkey
e-mail: nazlicantezgel@marun.edu.tr; nazlican.tezgel@abdiibrahim.com.tr

O. Kırtel · E. Toksoy Oner (✉)
Bioengineering Department, Industrial Biotechnology and Systems Biology Research Group, Marmara University, Istanbul, Turkey
e-mail: onurkirtel@marmara.edu.tr; ebru.toksoy@marmara.edu.tr

W. Van den Ende
Laboratory of Molecular Plant Biology, KU Leuven, Leuven, Belgium
e-mail: wim.vandenende@kuleuven.be

N. K. Arora et al. (eds.), *Microbial Enzymes: Roles and Applications in Industries*, Microorganisms for Sustainability 11,
https://doi.org/10.1007/978-981-15-1710-5_1

Gram-positive bacteria. This chapter first introduces fructans and its microbial and enzymatic production processes followed by the discussion on different classes and structure-functional features of FT enzymes.

1.1 Introduction

Fructans are one of the most widespread polysaccharides in nature. They are multi-functional carbohydrates attracting significant attention since their discovery in the 1800s. For centuries, the ancient people gained benefit from fructan sources without knowledge of their existence in many staple foods. The chemistry of fructan was established in the mid of 1900s by many acclaimed carbohydrate chemists (Suzuki 1993). Since then, the occurrence of fructose-based oligo- and polysaccharides has been studied extensively in plants and microorganisms. These polysaccharides are also considered as precursors for other metabolites, which formed the origins of prebiotic Earth (Stern and Jedrzejas 2008). Apart from these, there are various other properties of fructans that make them potential biopolymers to be used in biotechnological applications (Penadés et al. 2017).

Fructans have the largest market share among the natural functional additives in the food industry which comprised more than 65% of the total volume in 2014 (https://www.gminsights.com/industry-analysis/inulin-market). They are indigestible, water-soluble, and low-calorie sweeteners used in food products as an alternative to sucrose and thereby can be used safely by diabetic patients (Zambelli et al. 2016). Fructans from both the plant and microbial origins have become the focus of interest for their prebiotic properties by which they promote the growth of beneficial microorganisms in the gastrointestinal tract (Van Laere and Van den Ende 2002; Apolinario et al. 2014; Toksoy Oner et al. 2016; Tamura et al. 2017). Moreover, fructans are proposed to be associated with native immune responses both in plants and in animals (Peshev and Van den Ende 2014), which in turn makes them important molecules that function as immunomodulators (Vogt et al. 2014; Tamura et al. 2017). Escalating number of evidences showed that fructans can also be used in a variety of applications such as drug carriers for peptides and proteins (Sezer et al. 2015), bandages in wound-burned tissue healing (Costa et al. 2013), anti-tumorigenesis patterns (Sarilmiser and Toksoy Oner 2014), cryoprotectants in cereals like oat and wheat (Livingston and Henson 1998), and stress tolerance markers in fructan-accumulating plants (Valluru and Van den Ende 2008). Future studies are expected to further boost up their uses in high-value biotechnological applications.

1.2 Fructan Diversity and Function

Fructans are nonstructural carbohydrates holding high ground after glucose-based oligosaccharides in nature. Fructans occur as water-soluble fructose-based biopolymers and are naturally found in about 15% of flowering plants and many bacteria while their occurrence is restricted to certain fungi and scarce amount of algal species such as *Dasycladales* and *Cladophorales* (Hendry 1993). In spite of their existence in almost all biological kingdoms, fructans do not participate in the physiology of animals (Versluys et al. 2018). Then what makes these multifunctional carbohydrates to be regarded as promising biopolymers for human use? Indeed, fructans diversify in a way that promises great opportunities not only to the microbial world but also to higher organisms by many different aspects. Beacuse of the biotechnological applications, fructan research is gaining speed so as to improve production in a more energetic and cost-efficient way (Toksoy Oner et al. 2016). In this sense, occurrence, structural-functional relationships, and routes of biosynthesis of fructans gain more importance to understand the mechanism behind their production.

Fructans are structurally diverse among biological taxa they occur in. One of the notable difference is in glycosidic bonds between fructosyl moieties in their structures. Generally, monomeric fructose units of fructans are attached via β-2,1 and/or β-2,6 linkages to form higher order fructan oligosaccharides (FOS) and polysaccharides. The levan-type polymers having β-2,6-linked fructosyl residues on the backbone with occasional β-2,1-type branching constitute the majority of microbial fructans (Han et al. 2016). This type of microbial fructans are widespread among both Gram-positive bacteria including the genera *Bacillus, Paenibacillus, Geobacillus, Streptococcus*, and *Microbacterium*, and Gram-negative bacteria such as *Acetobacter, Pseudomonas*, *Zymomonas*, and *Halomonas* (Toksoy Oner et al. 2016). On the other hand, microbial inulin-type fructans formed by substitution of fructosyl units with β-2,1 linkage type and β-2,6 branches occur in only a limited number of lactic acid bacteria belonging to Gram-positive bacterial genera, namely *Streptococcus, Lactobacillus, Leuconostoc, Weissella,* and *Bacillus agaradhaerens* (Anwar et al. 2010; Toksoy Oner et al. 2016; Kralj et al. 2018).

In contrast to microbes, fructans of plant origin show greater structural variety. Fructans are classified into five main groups in plants depending on glycosidic attachments between fructosyl moieties. The first group, inulin, is the most widely studied plant fructan and dominant in dicot species mainly in Asteraceae which includes chicory, sunflowers, daisies, and thistles (Hendry 1993; Vijn and Smeekens 1999; Van Laere and Van den Ende 2002). Inulin-type fructans are further subdivided into two different groups in plants: regular inulins (nonreducing sugars) and inulo-n-oses (reducing sugars). Fructosyl subunits linked to a starter glucosyl moiety (α-D-Glc) allow the formation of the building block of regular inulin-type fructans, 1-kestotriose. Absence of a starter glucosyl from inulin chain results in formation of the other type of inulin fructans, inulo-n-oses (Van Laere and Van den Ende 2002). Levan-type plant fructans (also called phleins) (group 2) are formed by substitution of fructosyl units to 6-kestotriose backbone and present in certain species from monocot families such as Poaceae, Asparagaceae, and Haemodoraceae (Vijn and Smeekens 1999). Graminans (group 3) are mixed fructans of plant origin representing both β-2,1 and β-2,6 linkages (Carpita et al. 1989; Van den Ende et al. 2011). Neo-inulin- and –levan-type fructans make up the fourth group and they are formed on a 6G-kestotriose backbone having an internal glucose moiety. β-2,1

Linkages predominate in neo-inulin fructans while neo-levan-type fructans contain β-2,6 linkages. Fructosyl elongations of the molecule occur at either end of the glucose unit in the backbone (Martínez-Padilla 2017). Combination of graminan and neo-type fructans forms more complex types (of fructans) of group 5, agavins, which have been found in some plant species such as *Agave* spp. and *Achyranthes bidentate* (López et al. 2003; Wang et al. 2015).

Although both microbes and plants share the same linkage types to synthesize fructan polymers, microbial fructans are characterized by their greater degree of polymerization (DP), which can exceed more than 20,000 fructosyl units (Toksoy Oner et al. 2016; Versluys et al. 2018) whereas plant fructans occur with a DP varying from 3 to a few hundred (Van den Ende 2013). However, occasional production of low DP microbial FOS is also possible and sometimes even dominates in some genera and under specific reaction conditions (van Hijum et al. 2006). The variations in DP and branching result mainly due to the action of enzymes playing a role in fructan biosynthesis.

1.3 Biosynthesis of Fructans

Fructans are biosynthesized by the action of different enzymes. The main mechanism of action comprises the transfer of a fructosyl unit from a sucrose molecule to an acceptor. Inulobiose (β-D-fructosyl-2,1-β-D-fructofuranoside) and levanbiose (β-D-fructosyl-2,6-β-D-fructofuranoside) are the simplest structures constructing fructan polymers in nature. Plants catalyze the cleavage and transfer of fructosyl unit to an acceptor by at least two different enzymes; however, microbes only require a single enzyme activity to synthesize fructan polymers (Velázquez-Hernández et al. 2009). Microbial fructans are produced generally by extracellular FTs. Bacteria use FTs belonging to glycosyl hydrolase family 68 (GH68) of glycosidase hydrolases to synthesize fructans. In nature, biosynthesis of microbial fructans requires action of a single enzyme, either inulosucrase (EC 2.4.1.9, sucrase:2,1-β-D-fructan:1-β-D-fructosyltransferase) or levansucrase (EC 2.4.1.10, sucrose:2,6-β-fructan:6-β-D-fructosyltransferase) which catalyzes the synthesis of inulin-type or levan-type fructans, respectively (Fig. 1.1) (Banguela and Hernández 2006; Velázquez-Hernández et al. 2009), and the fate of the biosynthesis varies greatly depending on the reaction conditions. However, dual action of those enzymes has been proposed in streptococci as well (van Hijum et al. 2002).

Levansucrases are the best studied FTs which can directly catalyze not only biosynthesis but also hydrolysis of microbial levans in the absence of substrate sucrose (Kralj et al. 2008). Crystal structures of levansucrases have been identified from *Bacillus subtilis* (Protein Data Bank [PDB] ID: 1OYG; Meng and Fütterer 2003), *Gluconacetobacter diazotrophicus* (PDB ID: 1 W18; Martínez-Fleites et al. 2005), *Bacillus megaterium* (PDB ID: 3OM2; Strube et al. 2011), and *Erwinia amylovora* (PDB ID: 4D47; Wuerges et al. 2015). In the most general perspective, levansucrase uses a donor sucrose molecule, hydrolyzes it, and transfers the resulting fructosyl moiety to an acceptor sucrose molecule with β-2,6 linkages, leading to the

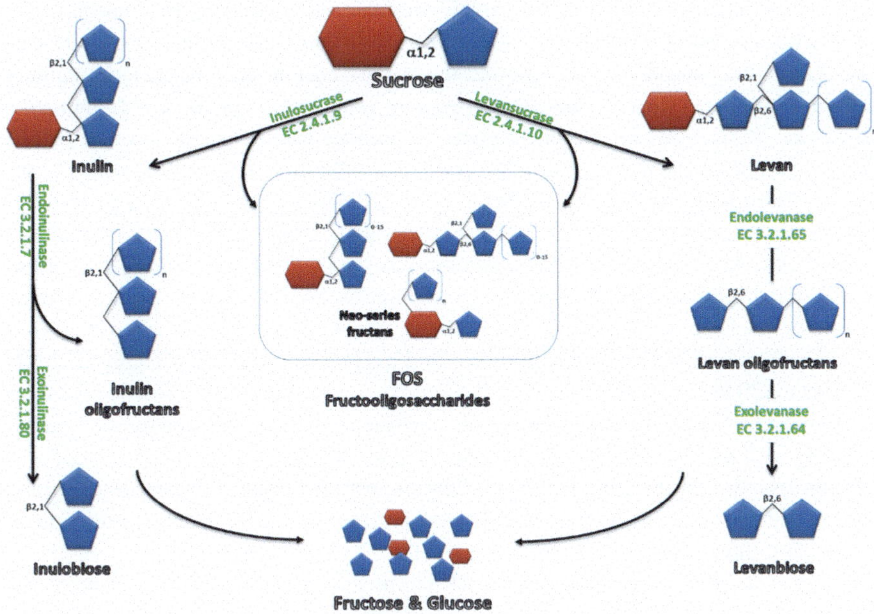

Fig. 1.1 Microbial fructan biosynthesis and degradation

formation of 6-kestotriose. Consequent transfers of fructosyl units to 6-kestotriose polymer backbone result in the construction of levan-type fructans. Depending on the reaction conditions, levansucrase can also use other molecules as acceptors, such as water in sucrose hydrolysis resulting in the release of glucose and fructose, with glucose leading to the formation of sucrose (Glc[α1-β2]Fru*f*) or a sucrose isomer, blastose (β-D-Fru*f*-[2↔6]-D-Glc*p*); fructose enabling the synthesis of inulobiose or levanbiose; and levan elongating the polymer chain (Li et al. 2015; Toksoy Oner et al. 2016; Miranda-Molina et al. 2017). As exceptions, *Leuconostoc mesenteroides* uses maltose as an alternative fructosyl acceptor to produce trisaccharide erlose (Kang et al. 2005) while *B. megaterium* synthesizes blastose (Homann et al. 2007). Blastose is a sucrose-derived neo-type disaccharide which is usually produced as a by-product of levansucrase reactions with sucrose as a substrate (Homann et al. 2007). Activity of levansucrase has also been reported on lactose leading to the production of lactosucrose (Chen and Gänzle 2017). Raffinose, a trisaccharide composed of galactose, glucose, and fructose, is also known to be used as a substrate by levansucrase (Trujillo et al. 2004). *B. subtilis* is able to produce both inducible and constitutive extracellular levansucrase, which use sucrose and glucose as carbon source, respectively (Abdel-Fattah et al. 2005; Srikanth et al. 2015). Recent studies on *Bacillus methylotrophicus* revealed an intracellular levansucrase action to produce levan in a more efficient way (Zhang et al. 2014; Wu et al. 2015; Li et al. 2015).

Inulosucrases catalyze bacterial inulin synthesis using mainly sucrose as an acceptor. Mechanism of action is similar to that of levansucrase; however, inulosucrase elongates the growing fructan chain by addition of fructosyl units with β-2,1 linkage.

While levansucrases have been intensively studied, the available information about inulosucrases is rather limited in the literature. Up to date, inulin biosynthesis by inulosucrases in microbes has been reported in species from lactic acid bacteria, namely *Lactobacillus*, *Leuconostoc*, and *Streptococcus* (van Hijum et al. 2002; Olivares-Illana et al. 2003); however, in recent years, existence of inulosucrase in other genera belonging to Gram-positive bacteria has been identified including *Streptomyces turgidiscabies* (Dilipkumar et al. 2011), *Actinobacteria* (Frasch et al. 2017), and *Bacillus agaradhaerens* (Kralj et al. 2018). On the other hand, crystal structure of an inulosucrase has only been identified from *Lactobacillus johnsonii* (Pijning et al. 2011; PDB ID: 2YFR). Inulosucrases of *Streptococcus mutans* are able to synthesize a trace amount of sucrose, 1-kestotriose, nystose (1,1-kestotetraose), and also raffinose-derived inulins in addition to inulin oligo- and polysaccharides (Ozimek et al. 2006; Anwar et al. 2010). Inulosucrase from *Leuconostoc citreum* has the ability to catalyze the synthesis of glucose-based polymers representing glycosyltransferase activity alongside with its FT activity (Olivares-Illana et al. 2003).

Fungi use fungal FTs classified in GH32 family glycosidase hydrolases which are evolutionarily closer to plant fructan-hydrolyzing enzymes, β-fructofuranosidases (invertases), than bacterial FTs (Lammens et al. 2009) to catalyze the biosynthesis of linear β-2,1-linked fructan polymers (Ritsema and Smeekens 2003). Indeed, the main mechanism of the enzyme comprises equimolar hydrolysis of sucrose into its monomers, glucose, and fructose. However, depending on the biological source and sucrose availability, reaction favors towards the synthesis of FOS when the enzyme uses sucrose (or FOS) as an acceptor rather than a water molecule, thus acting as a transfructosylase (Ritsema et al. 2006). Therefore β-fructofuranosidases show hydrolytic activity in low sucrose concentration and transfructosylating activity in high sucrose concentrations (Ganaie et al. 2014). Fungal species preferentially transfer a fructosyl group of sucrose, 1-kestotriose, nystose, or less preferably raffinose to an acceptor. Generally, FOS with DP ranging from 3 to 10 are predominant among fungal fructans; however, chain sizes show great variation between fungal species (Banguela and Hernández 2006). The occurrence of fungal FTs has been revealed in species of *Aspergillus, Aureobasidium, Kluyveromyces,* and *Penicillium* (Chuankhayan et al. 2010). The crystal structure of fungal FT enzymes has been reported in *Aspergillus japonicus* (PDB ID: 3LF7; Chuankhayan et al. 2010), *Schwanniomyces occidentalis* (PDB ID: 3KF5, Alvaro-Benito et al. 2010), *Aspergillus ficuum* (PDB ID: 3RWK, Pouyez et al., 2012), *Saccharomyces cerevisiae* (PDB ID: 4EQV, Sainz-Polo et al. 2013), *Xanthophyllomyces dendrorhous* (PDB ID:5ANN; Ramírez-Escudero et al. 2016), and *Aspergillus kawachii* (PDB ID: 5XH8, Nagaya et al. 2017).

Unlike microbes, plants synthesize fructans by the combined action of multiple enzymes. There are four FTs studied to date which take part in fructan production in the vacuolar compartment of higher plants. Plant FTs share GH-J clan of glycoside hydrolase enzymes with microbial FTs whereas they evolutionarily diverge from each other and belong to GH32 enzyme family (Lammens et al. 2009). Vacuolar sucrose, the central molecule in plant fructan biosynthesis, is the substrate used to initiate catalysis by sucrose:sucrose 1-fructosyltransferase (1-SST, EC

2.4.1.99) and sucrose:(sucrose/fructan) 6-fructosyltransferase (6-SST, EC 2.4.1.- / 6-SFT, EC 2.4.1.10) in order to form 1-kestotriose and 6-kestotriose, respectively. Formed FOSs are further elongated to inulin and levan polysaccharides by the action of fructan:fructan 1-fructosyltransferase (1-FFT, EC 2.4.1.100) and 6-SST/6-SFT, respectively. Fructan:fructan 6G-fructosyltransferase (6G-FFT, EC 2.4.1.243) synthesizes fructan neo-series from 1-kestotriose which is polymerized to neo-inulin type of fructans by 1-FFT, whereas 6-SFT is used to synthesize neo-levan-type fructans. 1-Kestotriose of plant origin is also prone to attacks of 6-SFT enzyme leading to formation of bifurcose, the smallest graminan type of fructans in plants. Bifurcose is elongated by either 1-FFT or 6-SFT forming graminans (Vijn and Smeekens 1999; Livingston III et al. 2009; Van den Ende 2013; Van den Ende et al. 2002). The crystal structure of 6-SST/6-SFT enzyme has been obtained from *Pachysandra terminalis* (Lammens et al. 2012; PDB ID: 3UGF).

1.4 Microbial Fructan Production

The demand for sucrose-derived fructans has increased in many industries. Alongside their natural occurrence by means of "biosynthesis," different types of fructans are also produced enzymatically by many industrial processes (Table 1.1). However, among them, there is a special interest on FOS rather than long-chain fructans since FOSs are generally recognized as safe (GRAS) molecules and they are commercially available for animal and human consumption since last few decades (Nobre et al. 2015).

Microbial FTs are the main enzymes used in industrial FOS production (Nobre et al. 2015). Commercially important FOSs are generally inulin-type fructans including 1-kestotriose (GF2), nystose (GF3), and 1-β-fructofuranosyl nystose (GF4) in which one to three fructosyl unit(s) are transferred to the sucrose donor at the β-2,1 position by releasing glucose into the reaction medium (Prapulla et al. 2000). These type of fructans can be produced chemically by glycosylation and *de novo* synthesis (Nobre et al. 2015). However, chemical FOS synthesis is not practical or economically feasible due to several reasons (Barreteau et al. 2006). The production requires many reaction steps which are laborious and use expensive/hazardous chemicals usually yielding low-quantity fructans (Palcic 1999; Prapulla et al. 2000). Extractions from plant sources are also not very applicable since the application of FOS biosynthetic enzymes of plant origins is limited to seasonal changes.

On the other hand, in vitro enzymatic production of fructans has been reported for many organisms using whole-cell synthesis and isolated enzymes besides the promising benefits of immobilization technology (Bali et al. 2015).

Table 1.1 Microbial fructan production

Microorganism	Enzyme	Fructan production method	Product	Reference
Aureobasidium sp. ATCC 20524	Inulosucrase	Immobilized enzyme	Production of 1-kestose from sucrose by immobilization on a shirasu porous glass	Hayashi et al. (1991)
Aspergillus japonicus	β-Fructofuranosidase	Immobilized enzyme	Production of 127.7 g/L and 71.3 U/mL FOS by immobilization on coffee silverskin	Mussatto and Teixeira (2010)
A. japonicas	β-Fructofuranosidase	Immobilized enzyme	Production of 61% FOS of the w/w of total sugars in the reaction mixture by immobilizing on gluten	Chien et al. (2001)
Aspergillus japonicus ATCC 20236	β-Fructofuranosidase	Immobilized enzyme	Production of 6.61 g/L/h FOS (0.66 g/g based on total substrate sucrose and 0.73 g/g based on consumed substrate sucrose) by 1.49 g/g enzyme immobilization on corncobs	Mussatto et al. (2009b)
Aspergillus japonicus CCRC 93007	β-Fructofuranosidase	Whole-cell synthesis	Production of more than 160 g 6 L FOS with *Gluconobacter oxydans* ATCC 23771 (with glucose dehydrogenase activity) in a 7-day continuous bioreactor system equipped with microfiltration module	Sheu et al. (2002)
Aspergillus oryzae	β-Fructofuranosidases	Isolated enzyme	Production of 53% FOS with 60% sucrose at 55 °C and pH 5.15 in a submerged fermentor	Sangeetha et al. (2004)
Aureobasidium pullulans	β-Fructofuranosidases	Whole-cell synthesis	166 g 6 L FOS yield from 360 g 6 L molasses sugar	Shin et al. (2004)
Aureobasidium pullulans	β-Fructofuranosidases	Immobilized enzyme	Continuous production of 180 g FOS/L/h for more than 100 days by immobilization on calcium alginate	Jung et al. (2011)
Aureobasidium pullulans CFR 77	Intracellular FTase	Whole-cell synthesis	Ultrasonication has been used to release intracellular FTase from *A. pullulans* CFR 77 for the subsequent production of FOS	Lateef et al. (2007)
Bacillus subtilis	Levansucrase	Immobilized enzyme	Production of 85% levan of total polymer production by immobilization on calcium phosphate gel	Chambert and Petit-Glatron (1993)
Bacillus subtilis (natto) Takahashi	Levansucrase	Whole-cell synthesis	Production of 49.4 g/L levan in the first 21 h of a submerged fermentation process	Shih et al. (2005)

Bacillus subtilis (natto) Takahashi	Levansucrase	Immobilized enzyme	Production of 70.6 g/L levan by immobilization on alginate beads in 72 h	Shih et al. (2010)
Bacillus subtilis (natto) Takahashi	Levansucrase	Whole-cell synthesis	Production of 61 g/L and 100 g/L of levan in batch and fed-batch suspension cultures of *B. subtilis* (natto) Takahashi using 10-L stirred bioreactors	Wu et al. (2013)
Bacillus subtilis NCIMB11871	Fructosyltransferase	Immobilized enzyme	Production of galactosyl-fructoside (sucrose analogue) by immobilization on Eupergit® C 250 L and Trisopor®-Amino at 50 °C	Baciu et al. (2005)
Bacillus subtilis NRC33a	Levansucrase	Immobilized enzyme	Production of levan (up to 86%) by immobilization on chitosan	Esawy et al. (2008)
Bacillus amyloliquefaciens NK-1	Levansucrase	Whole-cell synthesis	Co-production of levan and PGA leading to up to a tenfold increase in levan yields	Feng et al. (2015)
Halomonas smyrnensis AAD6^T	Levansucrase	Whole-cell synthesis	Production of high level of levan from sucrose when boric acid is supplemented in the reaction medium	Kazak Sarilmiser et al. (2015)
Leuconostoc citreum CW 28	Inulosucrase	Isolated enzyme	Production of inulin polymers with a molecular weight between 2.6 and 3.4 kDa	Ortiz-Soto et al. (2004)
Microbacterium laevaniformans PTCC 1406	Levansucrase	Whole-cell synthesis	Increase in yield more than four times when growth with sucrose syrup	Moosavi-Nasab et al. (2010)
Paenibacillus polymyxa NRRL B-18475	Levansucrase	Whole-cell synthesis	Production of 2000 kDa levan with 71% β(2–6) linkages, 13% terminal groups, and 12% branching at the C1 position with a β(1–2) linkage	Han and Clarke (1990)

(continued)

Table 1.1 (continued)

Microorganism	Enzyme	Fructan production method	Product	Reference
Penicillium citrinum KCCM 11663	β-Fructofuranosidases	Whole-cell immobilization	Continuous production of neo-FOS yielding 49 g/L in packed bed reactor	Park et al. (2005)
Penicillium expansum	β-Fructofuranosidases	Whole-cell synthesis	Production of 0.58 g FOS and 6 g sucrose in shake flasks	Prata et al. (2010)
Scopulariopsis brevicaulis	β-Fructofuranosidases	Whole-cell synthesis	Production of 1-kestose with theoretical yield of 85%, whereas crystallization and recrystallization yield of 71.0% and 78.0% with 99.9% purity	Takeda et al. (1994)
Schwanniomyces occidentalis	β-Fructofuranosidases	Whole-cell synthesis	Production of 6-kestose yielding 16.4% (w/w) of the total carbohydrates in the reaction mixture (sucrose, 1-kestose, nystose, and raffinose)	Alvaro-Benito et al. (2007)
Zymomonas mobilis	Levansucrase	Isolated enzyme	Production of 1-kestostriose, 6-kestotriose, neokestose, nystose, and some other non-identified fructooligosaccharides from sucrose syrup	Bekers et al. (2002)
Zymomonas mobilis	Levansucrase	Immobilized enzyme	Production of levan by immobilization on titanium-activated magnetite	Jang et al. (2001)

1.4.1 Enzymatic Production

1.4.1.1 Whole-Cell Synthesis

The yeast *Xanthophyllomyces dendrorhous* (*Phaffia rhodozyma*) is able to produce trisaccharide neokestose in the presence of sucrose. The highest yield obtained from this organism has been reported when the cells were harvested at the exponential phase while recycling of the cells caused a reduction in the final yield (Kritzinger et al. 2003). *Aureobasidium pullulans* is able to produce both intracellular and extracellular FTs. Lateef et al. (2007) used ultrasonication to release intracellular FTs of *A. pullulans* CFR 77 to produce FOS. In another study, 166 g of FOS was recovered from *A. pullulans* when 360 g of the substrate was fed into a reaction medium (Shin et al. 2004). Substrate feeding of *A. pullulans* has also been reported to increase the production of FTs up to 47% and also enhance enzyme specificity and final concentration (Yun et al. 1997a, b).

The yeast *Schwanniomyces occidentalis* can synthesize 6-kestotriose by using a substrate mixture containing sucrose, 1-kestotriose, nystose, and raffinose (Alvaro-Benito et al. 2007). *Aspergillus oryzae* KB is known to produce two types of β-fructofuranosidases having the ability to synthesize either 1-kestotriose, nystose, and fructosyl nystose or glucose and fructose (Kurakake et al. 2008). Enzymatic production of 1-kestotriose by *Scopulariopsis brevicaulis* has been achieved with 99.9% purity (Takeda et al. 1994). In a shake flask system, 3.25 g FOS/L/h was produced by *Penicillium expansum* (Prata et al. 2010) whereas, 160.0 gFOS/L/h was produced by β-fructofuranosidases from *A. japonicus* and *A. pullulans* in a continuous bioreactor (Sheu et al. 2002). Inulin-producing *Lactobacillus* strains have also been reported. *L. citreum* CW 28 inulosucrase was demonstrated to produce inulin polymer with a molecular weight between 1.35 and 1.60 × 10^3 kDa (Ortiz-Soto et al. 2004).

Microbes producing long-chain fructan polymers have been isolated from several species such as *Acetobacter xylinum* (Tajima et al. 1997), *Bacillus polymyxa* (Han and Clarke 1990; Han and Watson 1992), *B. subtilis* (natto) Takahashi (Shih et al. 2005), and *Microbacterium laevaniformans* (Bae et al. 2008). Gram-positive bacteria *B. subtilis* (natto) Takahashi was reported to produce 49.4 g/L levan in the first 21 h of a submerged fermentation process (Shih et al. 2005). In another study, 61 g/L and 100 g/L of levan productions have been achieved by fermentation of batch and fed-batch suspension cultures of *B. subtilis* (natto) Takahashi in a 10-L stirred bioreactor (Wu et al. 2013). *M. laevaniformans* PTCC 1406 cultures are known to synthesize levan-type fructans when they are grown on sucrose under controlled fermentation conditions. The yield was reported to increase more than four times when the microbes were fed on sucrose syrup (Moosavi-Nasab et al. 2010). The first levan biosynthesis in Gram-negative bacteria was reported in *G. diazotrophicus* (*Acetobacter diazotrophicus*) (Hernandez et al. 1995). With a different strain from the same microorganism, *G. diazotrophicus* PAl 5, 24.7 g/L levan was produced from 100 g/L sucrose feed without extra nitrogen supplementation (Molinari and Boiardi 2013). Levan production is also possible by extremophiles. Gram-negative *Halomonas* sp. has been reported to be the first halophilic bacterium that produces levan (Poli et al. 2009). High level of levan biopolymer was yielded by *Halomonas smyrnensis* AAD6^T levansucrase from sucrose when boric acid is supplemented in the reaction medium (Kazak Sarilmiser et al. 2015). The same *Halomonas* strain fed on pretreated starch and sugar beet molasses was also reported to produce high amounts of levan biopolymer (Küçükkaşık et al. 2011).

1.4.1.2 Production with Isolated Enzymes

The enzymes involved in FOS production process are also cultivated and isolated to be used in biotransformation phenomena yielding desired fructan products under controlled reaction conditions (Ganaie et al. 2014). Microbial enzymes isolated from organisms are purified by different methods such as centrifugation, chemical precipitation, and column chromatography allowing the recovery of maximum biopolymer products compared to whole-cell synthesis (Bali et al. 2015).

Inulosucrase isolated from *L. citreum CW* 28 was shown to yield inulin polymers with a molecular weight between 2.6 and 3.4 kDa (Ortiz-Soto et al. 2004). Expression of the fructosyltransferase gene of *S. mutans* in an *Escherichia coli* strain was also reported to lead to the production of very-high-molecular-weight inulins (Heyer et al. 1998). *Bacillus cereus* was reported to produce 37.40 U/mL extracellular FTs when fed on 16% (w/v) sucrose in submerged fermentation while 29.1 U/g enzyme was produced in solid-state fermentation (SSF) (El-Beih et al. 2009). FTs isolated from *A. oryzae* have been reported to yield 53% FOS by production with 60% sucrose (at 55 °C and pH 5.15) in a submerged fermentor (Sangeetha et al. 2004). Isolated *Zymomonas mobilis* levansucrase was reported to be used in the production of not only 1-kestotriose, 6-kestotriose, neokestose, and nystose, but also other non-identified fructo-oligosaccharides from sucrose syrup (Bekers et al. 2002).

During microbial fructan production, glucose released as a by-product, is utilized by microbes for growth; however, a part of it is non-metabolized and accumulated in the reaction medium possessing an inhibitory effect on microbial FTs. The strategies to utilize the remaining glucose in the reaction medium have been proposed by different methods during fructan production. One of them is the usage of enzyme mixtures. Mixed-enzyme systems are used in industrial FOS production processes (Table 1.2). β-Fructofuranosidase and glucose oxidase were used together to produce FOS from sucrose and glucose feed in a batch system (Yun et al. 1994). By this way, the amount of free glucose in the fermentation medium was completely consumed by glucose oxidase with 98% FOS recovery. Also, enzymes from different microbial sources produced high-content FOS from sucrose-derived substrates (Maugeri and Hernalsteens 2007).

1.4.1.3 Production by Hydrolysis of Fructans

In contrast to fructan biosynthesis, commercial FOS production is also possible via biodegradation of fructans by fructan hydrolases (Sangeetha et al. 2004). Microbial fructans can be hydrolyzed by inulinases (2,1-β-D-fructan fructanohydrolase, EC 3.2.1.7) and levanases (2,6-β-D-fructan fructanohydrolase, EC 3.2.1.65) in an endo- or exo-manner (Cote and Ahlgren 1993; Versluys et al. 2018). The reaction mechanism differs depending on the enzyme functioning on polymer backbone (Fig. 1.1).

Table 1.2 Mixed-enzyme systems employed in microbial fructan production

Mixed enzyme system	Enzymes	Product	Reference
Candida sp. (LEB-I3) *Rhodotorula* sp. (LEB-U5) *Cryptococcus* sp. (LEB-V2) *Rhodotorula* sp. (LEB-V10)	Fructosyltransferases	Production of 100 g/L of FOS from 500 g/L sucrose solution	Maugeri and Hernalsteens (2007)
Aureobasidium pullulans KFCC *Aspergillus niger*	β-Fructofuranosidase Glucose oxidase	Production of 98% FOS with complete consumption of sucrose and glucose by 10 units of β-fructofuranosidase with the combination of 15 units of glucose oxidase per gram sucrose at pH 5.5, 40 °C with 400 g/L initial sucrose concentration, 550 rpm agitation speed, and 0.71 L/min oxygen flow rate	Yun et al. (1994)
Aspergillus aculeatus Aspergillus niger	Transfructosylating activity of pectinases (Pectinex Ultra SP-L and Rapidase TF)	Production of 61.5% w/w (387 g/L) FOS from total carbohydrate in the reaction mixture by immobilization on polymethacrylate-based polymer (Sepabeads® EC)	Ghazi et al. (2005)
Aspergillus niger ATCC 20611 *Aspergillus japonicus* TIT-KJ1	β-Fructofuranosidases	Production of 60% FOS of the total sugars in the reaction mixtures from 50% (w/w) sucrose solution by immobilization of enzyme mixture on methacrylamide-based polymeric beads	Chiang et al. (1997)

Microbial fructan exohydrolases remove one or two terminal fructosyl residues from the fructan chain whereas endohydrolases split β linkages in oligofructosyl chain at random sites leading to the production of oligofructans in various sizes (Murakami et al. 1990).

Microbial inulinases specifically hydrolyze inulin oligo- and polysaccharides by the actions of endo-inulinases (EC 3.2.1.7) (Naidoo et al. 2015). Levanases show a similar substrate specificity in the hydrolysis of levan fructans by endo-levanases (EC 3.2.1.65) while exo-levanases (EC 3.2.1.64) are able to degrade sucrose, raffinose, and inulin other than levan biopolymers (Menendez et al. 2004). Inulinase production has been reported in yeast *Kluyveromyces marxianus* var. *bulgaricus* fed on Yacon (*Polymnia sonchifolia*) (Cazetta et al. 2005). Levan FOSs were also reported to be produced by hydrolysis of levan from *Zymomonas mobilis* (de Paula et al. 2008).

Under certain conditions, levansucrases, inulosucrases, and invertases (β-fructofuranosidases; EC 3.2.1.26) can also function as exo-levanases in some microbes showing both inulin- and levan-type hydrolyzing activity (Uchiyama 1993; Versluys et al. 2018).

1.4.2 Immobilization Technology

Many microbes or the enzymes isolated from those microorganisms are immobilized on different materials to produce catalytically and economically feasible fructans under optimum reaction conditions. Calcium alginate beads were reported to be used in FOS production by *A. japonicus* (Cruz et al. 1998), *A. foetidus* (Markosian et al. 2007), and *Aspergillus niger* (Lin and Lee 2008). The maximum yield from sucrose was obtained when the organisms were immobilized on calcium alginate beads.

Interestingly, different materials such as stainless steel, polyurethane foam, foam glass, and vegetable fibers were reported to be used as immobilizators for industrial FOS production (Mussatto et al. 2009a). Moreover, several lignocellulosic materials such as corncobs, cork oak, coffee husks, and wheat straw and grain were also used as support materials during FOS and β-fructofuranosidase production by immobilized *A. japonicus* cells (Mussatto et al. 2009b). Among them, corncobs were proposed to serve the most effective immobilizing matrix for FOS production when compared to non-immobilized cell systems (Bali et al. 2015).

Levan-producing *B. subtilis* (natto) Takahashi cells were immobilized on alginate beads to produce 70.6 g/L levan in 72 h (Shih et al. 2010). Immobilization of *B. subtilis* levansucrase enzyme on calcium phosphate gel was demonstrated to yield 85% levan of total polymer production (Chambert and Petit-Glatron 1993). *B. subtilis* NRC33a strain was reported to yield high amounts of levan (up to 86%) when the enzyme is immobilized on chitosan. The 51.13% of enzyme was stated to retain on chitosan beads after 14 repeats of usage (Esawy et al. 2008). *Z. mobilis* levansucrase was revealed to produce levan biopolymer by enzyme binding on titanium-activated magnetite (Jang et al. 2001).

Even though immobilization increases the rate of FOS production, many studies have reported that the higher amount of FOS yields can also be achieved by free cells than it has done with immobilized cells (Ning et al. 2010). Therefore, the main driving force in FOS production can be considered as the catalytic properties of the enzymes specialized for microbial fructan synthesis.

1.5 Microbial Fructosyltransferases

1.5.1 Mechanism of Action

1.5.1.1 Enzyme Structure and Function

Majority of microbial FTs which are responsible for transfructosylation reactions in microbes belong to GH68 family of enzymes. However, a few microbial fructan biosynthetic enzymes are involved in GH32 enzyme family. Bacterial FTs, namely inulosucrases, levansucrases, and a few β-fructofuranosidase-like enzymes, are classified in GH68 enzyme family, sharing a common β-propeller domain in the active site with the enzymes of GH-J clan. The propeller is formed by four antiparallel β-strands, each located in fivefold blades, hiding the catalytic triad in the core of the enzyme (Lammens et al. 2009). The active site is superimposed in a negatively charged funnel cavity enabling proper bonding with fructosyl donor substrates. By this way, sucrose has been stated to position similarly in the catalytic pocket of all GH68 family enzymes. In addition to general structure, fungal FTs (β-fructofuranosidases) involved in GH32 enzyme family hold a six-β-stranded β-sandwich structure in their C-terminal domain (Lammens et al. 2009) which was suggested to play a role in recognition of microbial fructans with high DPs (Le Roy et al. 2007). Studies concluded that GH68 and GH32 families show sequence homology sharing several conserved motifs (Naumoff 2001). Nevertheless, they are not structurally homologous proteins and have several enzyme family-specific properties proposed to be involved in the catalytic activity.

The catalytic triad of GH-J clan is fully conserved among the enzyme families. This set of three acidic amino acid residues (Asp-Glu-Asp), involved directly in the enzyme-substrate-binding activity, is located in the N-terminal of the propeller. Amino acid residues serve essential functions in catalysis and each one is conserved in different motifs. One aspartate residue (WMNDPNG motif) has been identified to function as the catalytic nucleophile while glutamate (EC motif) was revealed to act as the acid/base catalyst. The other aspartate residue (RDP motif) involves in substrate binding, thus functioning as a transition-state stabilizer (Meng and Fütterer 2003). Next to glutamate, the acid/base catalyst, EC motif possesses a conserved cysteine residue in GH32 enzyme family; however, an arginine residue is located in the EC motif of GH68 enzymes proposing enzyme family specificity (Lammens et al. 2009). Therefore, sequence variation in the motifs involved in catalysis results in the formation of enzyme products with different linkage types and chain length among GH-J clan. However, sequential differences are also valid among each GH family although member enzymes shared conserved domains.

Bacterial FTs are extracellular proteins belonging to GH68 enzyme family which have five main domains: a signal peptide (1), an N-terminal domain (2), a catalytic domain (3), a cell wall-binding domain (4), and a region in the C-terminal domain (5) (Velázquez-Hernández et al. 2009). The signal peptide is a widely shared domain among Gram-positive bacteria while the majority of FTs of Gram-negative bacteria are secreted as signal-peptide-independent proteins with an exception of levansucrase of *G. diazotrophicus*. *G. diazotrophicus* levansucrase is secreted by type II secretion system which requires cleavage of a precursor signal peptide located in the first 30 amino acid residues of the organism (Arrieta et al. 2004). The N-terminal of microbial FTs is generally 90 amino acids long containing several conserved residues or repetitive sequences (Lammens et al. 2009). FTs of lactic acid bacteria share no homology in sequence but in length, however, they contain some residues in common. *Lactobacillus sanfranciscensis* TMW1.392 levansucrase contains a repetitive 16-amino-acid sequence in the N-terminal of the enzyme (Tieking et al. 2005). FTs of Gram-negative bacteria also share a motif (WT[R/I]ADA[L/M]) in the N-terminal domain (Tajima et al. 2000). The catalytic domains of microbial FTs possess 11 common regions in their sequences. Among microbes, only Gram-positive bacteria have been shown to conserve aspartate residue in the catalytic domain (Ozimek et al. 2005). The cell wall-binding domain in the C-terminal of microbial FTs varies sequentially depending on the microbial origin. The cell wall-binding protein of Gram-positive bacteria typically harbors a 6-repeat 12-amino-acid region containing PXX-LPXTG motives and a hydrophobic region (Tieking et al. 2005) whereas a glycine-rich region predominates in the cell wall-binding proteins of Gram-negative origin (Song et al. 1998).

Microbial FTs operate via a double-displacement (ping-pong) mechanism allowing formation and hydrolysis of fructosyl-enzyme intermediate. The enzyme mechanism involves a two-step reaction. In the first step, a general glycosylation reaction takes place where a covalent fructosyl-enzyme intermediate is formed by a nucleophilic attack on the anomeric carbon of the substrate molecule (generally sucrose). The second step proceeds as a deglycosylation reaction leading to hydrolysis of the fructosyl-enzyme intermediate by removal of a proton from a fructosyl acceptor (Antošová and Polakovič 2001). The enzyme affinity towards a water molecule as an acceptor is considerably low reducing the hydrolysis activity of microbial FTs.

Most microbial levansucrases represent a simple Michaelis-Menten kinetics retaining substrate saturation conditions. As exceptions, levansucrases from *Lactobacillus reuteri* and *L. sanfranciscensis* do not obey saturation kinetics under normal operating conditions (van Hijum et al. 2006). Levansucrases function in a way in which the synthesized fructan polymer remains bound to the enzyme during chain elongation (processive or proportionate reaction) and is used as a fructosyl donor in substrate depletion (Kralj et al. 2008), while synthesis of short-chain fructans requires release of fructan chain after each transfructosylation reaction (non-processive or disproportionate reaction) (Ozimek et al. 2006). When the available substrate is limited in the environment, microbial levansucrases are able to use the synthesized polymer as a substrate by cleaving the backbone at the β-2,6 linkages (exo-hydrolysis). Then the reaction stops when the enzyme encounters a β-2,1 linkage at the branching points. In this manner, levan chains are prevented from extensive degradation by levansucrase enzyme. Therefore, levan production often requires a prompt enzyme inhibition step at the end of biosynthesis (Toksoy Oner et al. 2016). In contrast to many levansucrases, inulosucrases do not follow Michaelis-Menten kinetics (Anwar et al. 2008).

β-Fructofuranosidases (invertases) are involved in the reversible biosynthesis of several microbial FOS. Fructan production by β-fructofuranosidases depends on two mechanisms: either reverse hydrolysis or transfructosylation (Antošová and Polakovič 2001). β-Fructofuranosidases obey Michaelis-Menten kinetics (Ritsema et al. 2006); therefore the yield depends on the equilibrium between fructosyl-enzyme complex to a fructose (or kestose) and a free enzyme. Transfructosylation allows transfer of a fructosyl unit from a donor (sucrose or FOS) to an acceptor, allowing the formation of the product (a FOS having one more fructosyl unit than the acceptor) which is also a potential donor molecule to β-fructofuranosidases (Antošová and Polakovič 2001). The final FOS yield can be increased accordingly with the increasing substrate concentration. However, FOS produced by a fructofuranosidase action is usually low even when the microbe is exposed to high substrate concentrations whereas high FOS production is possible even in dilute substrate concentrations by a transfructosylase action (Antošová and Polakovič 2001). Therefore, β-fructofuranosidases having high transfructosylase activity often yield higher amounts of FOS. Likewise, β-fructofuranosidases with transfructosylase activity show catalytically different profile than the fungal FTs (β-fructofuranosidases) which synthesize FOS by means of reverse hydrolysis (Antošová and Polakovič 2001).

1.5.1.2 Physicochemical Properties

Bacterial FTs are generally considered as extracellular monomeric enzymes bearing a single domain with some exceptional cases (Olivares-Illana et al. 2003; Ortiz-Soto et al. 2004; Velázquez-Hernández et al. 2009). One of them is an inulosucrase from *L. citreum* CW28 that is a cell wall-associated FT possessing three domains in its crystal structure. The enzyme shows unusually high sequential similarity to glucosyltransferases, which are responsible for catalyzing the transfer of a glucosyl moiety to a growing polysaccharide chain, by having an identical glucan-binding domain. Correspondingly, *L. citreum* CW28 inulosucrase is known as the only bacterial inulosucrase having a molecular weight of 165 kDa, being the highest reported so far (Olivares-Illana et al. 2003). However, most inulosucrases are relatively smaller enzymes having a molecular weight ranging from 45 to 95 kDa (van Hijum et al. 2002; Dilipkumar et al. 2011; Frasch et al. 2017; Kralj et al. 2018). Similar to inulosucrases, there have been several dimeric levansucrases reported in *Actinomyces viscosus* T-14 V (Pabst 1977), *Rahnella aquatilis* JMC-1683 (Ohtsuka et al. 1992), and *L. mesenteroides* B-512 FMC (Kang et al. 2005). Among them *A. viscosus* T-14 V levansucrases have the highest molecular weight, 250 kDa, also being the highest mass reported for levansucrases (Pabst et al. 1979) while levansucrase of *G. diazotrophicus* possesses the highest molecular mass within monomeric levansucrases with 200 kDa (Hernandez et al. 1995). The smallest levansucrase having molecular weight of 20 kDa was reported to occur in *B. subtilis* (Tanaka et al. 1978). In contrast to bacterial FTs, microbial β-fructofuranosidases are multimeric enzymes whereas a monomeric β-fructofuranosidase was reported in *Bifidobacterium adolescentis* G1 (Muramatsu et al. 1993). Consequently, microbial β-fructofuranosidases possess much higher molecular weights compared to bacterial FTs.

The reported optimum pH for inulosucrase activity is from 5.0 to 6.5 (van Hijum et al. 2002; Ortiz-Soto et al. 2004; Del Moral et al. 2008) while pI for inulosucrases ranges from 4.5 to 5.0 (Velázquez-Hernández et al. 2009). Levansucrases represent a broader optimum pH range for their activity. *Z. mobilis* levansucrase (Han et al. 2009) operates at pH 7.0 while optimal enzymatic activity was observed at pH 4.6 in *Lactobacillus panis* levansucrase (Waldherr et al. 2008). pI for levansucrases was reported to range from 2.6 to 5.5 (Velázquez-Hernández et al. 2009). Bacterial β-fructofuranosidases occur to be active at acidic and neutral-to-alkali pH conditions (Lincoln and More 2017). However, enzymes from several *Aspergillus* species have been recorded to be active at highly alkaline pH (Antošová and Polakovič 2001).

Bacterial inulosucrases operate best at temperatures ranging from 20 °C to 50 °C. However, Ozimek et al. (2005) have revealed that *Lactobacillus reuteri* strain 121 has the ability to operate at 55 °C in the presence of Ca^{2+}. Besides, recombinant inulosucrase from *Lactobacillus gasseri* DSM 20604 (Díez-Municio et al. 2013) and inulosucrase from *L. johnsonii* NCC 533 (Anwar et al. 2008) were also reported to show optimum activity at 55 °C. Interestingly, levansucrase of *B. subtilis* has been stated to be more productive over a period of 10 h at 0 °C; however the enzyme activity is higher at 37 °C (Tanaka et al. 1978), while 60 °C is just enough for the activity of levansucrases from *R. aquatilis* JCM-1683 (Ohtsuka et al. 1992) and *Bacillus* sp. TH4-2 (Ammar et al. 2002). Fungal FTs can work at higher temperatures, bacterial inulosucrases do such as FTs from *A. oryzae* which has been declared to function optimally at 60 °C (Maiorano et al. 2008).

Substrate affinity differs remarkably among kinetically characterized microbial FTs. Km for sucrose in wild-type bacterial inulosucrases was obtained as 12 mM and 99.2 mM in *L. reuteri* 121 (Ozimek et al. 2006) and *L. citreum* (Ortiz-Soto et al. 2004), respectively. Levansucrases show a wide spectrum of sucrose affinity. The lowest levansucrase affinity to sucrose has been reported in *Pseudomonas syringae* pv. *phaseolicola* (Km: 160 mM) (Hettwer et al. 1995) while levansucrase from *B. megaterium* has more preference to sucrose as a substrate (Km: 4.1 mM) (Homann et al. 2007). Reported information about β-fructofuranosidase affinity to different substrates abounds in the literature more than microbial FTs. Other than sucrose, raffinose has been revealed as the most preferred substrate by microbial β-fructofuranosidases with Km values ranging from 1.3 to 392 mM (Alvaro-Benito et al. 2007; Martel et al. 2010).

1.5.2 Substrate Specificity

1.5.2.1 Role of Active Site in Substrate Specificity

Substrate specificity of microbial FTs directly depends on the enzyme origin and reaction conditions. In general, the enzyme-substrate complexes conserve certain amino acid residues which are essential for the polymerization of fructan chains in microbes; however, the enzyme topology, mainly distribution of amino acids in the substrate-binding site and shape of the catalytic pocket, is the main factor affecting proper binding. Therefore, the binding ability of an enzyme to a certain substrate is prone to differentiate depending on the conditions of the reaction medium, the natural habitat of the organism, or the position and the type of amino acids involved in the catalysis.

Members of GH-J clan, levansucrase *Bs*SacB of *B. subtilis* and inulosucrase InuJ of *L. johnsonii*, share the highest sequence homology among GH68 enzymes; however, their catalytic site differs by size due to substitution of a lysine residue (K363) of *Bs*SacB by an arginine amino acid (R545) in InuJ making entrance of the catalytic pocket narrower for substrate binding, thus affecting the preference on fructosyl donors and acceptors in respective organisms (Pijning et al. 2011).

On the other hand, the catalytic triad is fully conserved among microbial FTs, located in the N-terminal of the enzyme and directly involved in enzyme-substrate binding. However, their activity is more directing and stabilizing the catalysis once the substrate binding occurs rather than recognizing the binding regions on the substrate. Similarly, subsite −1 (according to Davies et al. 1997 nomenclature) which is highly specific within microbial FTs (Ozimek et al. 2006; Chuankhayan et al. 2010), located at the very bottom of the catalytic funnel but different than catalytic triad, is directly involved in the recognition of substrate, thereby being accepted as the fructosyl-binding site for sucrose and sucrose analogues. Subsite −1 is the only sugar-binding donor site in microbial FTs where the first attachment between the donor substrate and the enzyme occurs (Davies et al. 1997; Ozimek et al. 2006). Further donor subsites (negative) in the active site are occupied upon attachment of fructosyl units of longer chain donor substrates to the enzyme. When sucrose is used as a substrate, fructosyl unit covalently binds to subsite −1 (fructosyl site) whereas glucose moiety attaches to subsite +1 (glucosyl site). In contrast to subsite −1, subsite +1 has an affinity for both fructose and glucose making it more flexible for substrate recognition; therefore, this site also plays a role in transglycosylation where sucrose functions as an acceptor substrate (Ozimek et al. 2006; van Hijum et al. 2006).

Pijning et al. (2011) characterized the binding site of inulosucrase, InuJ of *L. johnsonii* NCC533, and suggested several binding modes of the enzyme for production of inulin-type FOS and inulin polymers. The R424 residue of InuJ was proposed to be involved in the formation of short inulin chains by interacting with the general acid/base E524 through a salt bridge which serves as a border for substrate binding at subsite −1 and blocks further donor subsites (−2, −3, etc.) and thus prevents the binding of longer chain donor substrates. The same type of binding was also demonstrated in between E342 and R246 of levansucrase, SacB of *B. subtilis* (Ozimek et al. 2004), and a similar one also occurs in the amylosucrase of *Neisseria polysaccharea* (Albenne et al. 2002). Residues D272, D425, and E524 were validated as the nucleophile, transition-state stabilizer, and general acid/base, respectively, and importantly they have the same positions in both InuJ and *Bs*SacB (Ozimek et al. 2006; Pijning et al. 2011). Besides, inulosucrase and levansucrases have a tendency to use the same acceptor substrates such as sucrose, kestotriose, and raffinose suggesting the active site mimicry in between these enzymes. However, they discretely differ from each other in terms of glycosidic bond specificity (regioselectivity), preference for the synthesis of FOS over fructan polymers, and hydrolysis/transglycosylation ratio (Ozimek et al. 2006; Kralj et al. 2008; Meng and Fütterer 2008; Visnapuu et al. 2011). Studies on substitutions of amino acid residues near or next to subsite −1 revealed specific substrate specificities in respective FTs indicating the presence of unidentified further substrate-binding subsites (Ozimek et al. 2006).

The fructosyl unit of a donor substrate binds to the nucleophile of the enzyme at subsite −1 to be transferred to an acceptor molecule for transfructosylation activity to be initiated. Sequence analyses have validated the conservation of subsite −1 among microbial FTs while there have been different scenarios proposed for positive subsites. Subsite +1 is occupied by an arginine residue in Gram-positive SacB

(*B. subtilis*), Inu (*Lb. reuteri*), InuJ (*Lb. johnsonii*), and SacB$_{Sm}$ (*S. mutans*) FTs; however this region is replaced by another basic amino acid histidine in Gram-negative LsdA (*G. diazotrophicus*), SucE2 (*Z. mobilis*), and Lsc1 (*P. syringae*) proteins. Other than arginine, lysine and asparagine residues also function in acceptor substrate binding at subsite +1 of Gram-positive FTs whereas acceptor site of Gram-negative FTs is dominated by polar amino acids such as threonine next to histidine, enabling proper hydrogen bond formation with the acceptor substrate (Frasch et al. 2017). In another study, substitution of H321 residue, which was proposed to be present at the equivalent position with A360 and H296, the amino acids known to have a key function in levansucrase activity in *B. subtilis* and *Z. mobilis*, respectively (Chambert and Petit-Glatron 1991; Li et al. 2008), by alanine (H321A), lysine (H321K), leucine (H321L), and serine (H321S) in levansucrase of *P. syringae* pv. tomato DC3000 (Lsc3) was acclaimed to have a decrease in enzyme affinity for sucrose suggesting a role in substrate preference (Visnapuu et al. 2011).

The structure of a recombinant FTs (AjFT) from *A. japonicus* CB05 with its substrate has revealed several amino acid residues which are essential for substrate recognition among fungal FTs (Chuankhayan et al. 2010). D60, D191, and E292 have been demonstrated to constitute the catalytic triad of the enzyme functioning in the stabilization of terminal fructosyl moiety of the substrate. The fructosyl or glucosyl unit at the +1 donor-binding subsite is stabilized by the side chains of R190, E292, E318, and H332; the carbonyl oxygen of I143; and the amide nitrogen of T145, while the side chain of E405 and the carbonyl oxygen of T404 of the enzyme stabilized the binding of substrate to +2 subsite through hydrophobic interactions. These residues are also known to lead to the formation of the inulin-type FOS in fungal FTs. Likewise, I143 and Q327 residues orient the binding of galactosyl moiety of substrate raffinose at +2 subsite, creating a longer distance between the glucosyl of galactosyl moiety and the general acid/base catalyst E292 (Chuankhayan et al. 2010).

1.5.2.2 Donor and Acceptor Specificities Among Microorganisms

Sucrose is the most preferential substrate in the reactions catalyzed by microbial FTs; however, these enzymes display wide substrate specificity including monosaccharides, disaccharides, aromatic and aliphatic alcohols, heterooligosaccharides, and fructosides (Velázquez-Hernández et al. 2009; Li et al. 2015). Acceptor substrate specificity has been widely studied in levansucrases of *B. subtilis* (Seibel et al. 2006), *M. laevaniformans* (Park et al. 2012; Kim et al. 2005), and *Bacillus licheniformis* (Lu et al. 2014). These enzymes have a preference to use lactose as a fructosyl acceptor to produce lactosucrose and they are able to use maltose and cellobiose to form erlose and fructosylated cellobiose, respectively. Additionally, melibiose can be utilized as an acceptor by certain strains of *B. subtilis* and *M. laevaniformans* to biosynthesize raffinose (Li et al. 2015). *B. subtilis* levansucrase is able to produce sucrose analogues by using several monosaccharide acceptors including D-galactose, D-xylose, and D-fucose and has a lesser preference to D-mannose and D-allose (Seibel et al. 2006) while *B. licheniformis* displays a transfer efficiency towards L-arabinose as an acceptor (Lu et al. 2014). On the other hand, inulosucrase from

L. gasseri DSM 20604 can catalyze the synthesis of novel maltosylfructosides in the presence of both sucrose and maltose, using donor sucrose molecule to transfer the fructosyl unit to an acceptor maltose molecule, besides its ability to synthesize inulin-type FOS solely with sucrose (Díez-Municio et al. 2013).

Next to sucrose, sucrose analogues, especially raffinose, are also used as donor substrates by *M. laevaniformans* levansucrase suggesting the enzyme preference to oligosaccharides with terminal sucrose residues (Kim et al. 2005). In addition to *M. laevaniformans,* levansucrase of *Bacillus amyloliquefaciens* has been reported to use raffinose as a donor substrate to produce levan-type fructans (Tian et al. 2011).

Interestingly, levansucrase from *R. aquatilis* has been found to use methanol as a fructosyl acceptor to produce methyl β-D-fructosides (Kim et al. 2000). Similarly, isopropanol and 1-pentanol are known to be used as acceptor substrates by *B. licheniformis* for the production of alkyl fructosidases (Lu et al. 2014).

1.5.2.3 Cofactor Requirements in Substrate Preference

Microbial FTs of Gram-positive origin require metal cofactors such as Ca^{2+} for FT activity (Ozimek et al. 2005) with an exception of HugO, inulosucrase from *Streptomyces viridochromogenes* DSM40736 (Tü494) (Frasch et al. 2017). A single aspartate residue was firstly revealed to make the most important contribution to Ca^{2+} ion binding to microbial inulosucrases and levansucrases but this evidence is not verified in any FTs from Gram-negative bacteria and fungi, and bacterial invertases as well (Meng and Fütterer 2003). Residues N310, D339, D241, and Q272 have been observed to function in $CaCl_2$ formation in SacB from *B. subtilis* (Frasch et al. 2017). Residues D520 of inulosucrase and D500 of levansucrase from *L. reuteri* were found to be analogous to D339 of SacB possessing high affinity to Ca^{2+} ion. Site-directed mutations of D520N, D520A, D500N, and D500A in *L. reuteri* 121 resulted in a prominent reduction in FT activity by causing a conformational change in the highly conserved catalytic DEIER motif and general acid catalyst, thus disrupting substrate binding (Ozimek et al. 2005). These findings suggest the requirement of a metal ion for FT activity in FTs from Gram-positive bacteria making them dissimilar from other microbial FTs in terms of substrate specificity.

1.5.2.4 Specificity in Structure

Several studies have revealed the structural differences between microbial FTs describing the molecular flexibility of FOS (French 1988; Mensink et al. 2015). According to French (1988), inulin-type FOSs are relatively flexible compared to levan-type FOS in terms of formation of left-handed or right-handed helical structures; however levan-type FOSs are more rigid and usually have preference to form left-handed helixes. These structural differences are considered as the limiting factors for such molecules having the ability to be an acceptor substrate during fructosylation, since they likely affect the proper binding mode depending on the enzyme shape and surface properties (Pijning et al. 2011).

1.5.3 Reaction Specificity

1.5.3.1 Effect of pH and Ionic Strength

pH and ionic strength of the reaction medium are important parameters that affect reaction selectivity of microbial FTs towards polymerization, transfructosylation, or hydrolysis. Studies on the effects of pH in microbial FT production have pointed out the importance of microbial strain. The course of the reaction mainly depends on intrinsic properties of the microbial source; therefore, optimal requirements in each microbial reaction catalyzed by FTs vary greatly suggesting the importance of steric effects in the active site that influence the enzyme conformation. Levansucrase from *Z. mobilis* exists in two distinctly active forms at different pH levels. At pH values above 7.0, the enzyme occurs in its natural conformation as a dimer and catalyzes the hydrolysis of sucrose or the synthesis of levan FOS whereas at pH values below 6.0, the enzyme undergoes conformational changes, precipitates, and forms microfibrils. Surprisingly, in its microfibril form, *Z. mobilis* levansucrase has been revealed to synthesize levan polymers with DP above 20,000 (Goldman et al. 2008).

Ionic strength has also been shown to be the sole source that determines the molecular weight of synthesized fructan polymers. Addition of 0.5 M NaCl to the reaction medium was reported to cause a sixfold decrease in the molecular size of *B. licheniformis* levansucrase (Nakapong et al. 2013). Besides, increase in ionic strength at different stages of reaction has been proposed to promote the preference on transfructosylation over hydrolysis in some microbes (Trujillo et al. 2004; Castillo and López-Munguía 2004) although pH variation has not been reported to change the rate of hydrolase over transferase activity (Toksoy Oner et al. 2016). These variations in pH and ionic strength of the enzyme direct the synthesis of FOS over high DP levan polymers, thereby implicitly affecting the product specificity.

1.5.3.2 Effect of Temperature

Temperature has a direct effect on microbial fructan production not only to initiate the reaction at optimal conditions but also to sustain the catalysis in favor of products at the desired molecular size and branching level. *B. licheniformis* RN-01 levansucrase tends to synthesize low-molecular-weight levan polymers (11 kDa) when the reaction temperature is decreased to 30 °C from its optimum operation temperature of 50 °C (Nakapong et al. 2013). In a similar manner, polymerization can be dominant over hydrolysis or vice versa when the reaction temperature is optimized (lowered or increased) in a way that it is not naturally favored by the enzyme (Toksoy Oner et al. 2016). Normally, elevated temperatures positively affect the fate of fructosylation reactions due to stabilizing the activity of substrate sucrose on microbial FTs at high temperatures since the viscosity of sucrose solution decreases with increasing temperatures making their processing easier (Antošová and Polakovič 2001).

1.5.3.3 Effects of Substrate and Enzyme Concentrations

Substrate concentration exhibits as a strong determinant of product molecular weight as well (Abdel-Fattah et al. 2005). Generally, increased substrate concentrations favor hydrolysis while low substrate concentrations promote polymerization activity in bacterial FTs. Additionally, sucrose concentration has been reported to have an effect on proportions of synthesized fructan oligo- and polysaccharides. FOS production is normally promoted at high substrate concentrations (Li et al. 2015; Ni et al. 2018). However, the reverse is also possible and thermodynamically more favorable in several microbial FTs. β-Fructofuranosidases are able to produce FOS at high concentrations even in dilute substrate concentrations. On the other hand, the yields of FOS which are produced by invertases having high FT activity are usually low even in high substrate concentrations (Antošová and Polakovič 2001).

Another determinant of product molecular weight is enzyme concentrations. High-molecular-weight levan polymers were able to be produced in a processive manner when the reaction was catalyzed by a lower amount of levansucrase in *B. subtilis* (Raga-Carbajal et al. 2016).

1.5.3.4 Effects of Solvents

Low-molecular-weight alcohols have been reported to change the dielectric constant of the reaction medium preserving levansucrase activity (Steinberg et al. 2002). The activity of *B. subtilis* levansucrase was enhanced for the recovery of high-molecular-weight levan in the presence of ethanol, polyethylene glycol, or acetonitrile. When the enzyme was incubated with 70% acetonitrile, transfructosylation activity was reported to be favored over hydrolysis (Chambert and Petit-Glatron 1989).

1.5.3.5 Effect of Inhibitors

The productivity of microbial FTs is competitively inhibited by glucose molecules (Jung et al. 1989). Glucose mainly occurs as a nonterminal reducing residue in fructan chains, is formed as a by-product during transfructosylation, and has a great potential to decrease the overall activity. Therefore, free glucose in the reaction medium is always eliminated by oxidative agents in commercially fructan production (Ates 2015).

The activity of microbial FTs can be reduced or even inactivated by several cationic metal ions and some sulfhydryl reagents. Cu^{2+}, Zn^{2+}, Hg^{2+}, Fe^{2+}, Pb^{2+}, and Ag^{+} have been reported to non-competitively inhibit the microbial FTs causing a decrease in their activity (Antošová and Polakovič 2001). The existence of *p*-chloromercuribenzoate (Sangiliyandi et al. 1999) and *N*-bromosuccinimide (Chang et al. 1994) in the reaction medium has resulted in enzyme inactivation suggesting the presence of thiol groups at or near the active site, thus destroying the catalytic activity. Besides, β-mercaptoethanol and dithiothreitol were found to inhibit the activity of microbial β-fructofuranosidases proposing the existence of disulfide bridges in the active site (Muramatsu et al. 1993) while EDTA has been reported to enhance enzyme activity (Gines et al. 2000).

1.5.3.6 Effect of Reaction Time

The duration of reaction also exhibits as a relatively strong determinant of proportions of short- and long-chain fructan polymers synthesized by microbial FTs. The significant impact of a levansucrase was reported in *B. amyloliquefaciens*. According to this study, FOSs were shown to be synthesized after 36 h while long-chain levan polymers were synthesized at a very early stage of the reaction (Tian et al. 2011). The reverse scenario has been demonstrated in another study with *Acetobacter diazotrophicus* levansucrase (Hernandez et al. 1995).

Inulosucrase from *L. gasseri* DSM 20604 showed the maximum FOS production after 24 h in the presence of sucrose substrate (Díez-Municio et al. 2013). 1-Kestotriose and nystose were reported to be the sole products of *L. reuteri* 121 inulosucrase, after 17 h of reaction (van Hijum et al. 2002). Furthermore, inulosucrases from *L. johnsonii* NCC 533 and *L. gasseri* DSM 20143 have been revealed to synthesize both inulin and FOS with various DP values after long reaction times (Anwar et al. 2008, 2010).

1.6 Industrial Applications of Microbial Fructosyltransferases

Fructans produced by the action of microbial FTs hold great potential with unique properties to be used in various industrial applications (Fig. 1.2). This section summarizes the promising features of inulin and levan fructans produced by microbial FTs.

1.6.1 Applications of Inulin Produced by Microbial Inulosucrases

1.6.1.1 Food Industry

Inulin oligosaccharides are extensively used in the food industry as food ingredients functioning as fat, sugar, or flour substitutes with great health benefits (Stevens et al. 2001; Barclay et al. 2010). Inulin FOS polymers produced by certain microorganisms provide texture and stability to dairy products, moisture to baked foods, crispness to low-calorie cookies and reduce the freezing point of frozen desserts (Kaur and Gupta 2002).

β-2,1 Linkage character makes inulin indigestible by humans and other animals; however, it is digestible by gut microbiota supporting the growth of beneficial microorganisms while repressing the growth of pathogens in the digestive system. These properties make inulin as a unique substance to be used as a prebiotic dietary fiber in dairy products to support the growth of bifidobacteria providing relief from constipation especially in elderly people (Barclay et al. 2010). Inulin, therefore, functions as a relatively low-calorie sweetener due to its indigestibility.

Another property of inulin biopolymers that food industries benefit from is the solubility level at different molecular weights and branching degrees. The solubility of inulin is closely dependent on the chain length of the fructan polymer which is defined with the action of inulosucrases in appropriate reaction conditions. Food processing takes advantage of the solubility of shorter chain inulins which can gelate the aqueous solution imitating the fat molecules when they are used as bulking

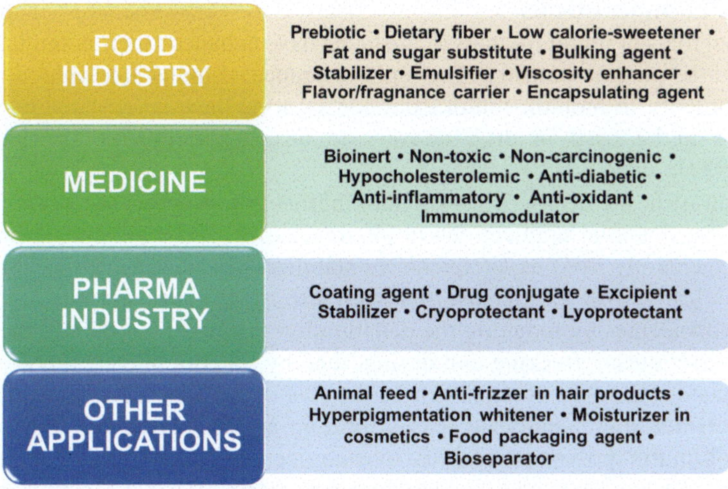

Fig. 1.2 Important features and potential applications of microbial fructosyltransferase products

agents (Blecker et al. 2001). Hence, along with its beneficial health effects, inulin is a promising agent in food processing improving the texture of several food products.

1.6.1.2 Medicine

Because inulosucrase is not naturally present in human physiology, presence of inulin polysaccharides in human plasma is only possible by uptake from external sources. This feature together with its non-toxicity and bioinertness has been used to measure the glomerular filtration rate (GFR) to monitor kidney functions which is usually called as "inulin clearance test" in medicine (Lascelles and Donaldson 1989).

Dietary inulin is known to decrease the risk of several cardiovascular diseases by reducing the triglyceride levels in serum. Inulin addition to daily diet has been investigated to cause a reduction in LDL cholesterol while there is an increase in HDL cholesterol (Russo et al. 2010). The risk of atherosclerosis was also reduced due to the lowered synthesis of fatty acids and triglycerides in the liver by inulin uptake (Kaur and Gupta 2002).

Inulin polymers have been reported to function as hormonal modifiers by regulating insulin and glucagon levels in plasma, thus playing an important role in both carbohydrate and lipid metabolisms (Roberfroid 1993; Luo et al. 1996). Furthermore, inulin was proposed to reduce the incidence of osteoporosis and colon cancer by enhancing Ca^{2+} absorption (Ohta et al. 1994) and suppressing the formation of early preneoplastic markers in potential malignant cells (Kulkarni and Reddy 1994; Wargovich et al. 1996), respectively. Additionally, inulin has been reported to possess immunomodulatory effects by inducing cell differentiation in intestine besides its function of enhancing gut flora, thus suggesting a role in tumor suppression in the colon (Roller et al. 2004; De Medina et al. 2010).

1.6.1.3 Pharmaceuticals

Functional roles of inulin fructans in medicine have enhanced their potential applications in the pharma industry. The rapid clearance of inulin from blood plasma and opportunity to monitor inulin excretion in urine have been stated as unique properties to be used in drug delivery to design meritorious drug vehicles (Fuchs 1987).

In contrast to the utility of short-chain inulins in food processing, long-chain inulins have potential applications in pharmaceuticals. Long-chain inulin polymers are generally used as excipients or stabilizers in pharmaceutical products (Barclay et al. 2010) to increase the shelf life of proteins, protein-containing drugs, and vaccines by reducing the denaturation rate and maintaining the structural integrity during freezing and drying, thereby functioning as both cryo- and lyoprotectants (Hinrichs et al. 2005). As the chain length increases, the solubility of the polymer significantly decreases forming a glassy structure. This unique feature of inulin polymers enables to stabilize the protein-based molecules, replacing the water molecules surrounding the protein by sugar moieties in polysaccharide chain diminishing the effects of denaturation during handling and storage (Hinrichs et al. 2001).

1.6.1.4 Animal Feed

Dietary inulin has been used as a low-dose antibiotic to promote the growth of rabbits. In an experiment 60 rabbits were treated with an antibiotic growth promoter and inulin. Rabbits fed on the inulin-based diet had increased bone and serum calcium, magnesium, and phosphorous levels with significantly lower blood glucose and triglyceride levels. Results revealed the possible growth promoter role of inulin biopolymer in rabbits suggesting the utilization of fructooligosaccharides as dietary fibers in animal diet as well (Montiel et al. 2013).

1.6.2 Applications of Levan Produced by Microbial Levansucrases

1.6.2.1 Food Industry

Similar to inulin, levan is a nontoxic and noncarcinogenic dietary fiber used in various processes as an emulsifier, a stabilizer, a viscosity enhancer, a flavor and/or a fragrance carrier, and an encapsulating agent (Han 1990; Toksoy Oner et al. 2016). Levan polymers synthesized by the action of levansucrases hold fatlike properties which have been suggested to replace the fat molecules in food products when they occur as levan phosphate (Roberts and Garegg 1998).

Besides, the high water-holding capability of microbial levan biopolymers improves rheological properties of foods and beverages. Levan-producing lactic acid bacteria are generally used in dairy products as starter cultures since microbial levans aid to improve the viscosity and prevent the whey separation during fermentation of milk products such as cheese and yogurt (Marshall and Rawson 1999).

Like inulin, levan has the potential to be used as a low-calorie sweetener in food products due to the absence of biosynthetic activity in the human body. As a low-calorie sweet-taste dietary fiber, microbial levan polymers have been suggested as ultrahigh-fructose syrups (UHFS) to replace the usage of sucrose in many foods and beverages (Han 1990).

1.6.2.2 Medicine

Levan polysaccharides are biocompatible biopolymers serving as antitumor agents by directly suppressing proliferation of tumor cells. However antitumor activity is restricted with a branching degree of the polymer. Studies with levans isolated from the cultures of *Gluconacetobacter xylinus*, *M. laevaniformans*, *R. aquatilis*, and *Z. mobilis* have shown that high-molecular-weight levans with enhanced DP values induce the effectiveness of the polymer against tumorigenesis (Yoo et al. 2004). Levan and levan analogs have been reported to be used as hypocholesterolemic agents to lower cholesterol levels and therefore, they are also considered as anti-obesity agents (Combie 2006). Levan reduces the level of leukocyte adhesion to blood vessels during inflammation promising a possible usage as an anti-inflammatory agent (Sedgwick et al. 1984). Likewise, microbial levan polymers have been proposed to achieve 81% antioxidant activity of ascorbic acid in a study with *A. xylinum* levan (Srikanth et al. 2015).

The antiproliferative activity of levan polymers produced by *Halomonas smyrnensis* AAD6^T has been indicated in MCF-7 breast cancer cells. Polymers have been revealed as a way of suppression of cancer mediated by an increase in apoptosis and oxidative stress (Queiroz et al. 2017).

Alongside immune-stimulating, antitumor, antidiabetic, and anti-inflammatory effects as many other polysaccharides, microbial levan polymers possess a unique property as metalloproteinase activators during tissue-healing process (Sturzoiu et al. 2011).

1.6.2.3 Pharmaceuticals

Due to their high water-holding capacity, microbial levans are preferred as coating agents in the delivery of essential nutrients such as selenium, cobalt, and iron, enabling safer travelling of the elements to the intestine while stabilizing their activity throughout the process (Bondarenko et al. 2016). Therefore they also serve as possible drug carriers of peptides and protein drug products that can be used in the pharma industry. A possible application has already been reported by Sezer et al. (2015) suggesting a levan-based microparticulate system for the delivery of vancomycin antibiotic (Sezer et al. 2015).

1.6.2.4 Cosmetics

Cosmeceutical properties of levan produced by *Z. mobilis* have been reported (Kim et al. 2006). Levan polymers are able to be used as excellent moisturizing agents possessing a transepidermal water loss (TEWL) which is almost equivalent to hyaluronic acid (Kim et al. 2005). Application of levan biopolymers has also been reported in beauty products such as shampoos, hair sprays, and moisturizers.

The main usage comprises film-forming ability of microbial levans enabling enhanced strength in damaged hair, and hair holding and anti-freezing properties (Gunn et al. 2009).

1.6.2.5 Bioseparation

Purification of several biological substances was achieved by a PEG/levan two-phase liquid system which operates based on a macroscopic phase separation. The system was designed to contain 60% of PEG and 6.77% of levan polymer enabling purification of bovine serum albumin, hen egg lysozyme, horse heart cytochrome c, horse heart hemoglobin, and horse heart myoglobin (Chung et al. 1997).

1.6.2.6 Packaging

Microbial levan is able to form an oxygen barrier when it is used as an ingredient in packaging films. This unique property makes them attractive for food packaging in a healthier and safer way. However, levan polymers are brittle for practical usage without the addition of plasticizers such as clay (Chen et al. 2014). Oxygen permeability has been reported in commercially available biofilms formulated with montmorillonite and 5% polyethylene glycol as lower than 0.05 cm^3/m^2 per day (Montana Polysaccharides 2015).

1.6.2.7 Aquaculture

An interesting study with aquatic species *Cyprinus carpio* (common carp) has pointed out the possible immune-enhancing role of levan polymers in aquatic animals. The study was conducted to explore the role of levan against a fish pathogen *Aeromonas hydrophila* (Rairakhwada et al. 2007). After a 75-day feeding period, 100% of *Cyprinus carpio* survivng on 0.5% levan diet were observed to survive. Another study with *Labeo rohita* (Hamilton) revealed a possible immunomodulatory function of levan diet in enhancing heat tolerance of aquatic species by causing an increase in heat-shock proteins in fish exposed to elevated temperatures (Gupta et al. 2010).

1.7 Conclusion

Microbes are versatile organisms with fascinating features applicable in many industries. They are the powerhouses of a variety of biopolymers providing convenience to our daily lives. Microbial enzymes accelerate the production of high-value substances directing practical utilization of microbe-based biopolymers. Among them, microbial FTs are one of the main players in today's health booster beneficial foods, disease-preventing agents, intelligent drugs, beauty tools, eco-friendly materials, and growth promoters in animals. As unique gifts of nature, fructan polymers have been attracting significant attention of researchers while opening new opportunities to the industry in different aspects. Their fast and simple biosynthesis from cheap carbon sources such as sugar beet molasses together with their structural redundancy enables ad hoc applications in many fields.

In light of these, understanding the mechanism of fructan biosynthesis and their possible roles in different processes has been an emerging area in since decades. In spite of ongoing projects and commercialized fructan products, structure-functional features of microbial FTs have not been fully understood. So far, microbial fructans and microbial fructan-producing enzymes were nominated by bioactive agents especially in human and animal health. As a given role in medicine, immunomodulatory effects of microbial fructans have been studied for their potential immune-enhancing functions in plants against several stress factors (Wang et al. 2009; Bolouri-Moghaddam and Van den Ende 2013; Van den Ende and El-Esawe 2014; Conrath et al. 2015; Ceusters et al. 2016). Very recently, evolutionary aspects and similarities in "fructan syndrome" between the microorganisms and plant worlds have been reviewed, pointing out to the possibility of a water scarcity-mediated fructan accumulation, or fructans playing a role in signaling events between plants and their host microbes (Versluys et al. 2018).

However, more efforts are required to overcome several process-related bottlenecks, such as optimizing the downstream processes to obtain biopolymers at desired purity levels. Studying fructan world and FTs is expected to enlighten hidden details regarding their synthesis, resulting in exploring for further futuristic industrial applications. As the information for fructans increases, their high-potential biotechnological applications are expected to enhance leading to sustainability of the fructan market and the industry.

References

Abdel-Fattah AF, Mahmoud DAR, Esawy MAT (2005) Production of levansucrase from *Bacillus subtilis* NCR 33A and enzyme synthesis of levan and fructo-oligosaccharides. Curr Microbiol 55:402–407

Albenne C, Skov LK, Mirza O, Gajhede M, Potocki-Véronèse G, Monsan P, Remaud-Simeon M (2002) Maltooligosaccharide disproportionation reaction: An intrinsic property of amylosucrase from *Neisseria polysaccharea*. FEBS Lett 527(1–3):67–70

Alvaro-Benito M, de Abreu M, Fernández-Arrojo L, Plou FJ, Jiménez-Barbero J, Ballesteros A, Polaina J, Fernández-Lobato M (2007) Characterization of a beta-fructofuranosidase from *Schwanniomyces occidentalis* with transfructosylating activity yielding the prebiotic 6-kestose. J Biotechnol 132:75–81

Alvaro-Benito M, Polo A, González B, Fernández-Lobato M, Sanz-Aparicio J (2010) Structural and kinetic analysis of *Schwanniomyces occidentalis* invertase reveals a new oligomerization pattern and the role of its supplementary domain in substrate binding. J Biol Chem 285:13930–13941

Ammar YB, MatsubaraTt IK, Iizuka M, Limpaseni T, Pongsawasdi P, Minamiura N (2002) Characterization of a thermostable levansucrase from *Bacillus* sp. TH4-2 capable of producing high molecular weight levan at high temperature. J Biotechnol 99(2):111–119

Antošová M, Polakovič M (2001) Fructosyltransferases: the enzymes catalyzing production of fructooligosaccharides. Chem Pap 55:350–358

Anwar MA, Kralj S, van der Maarel MJ, Dijkhuizen L (2008) The probiotic *Lactobacillus johnsonii* NCC 533 produces high-molecular-mass inulin from sucrose by using an inulosucrase enzyme. Appl Environ Microbiol 74:3426–3433

Anwar MA, Kralj S, Villar Piqué A, Leemhuis H, van der Maarel MJEC, Dijkhuizen L (2010) Inulin and levan synthesis by probiotic *Lactobacillus gasseri* strains: characterization of three novel fructansucrase enzymes and their fructan products. Microbiol 156:1264–1274

Apolinario AC, de Lima Damasceno BPG, de Macêdo Beltrão NE, Pessoa A, Converti A, da Silva JA (2014) Inulin-type fructans: A review on different aspects of biochemical and pharmaceutical technology. Carbohydr Polym 101:368–378

Arrieta JG, Sotolongo M, Menéndez C, Alfonso D, Trujillo LE, Soto M, Ramírez R, Hernández L (2004) A type II protein secretory pathway required for levansucrase secretion by *Gluconacetobacter diazotrophicus*. J Bacteriol 186:5031–5039

Ates O (2015) Systems biology of microbial exopolysaccharides production. Front Bioeng Biotechnol 3:200

Baciu IE, Jördening HJ, Seibel J, Buchholz K (2005) Investigations of the transfructosylation reaction by fructosyltransferase from *B. subtilis* NCIMB 11871 for the synthesis of the sucrose analogue galactosyl-fructoside. J Biotechnol 116(4):347–357

Bae IY, Oh IK, Lee S, Yoo SH, Lee HG (2008) Rheological characterization of levan polysaccharides from *Microbacterium laevaniformans*. Int J Biol Macromol 42:10–13

Bali V, Panesar PS, Bera MB, Panesar R (2015) Fructo-oligosaccharides: production, purification and potential applications. Cri Rev Food Sci Nutr 55(11):1475–1490

Banguela A, Hernández L (2006) Fructans: from natural sources to transgenic plants. Biotecnología Aplicada 23:202–210

Barclay T, Ginic-Markovic M, Cooper P, Petrovsky N (2010) Inulin—a versatile polysaccharide with multiple pharmaceutical and food chemical uses. J Excipients Food Chem 1(3):27–50

Barreteau H, Delattre C, Michaud P (2006) Production of oligosaccharides as promising new food additive generation. Food Technol Biotechnol 44:323–333

Bekers M, Laukevics J, Upite D, Kaminska E, Vigants A, Viesturs U, Pankova L, Danilevics A (2002) Fructooligosaccharide and levan producing activity of *Zymomonas mobilis* extracellular levansucrase. Process Biochem 38:701–706

Blecker C, Chevalier JP, Herck JC, van Fougnies C, Deroanne C, Paquot M (2001) Inulin: Its physicochemical properties and technological functionality. Recent Res Dev Agric Food Chem 5:125–131

Bolouri-Moghaddam MR, Van den Ende W (2013) Sweet immunity in the plant circadian regulatory network. J Exp Botany 64:1439–1449

Bondarenko OM, Ivask A, Kahru A, Vija H, Titma T, Visnapuu M, Joost U, Pudova K, Adamberg S, Visnapuu T, Alamäe T (2016) Bacterial polysaccharide levan as stabilizing, non-toxic and functional coating material for microelement-nanoparticles. Carbohydr Polym 136:710–720

Carpita NC, Kanabus J, Housley TL (1989) Linkage structure of fructans and fructan oligomers from *Triticum aestivum* and *Festuca arundinaceae* leaves. J Plant Physiol 134:162–168

Castillo E, López-Munguía A (2004) Synthesis of levan in water-miscible organic solvents. J Biotechnol 114:209–217

Cazetta ML, Martins PMM, Monti R, Contiero J (2005) Yacon (*Polymnia sonchifolia*) extract as a substrate to produce inulinase by *Kluyveromyces marxianus* var. *Bulgaricus*. J Food Eng 66(3):301–305

Ceusters N, Van den Ende W, Ceusters J (2016) Exploration of Sweet Immunity to Enhance Abiotic Stress Tolerance in Plants: Lessons from CAM. Prog. Botany 78:145–166

Chambert R, Petit-Glatron MF (1989) Study of the effect of organic solvents on the synthesis of levan and the hydrolysis of sucrose by *Bacillus subtilis* levansucrase. Carbohydr Res 191:117–123

Chambert R, Petit-Glatron MF (1991) Polymerase and hydrolase activities of *Bacillus subtilis* levansucrase can be separately modulated by site-directed mutagenesis. Biochem J 279:35–41

Chambert R, Petit-Glatron MF (1993) Immobilisation of levansucrase on calcium phosphate gel strongly increases its polymerase activity. Carbohydr Res 244:129–136

Chang CT, Lin YY, Tang MS, Lin CF (1994) Purification and properties of beta-fructofuranosidase from *Aspergillus oryzae* ATCC 76080. Biochem Mol Biol Int 32:269–277

Chen X, Gao H, Ploehn HJ (2014) Montmorillonite-levan nanocomposites with improved thermal and mechanical properties. Carbohydr Polym 101:565–573

Chen XY, Gänzle MG (2017) Lactose and lactose-derived oligosaccharides: More than prebiotics? Int J Dairy 67:61–72

Chiang CJ, Lee WC, Sheu DC, Duan KJ (1997) Immobilization of beta-fructofuranosidases from *Aspergillus* on methacrylamide-based polymeric beads for production of fructooligosaccharides. Biotechnol Prog 13(5):577–582

Chien CS, Lee WC, Lin TJ (2001) Immobilization of *A. japonicas* by entrapping cells in gluten for production of fructooligosaccharides. Enzym Microb Technol 29:252–257

Chuankhayan P, Hsieh CY, Huang YC, Hsieh YY, Guan HH, Hsieh YC, Tien YC, Chen CD, Chiang CM, Chen CJ (2010) Crystal structures of *Aspergillus japonicus* fructosyltransferase complex with donor/acceptor substrates reveal complete subsites in the active site for catalysis. J Biol Chem 285(30):23251–23264

Chung BH, Kim WK, Song KB, Kim CH, Rhee SK (1997) Novel polyethylene glycol/levan aqueous two-phase system for protein partitioning. Biotechnol Tech 11(5):327–329

Combie J (2006) Properties of levan and potential medical uses. Chapter: 13. Polysaccharides Drug Deliv Pharmaceut Appl:263–269

Conrath U, Beckers GJ, Langenbach CJ, Jaskiewicz MR (2015) Priming for enhanced defense. Annu Rev Phytopathol 53:97–119

Costa RR, Netol AI, Calgeris I, Correia CR, Pinho ACM, Fonseca J, Toksoy Oner E, Mano JF (2013) Adhesive nanostructured multilayer films using a bacterial exopolysaccharide for biomedical applications. J Mater Chem B 1:2367–2374

Cote GL, Ahlgren J (1993) Metabolism in microorganisms, Part I. Levan and levansucrase. In: Suzuki M, Chatterton NJ (eds) Science and technology of fructans. CRC Press, pp 141–168

Cruz R, Cruz VD, Belini MZ, Belote JG, Vieira CR (1998) Production of fructooligosaccharides by the mycelia of *Aspergillus japonicus* immobilized in calcium alginate. Bioresour Technol 65(1–2):139–143

Davies GJ, Wilson KS, Henrissat B (1997) Nomenclature for sugar-binding subsites in glycosyl-hydrolases. Biochem J 321:557–559

De Medina F, Martínez-Plata E, Romero I, Ortega M, Gonzalez R, Daddaoua AP, Martínez-Moya P, Zarzuelo A, Suárez MD, Martínez-Augustin O (2010) Non-absorbable glucids (active hexose correlated compound, inulin and fructooligosaccharides) exert immunomodulatory effects and induce differentiation in several intestinal cell types that are independent of their prebiotic actions. Proc Nutr Society 69(OCE3):E241

de Paula VC, Pinheiro IO, Lopes CE, Calazans GMT (2008) Microwave-assisted hydrolysis of *Zymomonas mobilis* levan envisaging oligofructan production. Bioresour Technol 99(7):2466–2470

Del Moral S, Olvera C, Rodriguez ME, Munguia AL (2008) Functional role of the additional domains in inulosucrase (IslA) from *Leuconostoc citreum* CW28. BMC Biochem 9:6–16

Díez-Municio M, de las Rivas B, Jimeno ML, Muñoz R, Moreno FJ, Herrero M (2013) Enzymatic synthesis and characterization of fructooligosaccharides and novel maltosylfructosides by inulosucrase from *Lactobacillus gasseri* DSM 20604. Appl Environ Microbiol 79(13):4129–4140

Dilipkumar M, Rajasimman M, Rajamohan N (2011) Optimization of inulinase production from garlic by *Streptomyces sp.* in solid state fermentation using statistical designs. Biotechnol Res Int:1–7

El-Beih FM, Abdel-Fattah AM, Hasanein DA, Mostafa FA, Abdel-Fatta AF (2009) Production and some properties of fructosyltransferase from *Bacillus cereus*. J Appl Sci Res 5:1132–1141

Esawy MA, Mahmoud DAR, Fattah AFA (2008) Immobilisation of *Bacillus subtilis* NRC33a levansucrase and some studies on its properties. Braz J Chem Eng 25:237–246

Feng J, Gu Y, Han L, Bi K, Quan Y, Yang C, Zhang W, Cao M, Wang S, Gao W, Sun Y, Song C (2015) Construction of a *Bacillus amyloliquefaciens* strain for high purity levan production. FEMS Microbiol Lett 362:1–7

Frasch HJ, van Leeuwen SS, Dijkhuizen L (2017) Molecular and biochemical characteristics of the inulosucrase HugO from *Streptomyces viridochromogenes* DSM40736 (Tü494). Microbiol 163:1030–1041

French AD (1988) Accessible conformations of the beta-D(2→1)-linked and beta-(2→6)-linked D-fructans inulin and levan. Carbohydr Res 176:17–29

Fuchs A (1987) Potentials for non-food utilization of fructose and inulin. Starch-Starke 39:335–343

Ganaie MA, Lateef A, Gupta US (2014) Enzymatic trends of fructooligosaccharides production by microorganisms. Appl Biochem Biotechnol 172(4):2143–2159

Ghazi I, de Segura AG, Fernández-Arrojo L, Alcalde M, Yates M, Rojas-Cervantes ML, Plou FJ, Ballesteros A (2005) Immobilisation of fructosyltransferase from *Aspergillus aculeatus* on epoxy-activated Sepabeads EC for the synthesis of fructo-oligosaccharides. J Mol Catal B Enzymatic 35(1–3):19–27

Gines SCD, Maldonado MC, Valdez GFD (2000) Purification and characterization of invertase from *Lactobacillus reuteri* CRL 1100. Curr Microbiol 40:181–184

Goldman D, Lavid N, Schwartz A, Shoham G, Danino D, Shoham Y (2008) Two active forms of *Zymomonas mobilis* levansucrase. An ordered microfibril structure of the enzyme promotes levan polymerization. J Biol Chem 283(47):32209–32217

Gunn E, Gabbianelli A, Crooks R, Shanmuganandamurthy K (2009) Derivatized polysaccharide polymer. US Patent 7,528,100, 5 May 2009

Gupta SK, Pal AK, Sahu NP, Dalvi RS, Akhtar MS, Jha AK, Baruah K (2010) Dietary microbial levan enhances tolerance of *Labeo rohita* (Hamilton) juveniles to thermal stress. Aquaculture 306:398–402

Han J, Xu X, Gao C, Liu Z, Wu Z (2016) Levan-producing *Leuconostoc citreum* strain BD1707 and its growth in tomato juice supplemented with sucrose. Appl Environ Microbiol 82(5):1383–1390

Han WC, Byun SH, Kim MH, Sohn EH, Lim JD, Um BH, Kim CH, Kang SA Jang KH (2009) Production of lactosucrose from sucrose and lactose by a levansucrase from *Zymomonas mobilis*. J Microbiol Biotechnol 19:1153–1160

Han YW (1990) Microbial levan. Adv Appl Microbiol 35:171–194

Han YW, Watson MA (1992) Production of microbial levan from sucrose, sugarcane juice and beet molasses. J Ind Microbiol 9:257–260

Han YW, Clarke MA (1990) Production and characterization of microbial levan. J Agric Food Chem 38:393–396

Hayashi S, Ito K, Nonoguchi M, Takasaki Y, Imada K (1991) Immobilization of a fructosyltransferring enzyme from *Aureobasidium sp.* on shirasu porous glass. J Ferment Bioeng 72(1):68–70

Hendry GAF (1993) Evolutionary origins and natural functions of fructans—a climatological, biogeographic and mechanistic appraisal. New Phytol J 123(1):3–14

Hernandez L, Arrieta J, Menendez C, Vazquez R, Coego A, Suarez V, Selman G, Petit-Glatron MF, Chambert R (1995) Isolation and enzymic properties of levansucrase secreted by *Acetobacter diazotrophicus* SRT4, a bacterium associated with sugar cane. Biochem J 309:113–118

Hettwer U, Gross M, Rudolph K (1995) Purification and characterization of an extracellular levansucrase from *Pseudomonas syringae* pv. *phaseolicola*. J Bacteriol 177:2834–2839

Heyer AG, Schroeer B, Radosta S, Wolff D, Czapla S, Springer J (1998) Structure of the enzymatically synthesized fructan inulin. Carbohydr Res 313:165–174

Hinrich WLJ, Sanders NN, De Smedt SC, Demeester J, Frijlink HW (2005) Inulin is a promising cryo- and lyoprotectant for PEGylated lipoplexes. J Control Release 103:465–479

Hinrichs WL, Prinsen MG, Frijlink HW (2001) Inulin glasses for the stabilization of therapeutic proteins. Int J Pharm 215(1–2):163–174

Homann A, Biedendieck R, Götze S, Jahn D, Seibel J (2007) Insights into polymer versus oligosaccharide synthesis: mutagenesis and mechanistic studies of a novel levansucrase from *Bacillus megaterium*. Biochem J 407:189–198

Jang KH, Song KB, Park BS, Kim CH, Chung BH, Choue RW, Lee KS, Lee C, Chun UH, Rhee SK (2001) Levan production by use of the recombinant levansucrase immobilized on titanium-activated magnetite. Process Biochem 37:339–343

Jung KH, Bang SH, Oh TK, Park HJ (2011) Industrial production of fructooligosaccharides by immobilized cells of *Aureobasidium pullulans* in a packed bed reactor. Biotechnol Lett 33(8):1621–1624

Jung KH, Yun JW, Kang KR, Lim JY, Lee JH (1989) Mathematical model for enzymatic production of fructo-oligosaccharides from sucrose. Enzym Microb Technol 11(8):491–494

Kang HK, Seo MY, Seo ES, Kim D, Chung SY, Kimura A, Day DF, Robyt JF (2005) Cloning and expression of levansucrase from *Leuconostoc mesenteroides* B-512 FMC in *Escherichia coli*. Biochim Biophys Acta 1727:5–15

Kaur N, Gupta AK (2002) Applications of inulin and oligofructose in health and nutrition. J Biosci 27(7):703–714

Kazak Sarilmiser H, Ates O, Ozdemir G, Arga KY, Toksoy Oner E (2015) Effective stimulating factors for microbial levan production by *Halomonas smyrnensis* AAD6T. J Biosci Bioeng 119:455–463

KazakSarilmiser H, Toksoy Oner E (2014) Investigation of anti-cancer activity of linear and aldehyde-activated levan from *Halomonas smyrnensis* AAD6T. Biochem Eng J 92:28–34

Kim KH, Chung CB, Kim YH, Kim KS, Han CS, Kim CH (2005) Cosmeceutical properties of levan produced by *Zymomonas mobilis*. J Cosmet Sci 56:395–406

Kim KH, Chung CB, Kim YH, Kim KS, Han CS, Kim CH (2006) Cosmeceutical properties of levan produced by *Zymomonas mobilis*. J Cosmet Sci 56(6):395–406

Kim MG, Kim CH, Lee JS, Song KB, Rhee SK (2000) Synthesis of methyl β-D-fructoside catalyzed by levansucrase from *Rahnella aquatilis*. Enzym Microb Technol 27:646–651

Kralj S, Buchholz K, Dijkhuizen L, Seibel J (2008) Fructansucrase enzymes and sucrose analogues: a new approach for the synthesis of unique fructo-oligosaccharides. Biocatal Biotransformation 26(1-2):32–41

Kralj S, Leeflang C, Sierra EI, Kempiński B, Alkan V, Kolkman M (2018) Synthesis of fructooligosaccharides (FosA) and inulin (InuO) by GH68 fructosyltransferases from *Bacillus agaradhaerens* strain WDG185. Carbohydr Poly 179:350–359

Kritzinger SM, Kilian SG, Potgieter MA, du Preez JC (2003) The effect of production parameters on the synthesis of the prebiotic trisaccharide, neokestose, by *Xanthophyllomyces dendrorhous* (*Phaffia rhodozyma*). Enzym Microb Technol 32(6):728–737

Küçükkaşık F, Kazak H, Güney D, Finore I, Poli A, Yenigün O, Nicolaus B, Toksoy Oner E (2011) Molasses as fermentation substrate for levan production by *Halomonas* sp. Appl Microbiol Biotechnol 89:1729–1740

Kulkarni N, Reddy BS (1994) Inhibitory effect of *Bifidobacterium longum* cultures on the azoxymethane-induced aberrant crypt loci formation and fecal bacterial β-glucuronidase. Proc Soc Exp Biol Med 207:278–283

Kurakake M, Ogawa K, Sugie M, Takemura A, Sugiura K, Komaki T (2008) Two types of beta-fructofuranosidases from *Aspergillus oryzae*. KB. J Agric Food Chem 56(2):591–596

Lammens W, Le Roy K, Schroeven L, Van Laere A, Rabijns A, Van den Ende W (2009) Structural insights into glycoside hydrolase family 32 and 68 enzymes: functional implications. J Exp Botany 60(3):727–740

Lammens W, Le Roy K, Yuan S, Vergauwen R, Rabijns A, Van Laere A, Strelkov SV, Van den Ende W (2012) Crystal structure of 6-SST/6-SFT from *Pachysandra terminalis*, a plant fructan biosynthesizing enzyme in complex with its acceptor substrate 6-kestose. Plant J Cell Mol Biol 70(2):205–219

Lascelles PT, Donaldson D (1989) Inulin clearance test. In: Diagnostic function tests in chemical pathology. Springer, Dordrecht, pp 96–97

Lateef A, Oloke JK, Prapulla SG (2007) The effect of ultrasonication on the release of fructosyltransferase from *Aureobasidium pullulans* CFR 77. Enzym Microb Technol 40(5):1067–1070

Le Roy K, Lammens W, Verhaest M, De Coninck B, Rabijns A, Van Laere A, Van den Ende W (2007) Unraveling the difference between invertases and fructan exohydrolases: a single

amino acid (Asp-239) substitution transforms Arabidopsis cell wall invertase1 into a fructan 1-exohydrolase. Plant Physiol 145(3):616–625
Li SY, Chen M, Li G, Yan YL, Yu HY, Zhan YH, Peng ZX, Wang J, Lin M (2008) Amino acid substitutions of His296 alter the catalytic properties of *Zymomonas mobilis* 10232 levansucrase. Acta Biochim Pol 55:201–206
Li W, Yu S, Zhang T, Jiang B, Mu W (2015) Recent novel applications of levansucrases. Appl Microbiol Biotechnol 99(17):6959–6969
Lin TJ, Lee YC (2008) High-content fructooligosaccharides production using two immobilized microorganisms in an internal-loop airlift bioreactor. J Chin Inst Chem Eng 39(3):211–217
Lincoln L, More SS (2017) Bacterial invertases: Occurrence, production, biochemical characterization, and significance of transfructosylation. J Basic Microbiol 57(10):803–813
Livingston DP, Henson CA (1998) Apoplastic sugars, fructans, fructan exohydrolase, and invertase in winter oat: responses to second-phase cold hardening. Plant Physiol 116(1):403–408
Livingston DP III, Hincha DK, Heyer AG (2009) Fructan and its relationship to abiotic stress tolerance in plants. Cell Mol Life Sci 66:2007–2023
López MG, Mancilla-Margalli NA, Mendoz-Diaz G (2003) Molecular structures of fructans from *Agave tequilana* Weber var. azul. J Agric Food Chem 51:7835–7840
Lu L, Fu F, Zhao R, Jin L, He C, Xu L, Xiao M (2014) A recombinant levansucrase from *Bacillus licheniformis* 8-37-0-1 catalyzes versatile transfructosylation reactions. Process Biochem 49:1503–1510
Luo J, Rizkalla SW, Alamovitch C, Boussairi A, Blayo A (1996) Chronic consumption of short chain fructooligosaccharides by healthy subjects decreased basal hepatic glucose production but had no effect on insulin-stimulated glucose metabolism. Am J Clin Nutr 63:639–645
Maiorano AE, Piccoli RM, da Silva ES, de Andrade Rodrigues MF (2008) Microbial production of fructosyltransferases for synthesis of pre-biotics. Biotechnol Lett 30:1867–1877
Markosian AA, Abelian LA, Adamian MO, Ekazhev ZD, Akopian ZI, Abelian VA (2007) Production of fructooligosaccharide syrup from sucrose in combination with palatinose and trehalose. Prikl Biokhim Mikrobiol 43(4):424–431
Marshall VM, Rawson HL (1999) Effects of exopolysaccharide producing strains of thermophilic lactic acid bacteria on the texture of stirred yoghurt. Int J Food Sci Technol 34:137–143
Martel CM, Warrilow AG, Jackson CJ, Mullins JG, Togawa RC, Parker JE, Morris MS, Donnison IS, Kelly DE, Kelly SL (2010) Expression, purification and use of the soluble domain of *Lactobacillus paracasei* beta-fructosidase to optimise production of bioethanol from grass fructans. Bioresour Technol 101:4395–4402
Martínez-Fleites C, Ortiz-Lombardia M, Pons T, Tarbouriech N, Taylor EJ, Arrieta JG, Hernández L, Davies GJ (2005) Crystal structure of levansucrase from the gram-negative bacterium *Gluconacetobacter diazotrophicus*. Biochem J 390:19–27
Martínez-Padilla LP (2017) Rheology of recent vegetal-based biopolymers. In: Masuelli M, Renard D (eds) Advances in physicochemical properties of biopolymers. Bentham Science Publishers, pp 184–207
Maugeri F, Hernalsteens S (2007) Screening of yeast strain for fructosylating activity. J Mol Catal B Enzym 49(1–6):43–49
Menéndez C, Hernández L, Banguela A, País J (2004) Functional production and secretion of the *Gluconacetobacter diazotrophicus* fructose-releasing exo-levanase (LsdB) in *Pichia pastoris*. Enzym Microb Technol 34(5):446–452
Meng G, Fütterer K (2008) Donor substrate recognition in the raffinose-bound E342A mutant of fructosyltransferase *Bacillus subtilis* levansucrase. BMC Struct Biol 88(16)
Meng GY, Fütterer K (2003) Structural framework of fructosyl transfer in *Bacillus subtilis* levansucrase. Nat Struct Biol 10:935–941
Mensink MA, Frijlink HW, van der Voort Maarschalk K, Hinrichs WL (2015) Inulin, a flexible oligosaccharide I: review of its physicochemical characteristics. Carbohydr Polym 130:405–419
Miranda-Molina A, Castillo E, López-Munguía A (2017) A novel two-step enzymatic synthesis of blastose, a β-d-fructofuranosyl-(2↔6)-d-glucopyranose sucrose analogue. Food Chem 227:202–210

Molinari ML, Boiardi JL (2013) Levans production by *Gluconacetobacter diazotrophicus*. Electron J Biotechnol 16(3):1–7

Montana Polysaccharides Corp (2015) Your source of levan. (Available from URL: www.polysaccharides.us)

Montiel RS, Acosta IT, Delgado EV, Juarez-Silva ME, Azaola A, Perez-Gil Romo F (2013) Inulin as a growth promoter in diets for rabbits. R Bras Zootec 42(12):885–891

Moosavi-Nasab M, Layegh B, Aminlari L, Hashemi MB (2010) Microbial production of levan using date syrup and investigation of its properties. WASET Int J Biol Biomol Agric Food Biotechnol Eng 4:603–609

Murakami H, Muroi H, Kuramoto T, Tamura Y, Mizutani K, Nakano H, Kitahata S (1990) Purification and some properties of a levanase from *Streptomyces sp*. No. 7-3. Agric Biol Chem 54(9):2247–2255

Muramatsu K, Onodera S, Kikuchi M, Shiomi N (1993) Purification and some properties of β-fructofuranosidase from *Bifidobacterium adolescentis* G1. Biosci Biotechnol Biochem 57:1681–1685

Mussatto SI, Aguilar CN, Rodrigues LR, Teixeira JA (2009b) Fructooligosaccharides and β-fructofuranosidase production by *Aspergillus japonicus* immobilized on lignocellulosic materials. J Mol Cat B Enzymatic 59(1–3):76–81

Mussatto SI, Rodrigues LR, Teixeira JA (2009a) Beta-fructofuranosidase production by repeated batch fermentation with immobilized *Aspergillus japonicus*. J Ind Microbiol Biotechnol 36(7):923–928

Mussatto SI, Teixeira JA (2010) Increase in the fructooligosaccharides yield and productivity by solid-state fermentation with *Aspergillus japonicus* using agro-industrial residues as support and nutrient source. Biochem Eng J 53(1):154–157

Nagaya M, Kimura M, Gozu Y, Sato S, Hirano K, Tochio T, Nishikawa A, Tonozuka T (2017) Crystal structure of a beta-fructofuranosidase with high transfructosylation activity from *Aspergillus kawachii*. Biosci Biotechnol Biochem 81:1786–1795

Naidoo K, Kumar A, Sharma V, Permaul K, Singh S (2015) purification and characterization of an endoinulinase from *Xanthomonas campestris* pv. *phaseoli* KM 24 mutant. Food Technol Biotechnol 53(2):146–153

Nakapong S, Pichyangkura R, Ito K, Iizuka M, Pongsawasdi P (2013) High expression level of levansucrase from *Bacillus licheniformis* RN-01 and synthesis of levan nanoparticles. Int J Biol Macromol 54:30–36

Naumoff DG (2001) Beta-fructosidase superfamily: Homology with some alpha-L-arabinases and beta-D-xylosidases. Proteins 42:66–76

Ni D, Zhu Y, Xu W, Bai Y, Zhang T, Mu W (2018) Biosynthesis of inulin from sucrose using inulosucrase from *Lactobacillus gasseri* DSM 20604. Int J Biol Macromol 109:1209–1218

Ning Y, Wang J, Chen J, Yang N, Jin Z, Xu X (2010) Production of neo-fructooligosaccharides using free-whole-cell biotransformation by *Xanthophyllomyces dendrorhous*. Bioresour Technol 101(19):7472–7478

Nobre C, Teixeira JA, Rodrigues LR (2015) New trends and technological challenges in the industrial production and purification of fructooligosaccharides. Cri Rev Food Sci Nutr 55(10):1444–1455

Ohta A, Ohtsuki M, Takizawa T, Inaba H, Adachi T, Kimura S (1994) Effects of fructooligosaccharides on the absorption of magnesium and calcium by cecectomized rats. Int J Vitam Nutr Res 64:316–323

Ohtsuka K, Hino S, Fukushima T, Osawa O, Kanematsu T, Uchida T (1992) Characterization of levansucrase from *Rahnella aquatilis* JCM-1683. Biosci Biotechnol Biochem 56(9):1373–1377

Olivares-Illana V, López-Munguía A, Olvera C (2003) Molecular characterization of inulosucrase from *Leuconostoc citreum*: a fructosyltransferase within a glucosyltransferase. J Bacteriol 185(12):3606–3612

Ortiz-Soto ME, Olivares-Illana V, López-Munguía A (2004) Biochemical properties of inulosucrase from *Leuconostoc citreum* CW28 used for inulin synthesis. Biocatal Biotransformation 22:275–281

Ozimek LK, Euverink GJ, van der Maarel MJ, Dijkhuizen L (2005) Mutational analysis of the role of calcium ions in the *Lactobacillus reuteri* strain 121 fructosyltransferase (levansucrase and inulosucrase) enzymes. FEBS Lett 579:1124–1128

Ozimek LK, Kralj S, van der Maarel MJEC, Dijkhuizen L (2006) The levansucrase and inulosucrase enzymes of *Lactobacillus reuteri* 121 catalyse processive and nonprocessive transglycosylation reactions. Microbiol 152:1187–1196

Ozimek LK, van Hijum SA, van Koningsveld GA, van Der Maarel MJ, van Geel-Schutten GH, Dijkhuizen L (2004) Site-directed mutagenesis study of the three catalytic residues of the fructosyltransferases of *Lactobacillus reuteri*121. FEBS Lett 560(1–3):131–133

Pabst MJ (1977) Levan and levansucrase of *Actinomyces viscosus*. Infect Immun 15:518–526

Pabst MJ, Cisar JO, Trummel CL (1979) The cell wall-associated levansucrase of *Actinomyces viscosus*. Biochim Biophys Acta 566:274–282

Palcic MM (1999) Biocatalytic synthesis of oligosaccharides. Curr Opin Biotechnol 10:616–624

Park MC, Lim JS, Kim JC, Park SW, Kim SW (2005) Continuous production of neo-fructooligosaccharides by immobilization of whole cells of *Penicillium citrinum*. Biotechnol Lett 27(2):127–130

Park J, Kim MI, Park YD, Shin I, Cha J, Kim CH, Rhee S (2012) Structural and functional basis for substrate specificity and catalysis of levan fructotransferase. J Biol Chem 287(37):31233–31241

Penadés S, Davis BG, Seeberger PH (2017) Chapter 58: glycans in nanotechnology. In: Varki A, Cummings RD, Esko JD (eds) Essentials of glycobiology, 3rd edn. Cold Spring Harbor Laboratory Press, Cold Spring Harbor, NY

Peshev D, Van den Ende W (2014) Fructans: prebiotics and immunomodulators. J Funct Foods 8(C):348–357

Pijning T, Anwar MA, Boger M, Dobruchowska JM, Leemhuis H, Kralj S, Dijkhuizen L, Dijkstra BW (2011) Crystal structure of inulosucrase from *Lactobacillus*: Insights into the substrate specificity and product specificity of GH68 fructansucrases. J Mol Biol 412:80–93

Poli A, Kazak H, Gürleyendağ B, Tommonaro G, Pieretti G, Toksoy Oner E, Nicolaus B (2009) High level synthesis of levan by a novel *Halomonas* species growing on defined media. Carbohydr Polym 78:651–657

Pouyez J, Mayard A, Vandamme AM, Roussel G, Perpete EA, Wouters J, Housen I, Michaux C (2012) First crystal structure of an endo-inulinase, INU2, from *Aspergillus ficuum*: Discovery of an extra-pocket in the catalytic domain responsible for its endo-activity. Biochimie 94:2423–2430

Prapulla SG, Subhaprada V, Karanth NG (2000) Microbial production of oligosaccharides: A review. Adv Appl Microbiol 47:299–343

Prata MB, Mussatto SI, Rodrigues LR, Teixeira JA (2010) Fructooligosaccharide production by *Penicillium expansum*. Biotechnol Lett 32(6):837–840

Queiroz EAIF, Fortes ZB, da Cunha MAA, Sarilmiser HK, Barbosa Dekker AM, Öner ET, Dekker RFH, Khaper N (2017) Levan promotes antiproliferative and pro-apoptotic effects in MCF-7 breast cancer cells mediated by oxidative stress. Int J Biol Macromol 102:565–570

Raga-Carbajal E, Carrillo-Nava E, Costas M, Porras-Dominguez J, López-Munguía A, Olvera C (2016) Size product modulation by enzyme concentration reveals two distinct levan elongation mechanisms in *Bacillus subtilis* levansucrase. Glycobiology 26(4):377–385

Rairakhwada D, Pal AK, Bhathena ZP, Sahu NP, Jha A, Mukherjee SC (2007) Dietary microbial levan enhances cellular non-specific immunity and survival of common carp (*Cyprinus carpio*) juveniles. Fish Shellfish Immunol 22:477–486

Ramírez-Escudero M, Gimeno-Pérez M, González B, Linde D, Merdzo Z, Fernández-Lobato M, Sanz-Aparicio J (2016) Structural analysis of beta-fructofuranosidase from *Xanthophyllomyces dendrorhous* reveals unique features and the crucial role of N-glycosylation in oligomerization and activity. J Biol Chem 291:6843

Ritsema T, Hernández L, Verhaar A, Altenbach D, Boller T, Wiemken A, Smeekens S (2006) Developing fructan-synthesizing capability in a plant invertase via mutations in the sucrose-binding box. Plant J 48:228–237

Ritsema T, Smeekens SC (2003) Engineering fructan metabolism in plants. J Plant Physiol 160(7):811–820

Roberfroid MB (1993) Dietary fibre, inulin and oligofructose: a review comparing their physiological effects. Crit Rev Food Sci Technol 33:103–148

Roberts EJ, Garegg PJ (1998) Levan derivatives, their preparation, composition and applications including medical and food applications. World patent appl. WO9803184. 29 Jan 1998

Roller M, Rechkemmer G, Watzl B (2004) Prebiotic inulin enriched with oligofructose in combination with the probiotics *Lactobacillus rhamnosus* and *Bifidobacterium lactis* modulates intestinal immune functions in rats. J Nutr 134:153

Russo F, Riezzo G, Chiloiro M, De Michele G, Chimienti G, Marconi E, D'Attoma B, Linsalata M, Clemente C (2010) Metabolic effects of a diet with inulin-enriched pasta in healthy young volunteers. Curr Pharma Design 16:825–831

Sainz-Polo MA, Ramírez-Escudero M, Lafraya A, González B, Marín-Navarro J, Polaina J, Sanz-Aparicio J (2013) Three-dimensional structure of *Saccharomyces* invertase: role of a non-catalytic domain in oligomerization and substrate specificity. J Biol Chem 288:9755–9766

Sangeetha PT, Ramesh MN, Prapulla SG (2004) Production of fructooligosaccharides by fructosyltransferase from *Aspergillus oryzea* CFR 202 and *Aureobasidium pullulans* CFR 77. Process Biochem 39:755–760

Sangiliyandi G, Raj GK, Gunasekaran P (1999) Elevated temperature and chemical modification selectively abolishes levan forming activity of levansucrase of *Zymomonas mobilis*. Biotechnol Lett 21:179

Sedgwick AD, Rutman A, Sin YM, Mackay AR, Willoughby DA (1984) The effects of levan on the acute inflammatory response. Br J Exp Pathol 65:215–222

Seibel J, Moraru R, Götze S, Buchholz K, Naamnieh S, Pawlowski A, Hecht HJ (2006) Synthesis of sucrose analogues and the mechanism of action of *Bacillus subtilis* fructosyltransferase (levansucrase). Carbohydr Res 341:2335–2349

Sezer AD, Kazak Sarilmiser H, Rayaman E, Cevikbas A, Toksoy Oner E, Akbuga J (2015) Development and characterization of vancomycin-loaded levan-based microparticular system for drug delivery. Pharm Dev Technol 22(5):627–634

Sheu DC, Duan KJ, Cheng CY, Bi JL, Chen JY (2002) Continuous production of high-content fructooligosaccharides by a complex cell system. Biotechnol Prog 18(6):1282–1286

Shih IL, Chen LD, Wang TC, Wu JY, Liaw KS (2010) Tandem production of levan and ethanol by microbial fermentation. Green Chem 12:1242–1247

Shih IL, Yu YT, Shieh CJ, Hsieh CY (2005) Selective production and characterization of levan by *Bacillus subtilis* (Natto) Takahashi. J Agric Food Chem 53:8211–8215

Shin HT, Baig SY, Lee SW, Suh DS, Kwon ST, Lim YB, Lee JH (2004) Production of fructo-oligosaccharides from molasses by *Aureobasidium pullulans* cells. Bioresour Technol 93(1):59–62

Song KB, Seo JW, Kim MG, Rhee SK (1998) Levansucrase of *Rahnella aquatilis* ATCC 33071: gene cloning, expression, and levan formation. Ann N Y Acad Sci 864:506–511

Srikanth R, Siddartha G, Sundhar Reddy CH, Harish BS, Janaki Ramaiah M, Uppuluri KB (2015) Antioxidant and anti-inflammatory levan produced from *Acetobacter xylinum* NCIM 2526 and its statistical optimization. Carbohydr Polym 123:8–16

Steinberg D, Rozen R, Bromshteym M, Zaks B, Gedalia I, Bachrach G (2002) Regulation of fructosyltransferase activity by carbohydrates, in solution and immobilized on hydroxyapatite surfaces. Carbohydr Res 337(8):701–710

Stern R, Jedrzejas MJ (2008) Carbohydrate polymers at the center of life's origins: the importance of molecular processivity. Chem Rev 108(12):5061–5085

Stevens CV, Meriggi A, Booten K (2001) Chemical modification of inulin, a valuable renewable resource and its industrial applications. Biomacromolecules 2:1–16

Strube CP, Homann A, Gamer M, Jahn D, Seibel J, Heinz DW (2011) Polysaccharide synthesis of the levansucrase SacB from *Bacillus megaterium* is controlled by distinct surface motifs. J Biol Chem 286:17593–17600

Sturzoiu C, Petrescu M, Galateanu B, Anton M, Nica C, Simionca S, Симйонка C, Iuri (Iu.) Y, Iurii G, Юрий DA, Stoian G (2011) *Zymomonas mobilis* levan is involved in metalloproteinases activation in healing of wounded and burned tissues. Lucrari Stiintifice: Zootehnie si Biotehnologii 44

Suzuki M (1993) History of fructan research: Rose to Edelman. In: Suzuki M, Chatterton NJ (eds) Science and technology of fructans. CRC Press, Boca Raton, pp 21–33

Tajima K, Tanio T, Kobayashi Y, Khono H, Fujiwara M, Shiba T, Erata T, Munekata M, Takai M (2000) Cloning and sequencing of the levansucrase gene from *Acetobacter xylinum* NCI 1005. DNA Res 7:237–242

Tajima K, Uenishi N, FujiwaraM ET, Munekata M, Takai M (1997) The production of a new water-soluble polysaccharide by *Acetobacter xylinum* NCI 1005 and its structural analysis by NMR spectroscopy. Carbohydr Res 305:117–122

Takeda H, Sato K, Kinoshita S, Sasaki H (1994) Production of 1-kestose by *Scopulariopsis brevicaulis*. J Ferment Bioeng 77(4):386–389

Tamura K, Hemsworth GR, De Jean G, Rogers TE, Pudlo AN, Urs K, Jain N, Davies GJ, Martens EC, Brumer H (2017) Molecular mechanism by which prominent human gut bacteroidetes utilize mixed-linkage beta-glucans, major health-promoting cereal polysaccharides. Cell Rep 21:417–430

Tanaka T, Oi S, Iizuka M, Yamamoto T (1978) Levansucrase of *Bacillus subtilis*. Agric Biol Chem 42:323–326

Tian F, Inthanavong L, Karboune S (2011) Purification and characterization of levansucrases from *Bacillus amyloliquefaciens* in intra- and extracellular forms useful for the synthesis of levan and fructooligosaccharides. Biosci Biotechnol Biochem 75:1929–1938

Tieking M, Ehrmann MA, Vogel RF, Ganzle MG (2005) Molecular and functional characterization of a levansucrase from the sourdough isolate *Lactobacillus sanfranciscensis* TMW 1.392. Appl Microbiol Biotechnol 66:655–663

Toksoy Oner E, Hernandez L, Combie J (2016) Review of levan polysaccharide: from a century of past experiences to future prospects. Biotechnol Adv 34:827–844

Trujillo LE, Gómez R, Banguela A, Soto M, Arrieta JG, Hernández L (2004) Catalytical properties of N-glycosylated *Gluconacetobacter diazotrophicus* levansucrase produced in yeast. Electron J Biotechnol 7:115–123

Uchiyama T (1993) Biosynthesis and degradation of fructans by microbial enzymes other than levansucrase. In: Chatterton J (ed) Suzuki M, Science and technology of fructans. CRC Press, pp 170–187

Valluru R, Van den Ende W (2008) Plant fructans in stress environments: emerging concepts and future prospects. J Exp Botany 59(11):2905–2916

Van den Ende W (2013) Multifunctional fructans and raffinose family oligosaccharides. Front Plant Sci 4:247

Van den Ende W, Coopman M, Clerens S, Vergauwen R, Le Roy K, Lammens W, Van Laere A (2011) Unexpected presence of graminan- and levan-type fructans in the evergreen frost-hardy eudicot *Pachysandra terminalis* (Buxaceae). Purification, cloning and functional analysis of a 6-SST/6-SFT enzyme. Plant Physiol 155:603–614

Van den Ende W, El-Esawe SK (2014) Sugar signaling pathways leading to fructan and anthocyanin accumulation: a dual function in abiotic and biotic stress responses? Environ Exp Botany 108:4–13

Van den Ende W, Michiels A, Roover JD, Van Laere A (2002) Fructan biosynthetic and breakdown enzymes in dicots evolved from different invertases. Expression of fructan genes throughout chicory development. Mini-Review Sci World J 2:1281–1295

van Hijum SA, Kralj S, Ozimek LK, Dijkhuizen L, van Geel-Schutten IG (2006) Structure-function relationships of glucansucrase and fructansucrase enzymes from lactic acid bacteria. Microbiol Mol Biol Rev 70:157–176

van Hijum SA, van Geel-Schutten GH, Rahaoui H, van der Maarel MJ, Dijkhuizen L (2002) Characterization of a novel fructosyltransferase from *Lactobacillus reuteri* that synthesizes high-molecular-weight inulin and inulin oligosaccharides. Appl Environ Microbiol 68(9):4390–4398

Van Laere A, Van Den Ende W (2002) Inulin metabolism in dicots: Chicory as a model system. Plant Cell Env 25:803–813

Velázquez-Hernández ML, Baizabal-Aguirre VM, Bravo-Patiño A, Cajero-Juárez M, Chávez-Moctezuma MP, Valdez-Alarcón JJ (2009) Microbial fructosyltransferases and the role of fructans. J Appl Microbiol 106:1763–1778

Versluys M, Kirtel O, Toksoy Oner E, Van den Ende W (2018) The fructan syndrome: Evolutionary aspects and common themes among plants and microbes. Plant Cell Environ 41:16–38

Vijn I, Smeekens S (1999) Fructan: more than a reserve carbohydrate? Plant Physiol 120:351–359

Visnapuu T, Mardo K, Mosoarca C, Zamfir AD, Vigants A, Alamäe T (2011) Levansucrases from *Pseudomonas syringae* pv. tomato and *P. chlororaphis* subsp. *aurantiaca*: Substrate specificity, polymerizing properties and usage of different acceptors for fructosylation. J Biotechnol 155(3):338–349

Vogt LM, Meyer D, Pullens G, Faas FM, Venema K, Ramasamy U, Schols HA, de Vos P (2014) Toll-like receptor 2 activation by β2→1-fructans protects barrier function of T84 human intestinal epithelial cells in a chain length–dependent manner. J Nutr 144(7):1002–1008

Waldherr FW, Meissner D, Vogel RF (2008) Genetic and functional characterization of *Lactobacillus panis* levansucrase. Arch Microbiol 190:497–505

Wang C, Hua D, Yan C (2015) Structural characterization and antioxidant activities of a novel fructan from *Achyranthes bidentata* blume, a famous medicinal plant in China. Ind Crop Prod 70:427–434

Wang F, Feng G, Chen K (2009) Defense responses of harvested tomato fruit to burdock fructooligosaccharide, a novel potential elicitor. Postharvest Biol Technol 52(1):110–116

Wargovich MH, Chen DD, Jimenez A, Steele VE, Velasco M, Stephens C, Price R, Gray K, Kelloff GJ (1996) Aberrant crypts as a biomarker for colon cancer, evaluation of potential chemopreventive agents in the rat. Cancer Epidemiol Biomark Prev 5:355–360

Wu C, Zhang T, Mu W, Miao M, Jiang B (2015) Biosynthesis of lactosylfructoside by an intracellular levansucrase from *Bacillus methylotrophicus* SK 21.002. Carbohydr Res 401:122–126

Wu FC, Chou SZ, Shih IL (2013) Factors affecting the production and molecular weight of levan of *Bacillus subtilis* natto in batch and fed-batch culture in fermenter. J Taiwan Inst Chem Eng 44:846–853

Wuerges J, Caputi L, Cianci M, Boivin S, Meijers R, Benini S (2015) The crystal structure of *Erwinia amylovora* levansucrase provides a snapshot of the products of sucrose hydrolysis trapped into the active site. J Struct Biol 191:290–298

Yoo SH, Yoon EJ, Cha J, Lee HG (2004) Antitumor activity of levan polysaccharides from selected microorganisms. Int J Biol Macromol 34(1–2):37–41

Yun JW, Kim DH, Kim BW, Song SK (1997a) Comparison of sugar compositions between inulo- and fructooligosaccharides produced by different enzyme forms. Biotechnol Lett 19(6):553–556

Yun JW, Kim DY, Song SK (1997b) Enhanced production of fructosyltransferase and glucosyltransferase by substrate-feeding cultures of *Aureobasidium pullulans*. J Ferment Bioeng 84(3):261–263

Yun JW, Lee MG, Song SK (1994) Batch production of high-content fructo-oligosaccharides from sucrose by the mixed-enzyme system of β-fructofuranosidase and glucose oxidase. J Ferment Bioeng 77(2):159–163

Zambelli P, Tamborini L, Cazzamalli S, Pinto A, Arioli S, Balzaretti S, Plou FJ, Fernandez-Arrojo L, Molinari F, Conti P, Romano D (2016) An efficient continuous flow process for the synthesis of a non-conventional mixture of fructooligosaccharides. Food Chem 190:607–613

Zhang T, Li R, Qian H, Mu W, Miao M, Jiang B (2014) Biosynthesis of levan by levansucrase from *Bacillus methylotrophicus* SK 21.002. Carbohydr Polym 101:975–981

Exploitation of *Saccharomyces cerevisiae* Enzymes in Food Processing and Preparation of Nutraceuticals and Pharmaceuticals

2

Elsa F. Vieira and Cristina Delerue-Matos

Abstract

Saccharomyces cerevisiae enzymes are well recognized for high biodiversity and industrial applicability. This chapter sketches the main research trends concerning the functionalities of *S. cerevisiae* enzymes in food processing, ethanol industry, and production of new nutraceuticals and pharmaceuticals. We review the key aspects of the overall production process for the enzyme production and the recent strategies to identify and improve its catalyst properties. Finally, we summarize the classical and latest applications of *S. cerevisiae* enzymes in the food industry, with specific attention given to their role in changing the nutritional, digestibility, and sensory properties in a variety of food sectors like dairy, bakery, beverages, brewing, wine, fish processing, and sweeteners. Additionally, the use of *S. cerevisiae* enzymes in the production of functional foods, namely protein hydrolysates/autolysates, with provided health benefits by reducing the risk of chronic diseases, as well as its use in the formulation of new cosmetic and pharmaceutical products, is explored.

2.1 Introduction

S. cerevisiae is the most useful yeast species, with participation in the fermentation processes of winemaking, baking, and brewing since ancient times. This eukaryotic model organism has been extensively used as the most suitable microbial resource for industrial enzyme production due to their wide biochemical diversity, feasibility

E. F. Vieira (✉) · C. Delerue-Matos
REQUIMTE-LAQV, Instituto Superior de Engenharia do Porto, Instituto Politécnico do Porto, R. Dr. António Bernardino de Almeida 431, Porto, Portugal
e-mail: elsa.vieira@graq.isep.ipp.pt

N. K. Arora et al. (eds.), *Microbial Enzymes: Roles and Applications in Industries*, Microorganisms for Sustainability 11,
https://doi.org/10.1007/978-981-15-1710-5_2

of mass culture, and simple genetic manipulation (Zhang and Kim 2012). In addition, the generally regarded as safe (GRAS) status given by Food and Drug Administration supports its use for human consumption (Fernandes and Carvalho 2017; Nandy and Srivastava 2018). Compared to animal and plant enzyme sources, *S. cerevisiae* enzymes are more active and stable and have higher yield (Liu and Kokare 2017).

The scientific term *S. cerevisiae* means "a mold which ferments the sugar into alcohol and carbon dioxide" (Saranraj et al. 2017). This yeast species is known by the common name "baker's yeast" in baking and confectionery fields, "wine's yeast" in the wine-like alcoholic beverage production, and "brewer's yeast" in the beer production (Faria-Oliveira et al. 2013). "Baker's yeast" can be grouped into three main types of commercial formulations: compressed yeast (also known as "granular yeast" or "instant dried yeast"), cream yeast, and dried yeast (available as "active dry yeast" and "instant dry yeast") (Saranraj et al. 2017). "Wine's yeast" participates in the fermentation of grape musts into wine (Marsit and Dequin 2015). Brewer's spent yeast (BSY) is the second major by-product from the brewing industry, after the brewer's spent grain (BSG) (Ferreira et al. 2010). BSY results from the cultivation of *S. cerevisiae* on malted barley, separated after the wort fermentation, debittered and dried, and presents a well-balanced amino acid profile, peptides, phenolic compounds, nucleotides, vitamins (B3, B6, B9), carbohydrates (mainly glucans and mannans), phospholipids, minerals, and trace elements (Kanauchi et al. 2005; Abbas 2006; Faria-Oliveira et al. 2013; Vieira et al. 2013; Vieira et al. 2016a, b, c). Its inner content is rich in several proteolytic enzymes (Roy et al. 1999, 2000; Hecht et al. 2014), which can be used in the production of several autolysates/hydrolysates with evidenced bioactivities, particularly antioxidant, antihypertensive, and anti-inflammatory properties (Mirzaei et al. 2015; Vieira et al. 2016a, b, c; Vieira et al. 2017a, b, c, d; Vieira and Ferreira 2017).

S. cerevisiae contains extracellular and intracellular enzymes, which are classified based on their catalytic functions: oxidoreductases, transferases, hydrolases, lyases, isomerases, and ligases (Liu and Kokare 2017). The isolation and purification of these enzymes and their application as useful biocatalysts for various industrial processes, namely in the production of several new products in the food, pharmaceutical, cosmetic, and biofuel industries, have increased over the last few years and revolutionized the market scenario of these industries.

This chapter aims to provide an updated overview of the use of *S. cerevisiae* enzymes in the production of new nutraceuticals with provided health benefits by reducing the risk of chronic diseases, as well as its use in the formulation of new pharmaceuticals, cosmeceuticals, and bioethanol production. Succinctly, a separate section is devoted to overview the key aspects of the overall production process for the *S. cerevisiae* enzyme production and the recent strategies to identify and improve its catalyst properties.

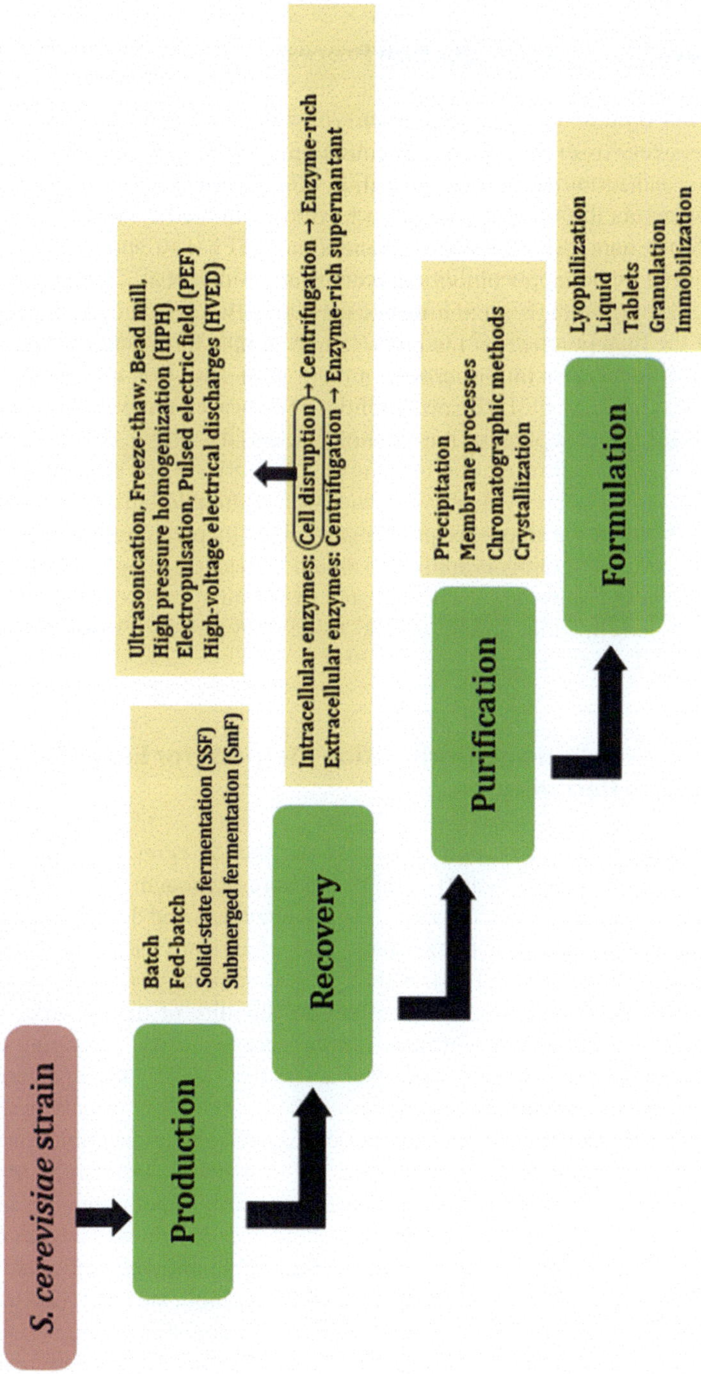

Fig. 2.1 A simplified process flowchart for *S. cerevisiae* enzyme production

2.2 Enzyme Production

2.2.1 Overall Enzyme Production Process

The production process of *S. cerevisiae* enzymes involves synthesis through fermentation, recovery from the fermentation medium, purification step (/s) to remove contaminants, and formulation procedure (Fig. 2.1). Commercial enzymes are produced in bioreactors through two possible cultivation methods, the submerged fermentation (SmF) and the solid-state fermentation (SSF) (Liu and Kokare 2017). SSF has acquired higher applicability in recent years (Singh et al. 2008) due to their recognized advantages: (1) high volumetric productivity, (2) relatively higher concentration of the final products, (3) requirement for simple fermentation equipment, and (4) less effluent generation, among others (Liu and Kokare 2017). When enzymes are produced by SmF, the extracellular enzymes are recovered by centrifugation (i.e., solid-liquid separation); in case of intracellular origin, cell disruption is required, followed by centrifugation to remove cell debris. When enzymes are produced by SSF, enzymes are typically excreted with a suitable buffer. In both types of cultivation, purification step is frequently carried out by precipitation followed by ultrafiltration; chromatography is applied to obtain high levels of purity. The enzyme concentrate is further submitted to a final step of formulation (i.e., lyophilization, granulation, immobilization, among others) to be delivered in a suitable form for food processing (Fernandes and Carvalho 2017).

2.2.2 Cell Disruption and Purification Methods for Enzymatic Fraction Recovery

Different treatments can be applied to recover the *S. cerevisiae* intracellular enzymes, such as (1) mechanical (i.e., high-pressure homogenization (HPH) and bead milling), (2) chemical (i.e., organic solvents, enzymes, and detergents), and (3) physical (i.e., ultrasonication, freeze-thaw, and electrically assisted treatment) (Ganeva et al. 2003; Liu et al. 2013). Although the chemical processes are applied on a large scale, the drastic procedures affect the stability of the enzymes and/or introduce additional impurities. Ultrasonication processes cause damaging effects on enzyme activities by altering their characteristics, substrates, and reactions between enzymes and substrates (Huang et al. 2017). Also, it is usually accompanied by temperature increase, high content of cell debris, and undesirable formation of sonochemical compounds (Shynkaryk et al. 2009). HPH is a frequently mechanical method employed for large-scale disruption of yeast cells. Although this method results in effective breakage of cells and high recovery of enzymes, it causes nonselective release of the products (i.e., the co-release of contaminants) and final products contain large quantity of cell debris, which complicates the downstream processing for purification. Thus, there is a growing interest in the application of more efficient physical/mechanical methods of cell disruption. Currently, application of electrotechnologies, such as the pulsed electric field (PEF) and high-voltage electrical discharges (HVED), are showing to be promising for intracellular enzyme extraction. HVED treatment results in an increase in the electrical conductivity and permeability of the whole sample while PEF treatment provides a clear extraction of intracellular enzymes with preservation of their functional activity (Ganeva et al. 2003). The methodology applied to purify the microbial enzymes (i.e., crystallization, electrophoresis, and chromatography) is based on their ionic properties, molecular size, and ability to get adsorbed (Liu and Kokare 2017).

2.3 Biotechnological Tools for Improving Enzyme Properties

Improvement or modification of *S. cerevisiae* enzymes may be achieved through protein engineering using genetic methods and/or adequate formulations (Fernandes and Carvalho 2017). Genetic engineering refers to the genetic material transference from one microorganism (donor) into another (host) that will produce the enzyme more efficiently (Liu and Kokare 2017). Enzyme immobilization confines or localizes it in a certain defined region of space while retaining the catalytic activity (Tavano 2013). This strategy overcomes some limitations of their free forms, namely (1) lack of long-term operational stability and shelf life, (2) recovery and reusability, (3) contamination of reaction products with the enzyme, and (4) performance in organic solvents, pH tolerance, selectivity, and heat stability (Ahmad and Sardar 2015; Madeira et al. 2017). Physical and chemical methods have been applied for enzyme immobilization; the first ones are characterized by a weak interaction between matrix and enzyme while chemical methods are based on the formation of covalent bond between the support and the enzyme. The challenge in the selection of the immobilization method is the prevention of enzyme activity loss; procedures such as adsorption, covalent coupling, entrapment, and cross-linking have been commonly used (Ahmad and Sardar 2015). When adsorption of enzymes onto insoluble supports is applied, enzymes are mixed with a suitable adsorbent under appropriate conditions of pH and ionic strength and further washed to obtain the immobilized enzyme in a directly usable form. In covalent immobilization, covalent bonds are formed between the enzyme and the support matrix. Entrapment is the restricted movement of enzymes in a porous gel and cross-linking involves attachment of biocatalysts to each other by bi- or multifunctional reagents or ligands. For instance, the cross-linked enzyme aggregates (CLEAs) involve a simple precipitation of the enzyme from aqueous solution by the addition of salts, or water-miscible organic solvents or nonionic polymers (Ahmad and Sardar 2015).

2.4 *S. cerevisiae* Enzymes

Generally, industrial enzymes are applicable in three main sectors: (1) 65% of the market in the detergent, starch, textile, leather, pulp and paper, and personal care industries; (2) 25% of the market in the food industry (baking, dairy, brewing, wine and juice, fats, and oils); and (3) 10% of the market in the animal feed industry (Cherry and Fidantsef 2003; Sharma et al. 2009; Adrio and Demain 2014; Ahmad and Sardar 2015). About 90% of the industrial enzymes are recombinant versions; proteases, amylases, and pectinases represent more than 60%, 30%, and 10%, respectively, of the enzyme market (Liu and Kokare 2017). Novozymes (Denmark) is the main supplier of industrial enzymes, followed by DSM (The Netherlands), DuPont (USA), and Roche (Switzerland). Although the United States, Western Europe, Japan, and Canada are still the largest consumers of industrial enzymes (Adrio and Demain 2014), developing countries in Asia Pacific, Eastern Europe, and Africa and the Middle East regions have emerged as promising industrial enzyme markets (Liu and Kokare 2017). The following section presents the main enzymes expressed by *S. cerevisiae* and their industrial applications; this information is summarized in Table 2.1.

2.4.1 Pectinases

Pectin is a hydrocolloid with great affinity for water and ability to form gels under certain conditions. Thus, the addition of pectinases, a group of enzymes that catalyze pectin degradation through depolymerization (hydrolases and lyases) and de-esterification (esterases) reaction, reduces viscosity and improves pectin gel pressability. These enzymes are highly advantageous in the development of wine, cider, and fruit juices (Pedrolli et al. 2009). Pectinases are divided into three major types: pectinesterases (PE), depolymerizing enzymes, and cleaving enzymes (Liu and Kokare 2017). Although *Aspergillus niger* is the most commonly used microbial species for the industrial production of pectinases (Gummadi and Panda 2003), few strains of *S. cerevisiae* have also been shown to be able to degrade pectin; polygalacturonases (PG) is the main pectinase reported (Alimardani-Theuil et al. 2011).

2.4.2 Chitinases

Chitin is the second most abundant polysaccharide in nature (after cellulose) and presents many important physiological functions and potential applications (i.e., control of phytopathogens, production of chitooligosaccharides, and treatment and degradation of chitin biowaste) (Karthik et al. 2017). Chitinases (EC 3.2.1.14) hydrolyze chitin into monomer N-acetyl-D-glucosamine (Liu and Kokare 2017). These enzymes are likely to be primarily responsible for wine haze formation (Marangon et al. 2011). According to Younes et al. (2011), *S. cerevisiae* PIR1, a wild yeast strain isolated from pinot noir grapes, secretes a class IV endochitinase with action in wine haze reduction. More recently, a patent for inhibiting or reducing haze in a liquid consumable (such as wine) was granted (Bauer and NDLOVU 2015). In this invention, suitable *S. cerevisiae* strains genetically modified or grown under environmental conditions for maximizing the cell-wall chitin and chitinase content are added to the liquid consumable, and chitinases are allowed to bind to the chitin in the *S. cerevisiae* cell walls.

Table 2.1 *S. cerevisiae* enzymes and their applications

Enzyme	Action	Applications	Reference
Pectinases (CECT 1389; ATCC52712 strains)	Catalyze the hydrolysis of α-1,4-glycosidic bonds in pectins or pectate	– In the fruit juice industry, pectinases are often used to (1) enhance juice extraction and clarification, (2) increase the yield of extracted juice, (3) reduce the processing time and turbidity, and (4) facilitate the extraction of color and aroma components from the cell into the juice – Enhance the starch (cassava, potato) recovery rates and yield	Blanco et al. (1997), Blanco et al. (1999), Dzogbefia et al. (2001), Djokoto et al. (2006), Dzogbefia et al. (2008), Ametefe et al. (2017), Agyepong and Barimah (2017)
Chitinases (PIR1 strain)	Catalyze the hydrolysis of chitin to its monomer N-acetyl-D-glucosamine	– Wine haze reduction/inhibition	Younes et al. (2011)
Invertases	Catalyze the hydrolysis of saccharose to glucose and fructose	– Manufacture of candy and jam	Akardere et al. (2010), Kumar and Kesavapillai (2012), AL-Sa'ady (2014)
Catalases (baker's yeast)	Catalyze the hydrolysis of hydrogen peroxide to water and dioxygen	– Food and textile processing (remove hydrogen peroxide that is used for sterilization or bleaching)	Seip and Di Cosimo (1992)
Tannases (CCMB 520 strain)	Catalyze the hydrolysis of tannins (tannic acid, methyl gallate, ethyl gallate, n-propyl gallate, and isoamyl gallate) to gallic acid	– Production of gallic acid and propyl gallate – Degradation of tannins present in the effluents of tanneries – Preparation of animal feeding – Pharmaceutical production (trimethoprim)	Belmares et al. (2004), Lopes et al. (2018)
Amylases (SDB strain)	Catalyze the hydrolysis of starch into sugars	– Food industry (production of syrups, reduction of viscosity of syrups, fruit juice clarification, starch solubilization) – Paper and pharmaceutical industries	Acourene and Ammouche (2012)

(continued)

Table 2.1 (continued)

Enzyme	Action	Applications	Reference
Alpha-glucosidases (baker's yeast)	Catalyze the hydrolysis of starch and disaccharides to glucose	– Production of bakery products – Starch processing – Flavoring production	Agrawal and Pandit (2003)
Inulinases	Catalyze the hydrolysis of inulin to fructose	– Production of fructose (sweetener and probiotic)	Kim et al. (1997), Onilude et al. (2012)
Cellulases, xylanases, and mannanases	Catalyze the conversion of lignocellulosic biomass to ethanol	– Bioethanol industry	van Zyl et al. (2015), Lane et al. (2018)
Lipases	Catalyze the hydrolysis of triglycerides into fatty acids and glycerol	– Food processing – Detergent, pharmaceutical, paper, cosmetic, and chemical synthesis industries – Biodiesel production	Schousboe (1976), Taketani et al. (1981), Degrassi et al. (1999), Białecka-Florjańczyk et al. (2010), Shi et al. (2012)
Laccases	Catalyze the oxidation of a wide range of phenolic compounds	– Textile industry, namely in hard-surface cleaning, in detergent formulations, and in tailoring processes – Beverage industry (juice clarification)	Bulter et al. (2003)

Further removal of the *S. cerevisiae* cells or cell-wall extract from the liquid consumable removes the chitinases, thereby inhibiting or reducing haze formation.

2.4.3 Invertases

Invertase (β-fructofuranosidases) (EC 3.2.1.26) hydrolyze the disaccharide into glucose and fructose (AL-Sa'ady 2014). This enzyme is crucial during the wine fermentation as it catalyzes the hydrolysis of saccharose present in grape must into glucose and fructose (Ribereau-Gayon et al. 2006). In the food industry, this enzyme is important in the preparation of jams and candies, where fructose is preferred over saccharose due to its high sweetness and crystallization properties (AL-Sa'ady, 2014). In the last few years, SSF has been applied to produce invertase from *S. cerevisiae* (Kumar and Kesavapillai 2012; AL-Sa'ady 2014). *S. cerevisiae* was reported to produce an extracellular β-D-fructofuranoside fructohydrolase (invertase) when grown on a medium containing β-fructofuranoside saccharose or raffinose (AL-Sa'ady 2014). Also, Akardere et al. (2010) purified invertase from baker's yeast (*S. cerevisiae*) with 363% of recovery.

2.4.4 Catalases

Catalases (EC 1.11.1.6) hydrolyze the hydrogen peroxide to water and molecular oxygen, protecting cellular proteins against reactive oxygen species modifications (Liu and Kokare 2017). These enzymes act through the protection of glucose-6-phosphate dehydrogenase against oxidative inactivation (Lushchak and Gospodaryov 2005). In addition, they have been applied in the food industry, acting in the removal of hydrogen peroxide that is used for sterilization or bleaching (Liu and Kokare 2017). An inexpensive and readily available commercial source of catalase is *S. cerevisiae* (baker's yeast) (Seip and Di Cosimo 1992). A culture medium was designed for maximal production of biomass and intracellular catalase by a potent enriched culture of *S. cerevisiae* CFR-201. This cell-free extract rich in catalase was reported to remove 99.9% of residual hydrogen peroxide from cold pasteurized milk at 45 °C for 1 h (Venkateshwaran et al. 1999).

2.4.5 Tannases

Tannase (EC, 3.1.1.20) is an extracellular enzyme that catalyses the hydrolysis of gallic acid esters and hydrolyzable tannins. They can be used in different industrial sectors, such as food (juice, beer, and wine) and pharmaceutical production (synthesis of the antibacterial drug, trimethoprim). In the food industry, tannase is commonly used as a clarifying agent in some wines, fruit juices, and refreshing drinks with coffee flavor, as well as in the production of gallic acid which is a potent antioxidant agent (Belmares et al. 2004; Lopes et al. 2018). Recently, a tannase was found in the crude extract of *S. cerevisiae* CCMB 520 subjected to SmF; the enzyme showed optimal biocatalytic performance at the tested temperature and pH ranges, as well as high thermal resistance (Lopes et al. 2018).

2.4.6 Amylases

Amylases hydrolyze starch into sugars (dextrins and oligosaccharides): α-Amylases (EC 3.2.1.1) catalyze the hydrolysis of internal α-1,4-*O*-glycosidic bonds in polysaccharides; β-amylases (EC 3.2.1.2) catalyze the hydrolysis of α-1, 4-glucan bonds to yield successive maltose units; and γ-amylases (EC 3.2.1.3) cleave α-(1–6) glycosidic bonds, as well as the last α-(1–4) glycosidic linkages at the nonreducing end of amylose and amylopectin (Sundarram and Murthy 2014, Liu and Kokare 2017). Amylases found applications in the starch saccharification, paper, food, and pharmaceutical industries. For instance, amylases are used in the (1) production of glucose/maltose syrups, (2) reduction of viscosity of sugar syrups, (3) clarification of fruit juice for longer shelf life, (4) solubilization of starch in the brewing industry, and (5) delaying of the staling of bread and other baked products (Liu and Kokare 2017). Extensive work has been done on the cloning of α-amylase genes in *S. cerevisiae*.

Acourene and Ammouche (2012) also reported the optimization of the cultural conditions in a stirred bioreactor for the extracellular secretion of α-amylase by the S. *cerevisiae SDB* strain.

2.4.7 Inulinases

Inulinase (EC 3.2.1.7) hydrolyzes the β-2,1 linkage of inulin into fructose or fructooligosaccharides, which have commercial applications in the food and pharmaceutical industries (Singh and Singh 2017). Fructose is a common alternative sweetener to saccharose, thus with a great importance in food industries; its medicinal properties are related to an increase of iron absorption in children, calcium absorption stimulation in postmenopausal women, stimulation of growth of *Bifidobacteria* in the large and small intestine, and prevention of colon cancer (Onilude et al. 2012). Recently, Onilude et al. (2012) reported inulinase production employing isolated inulinase-producing *S. cerevisiae* species from spontaneously fermented sugarcane in SSF.

2.4.8 Cellulases

Cellulases catalyze the hydrolysis of β-1,4-glycosidic bonds in cellulose chains; they are classified as endoglucanases (EC 3.2.1.4), exoglucanases (EC 3.2.1.91), and glucosidases (EC 3.2.1.21) (Liu and Kokare 2017). *S. cerevisiae* has been routinely used for ethanol production from several agricultural lignocellulosic feedstocks, such as corn (Białas et al. 2010), potato starch (Hashem and Darwish 2010), sugarcane bagasse (Martín et al. 2002), cassava (Nadir et al. 2009; Akaracharanya et al. 2011), sweet sorghum (Nuanpeng et al. 2018), oat and soybean (Cortivo et al. 2018), and citrus peel waste (Wilkins et al. 2007), among other crop sources.

2.4.9 Lipases

Lipases (EC 3.1.1.3) hydrolyze carboxyl ester bonds in triglycerides into fatty acids and glycerol; they are used in the food processing, detergent, pharmaceutical, paper, cosmetics, and chemical synthesis industries. Another promising application is the biodiesel production, a mixture of fatty acid alkyl esters, which is a potential alternative energy source derived from vegetable oils to petroleum-based products (Madeira et al. 2017). Several lipases, such as sterol ester hydrolase (Taketani et al. 1981), triacylglycerol lipase (Schousboe 1976), and carboxylic ester hydrolase (Degrassi et al. 1999), have been isolated from *S. cerevisiae*. More recently, Białecka-Florjańczyk et al. (2010) studied the hydrolysis of phenyl esters of alkane carboxylic acids by *S. cerevisiae* lipases. Also, the functional expression and characterization of five wax ester synthases in *S. cerevisiae* and their utility for biodiesel production were reported by Shi et al. (2012).

2.4.10 Laccases

Laccases catalyze the oxidation of phenols, polyphenols, and anilines and present several industrial applications, such as wastewater treatment, pulping, food preparation, textile manufacturing, and bioremediation (Mogharabi and Faramarzi 2014). In the textile industry, laccases are used in hard-surface cleaning, detergent formulations, and cork taint removal, whereas in the beverage industry, they are mostly used for fruit juice clarification (Cherry and Fidantsef 2003). Bulter et al. (2003) demonstrated the expression of laccase from *Myceliophthora thermophile* in *S. cerevisiae.* Although this work did not lead to a commercial production, this enzyme can be available for use in tailoring processes.

2.4.11 Proteases

Proteases (EC 3:4, 11–19, 20–24, 99) catalyze the hydrolysis of covalent peptide bonds. They are broadly classified based on optimum pH (acidic, neutral, or alkaline proteases), substrate specificity, site of action on protein substrates (endo- or exo-proteases), and their catalytic mechanisms (serine, aspartic, cysteine, or metalloproteases) (Liu and Kokare 2017). *S. cerevisiae* is a good source of vacuolar and cytoplasm proteases; the enzymes described up to now were reviewed by Hecht et al. (2014). Thus, due to their versatility, *S. cerevisiae* proteases can be used in the food, cleaning, and textile sectors. In the food processing, *S. cerevisiae* proteases have been used as valuable biocatalysts, particularly in the production of protein hydrolysates/autolysates with promoted biological properties, namely antioxidant, antihypertensive, and anti-inflammatory activities. These hydrolysates/autolysates present the potential to be used in the formulation of new nutraceuticals and pharmaceuticals. Furthermore, protease treatment can also potentiate and modulate several functional characteristics of the proteins, such as viscosity, dispersibility, solubility, foaming, and emulsifying, giving them advantages for use in various products in the food industry (dos Santos Aguilar and Sato 2018). For instance, Vieira et al. (2017b) reported that a protease extract obtained from brewer's spent yeast (*S. cerevisiae*) was efficient in the production of sardine protein hydrolysates (SPH) from muscle and viscera by-products with interesting functional properties (better emulsion, foaming, and oil binding properties). Results obtained with this enzymatic treatment were comparable to those prepared by the commercial enzymes Alcalase® and Neutrase® and suggested that muscle and viscera SPH can be used as functional food ingredients, such as pate and spread-texture food.

2.5 Recent Applications of *S. cerevisiae* Enzymes

The application of *S. cerevisiae* enzymes in the food and beverage industries, including dairy, bakery, beverage, fish processing, and protein hydrolysate, as well as in the formulation of new pharmaceuticals and cosmeceuticals and in the

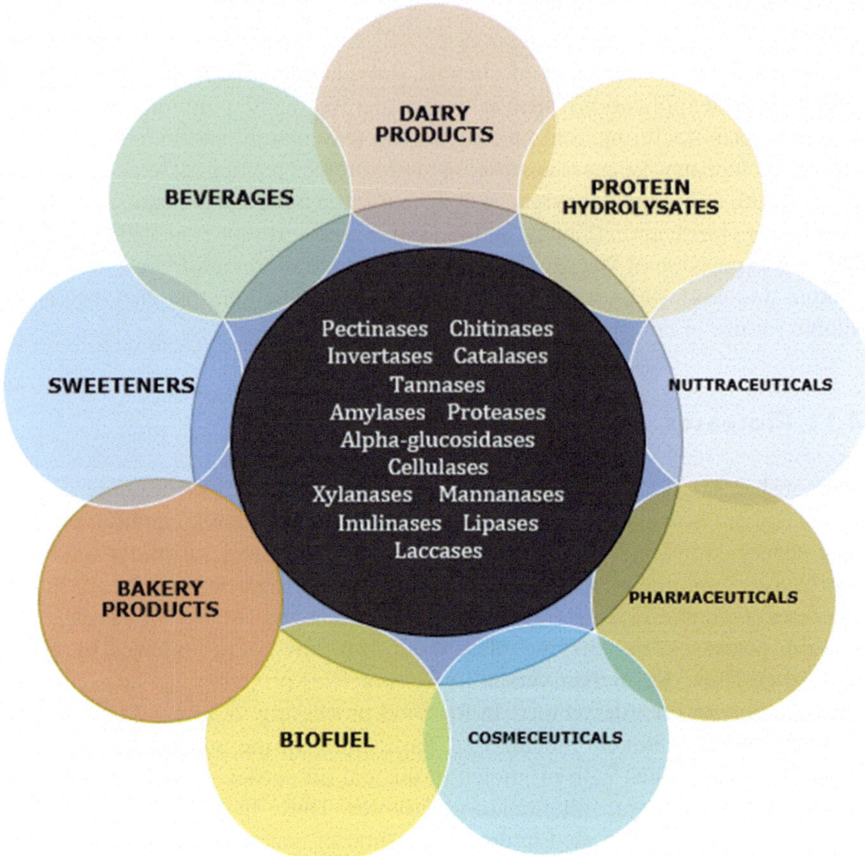

Fig. 2.2 Industrial application of *S. cerevisiae* enzymes

bioethanol industry, is well established (Fig. 2.2). This section aims to give some examples of application of *S. cerevisiae* enzymes in these fields.

2.5.1 Dairy Products

In the dairy industry, *S. cerevisiae* contributes substantially to the fermentation step by (1) supporting starter cultures, (2) inhibiting undesired microorganisms causing quality defects, (3) inhibiting occurrence of undesirable biochemical changes such as production of aromatic compounds, and (4) exerting lipolytic and proteolytic activities (Rasika et al. 2015). Acidophilus milk, a beverage containing a lactic acid content of 1.5–2.0% and used in the prevention of stomach disorder, uses *S. cerevisiae* to enhance the viability of bacterial strain and improve the antioxidant properties of fermented milk (Parrella et al. 2012). In addition, *S. cerevisiae* is able to hydrolyze milk proteins into angiotensin-converting enzyme (ACE) inhibitory peptides, which can be used in dietary supplements and/or as an alternative approach for antihypertensive medication. Roy et al. (2000) described the purification and characterization of *S. cerevisiae* K7 protease involved in the production of ACE inhibitory peptides from milk proteins (Roy et al. 2000). Moreover, the Calpis® sour milk, a product fermented with a starter containing *Lactobacillus helveticus* and *S. cerevisiae*, presents an antihypertensive effect due to the presence of the ACE inhibitory peptides Val-Pro-Pro and Ile-Pro-Pro (Vermeirssen et al. 2003). Recently, *S. cerevisiae* K7 was recommended as a potential yeast strain for the functional food industry due to its capacity to hydrolyze milk proteins into physiologically active peptides (Rasika et al. 2015).

2.5.2 Beverages (Wine, Beer, Juices)

S. cerevisiae is used in the beer fermentation; distinct yeast strains are used to produce ale or lager beer. Since the wort components (dextrins, β-glucan, and soluble proteins, among others) are not metabolized by natural strains of brewer's yeast, genetic manipulation is important to increase the capacity for synthesis of the extracellular amylases β-glucanases and β-glucosidase (Spier et al. 2016). *S. cerevisiae* is also the main strain used in wine fermentation. As previously mentioned, invertase catalyzes the hydrolysis of saccharose present in grape must into glucose and fructose molecules; β-1,3-glucanases are related to the cell-wall hydrolysis and release of mannoproteins into medium during autolysis of the yeast; β-glucanase enhances wine structure and stability; pectinases catalyze the degradation of pectic substances; while β-glucosidase improves the release of aromatic compounds in wine (Spier et al. 2016). *S. cerevisiae* PIR1, a wild yeast strain isolated from pinot noir grapes, was recently reported to secrete an acidic extracellular protease which was completely inhibited by pepstatin A (Younes et al. 2011). The same researchers showed that hydrolyzed proteins correspond to pathogenesis-related (PR) proteins, in particular thaumatin-like (TL) proteins and chitinases, which are implicated in the wine haze formation (Younes et al. 2013). Gainvors et al. (1994) demonstrated that the addition of a crude enzymatic extract from *S. cerevisiae* (SCPP 2180) with PE, PL, and PG activities to grape must had the same effect on the turbidity as the same quantity of the commercial preparation of Endozyme (Pascal Biotech SARL-Paris). Also, Blanco et al. (1997) showed that when wine fermentations are carried out using PG strains of *S. cerevisiae*, the clarification process is greatly facilitated, with the filtration time being reduced more than 50% in some cases. More recently, Dzogbefia et al. (2008) showed that a crude pectinase preparation from *S. cerevisiae* produced in Ghana could be effectively used for starch extraction from cassava, being a potential alternative to imported commercial enzymes. In the beverage industry, the biotechnological application of pectinase preparation from *S. cerevisiae* ATCC 52712 to pineapple and papaya juice extraction is also quite promising (Dzogbefia et al. 2001; Djokoto et al. 2006).

2.5.3 Sweeteners

Xylitol, mannitol, sorbitol, and erythritol are sugar alcohols that offer similar or better sweetness properties of saccharose (Park et al. 2016). The absence of xylose reductase (XR) and xylitol dehydrogenase (XDH) enzymes in *S. cerevisiae* has been encouraged to engineer xylitol-biosynthesizing *S. cerevisiae* recombinant strains (Lee et al. 2000; Costenoble et al. 2003; Bae et al. 2004; Oh et al. 2013; Park et al. 2016). For instance, Costenoble et al. (2003) demonstrated mannitol production by *S. cerevisiae* using elevated glucose concentrations in the presence of fructose.

2.5.4 Bakery

There has been an increasing interest in the use of *S. cerevisiae* lipolytic enzymes in the baking processes. For instance, phospholipases from *S. cerevisiae* have been used to degrade polar wheat lipids into emulsifying lipids, being a potential substitute or supplement of traditional emulsifiers. Also, current research has been dedicated to understand the mechanisms behind the enzymatic prevention of bread staling in the presence of *S. cerevisiae* α-amylases and xylanases (Liu and Kokare 2017).

2.5.5 Protein Hydrolysates/Autolysates

Enzymatic hydrolysis is an alternative approach for recovering protein fraction from different biomass sources, resulting in a soluble product with higher digestibility and presenting potential bioactive peptides with promising health benefits in the treatment of chronic diseases. *S. cerevisiae* proteases have been used in the production of several protein hydrolysates/autolysates (Table 2.2). For instance, protease extract from *S. cerevisiae* (as BSY) has been used to produce hydrolysates from sardine protein (muscle and viscera) by-products. Vieira and Ferreira (2017) reported the hydrolysis of sardine sarcoplasmic proteins by proteases extracted from BSY; the final hydrolysates produced using the substrate/enzyme ratio 1:0.27 (mg/U), 7 h, and 50 °C were showed to present antioxidant and ACE-I activities. The results also indicated that these protein hydrolysates could be used as a promising source of bioactive peptides.

BSG, the residual solid fraction of barley malt and other used grains remaining after filtration of wort, is the main brewing by-product. Recently, BSG proteins were effectively hydrolyzed using yeast extract rich in proteases obtained from *S. cerevisiae* (BSY). The final protein hydrolysate presented improved biological properties, and apparently the main BSY proteases responsible for the hydrolysis of BSG proteins under the optimum conditions belong to the class of serine peptidases and metallopeptidases (Vieira et al. 2016a). The same authors also described the optimization of BSY extract autolysis by response surface methodology (RSM) to produce a BSY autolysate comprising antioxidant and ACE inhibitory activities

Table 2.2 Production of nutraceuticals and pharmaceuticals through the action of *S. cerevisiae* proteases

Substrate	Enzyme	Conditions	Biological activity	Reference
Skimmed milk proteins	*S. cerevisiae* (cell-free extract rich in protease B)	37 °C for 3 h	ACE inhibitory activity (IC_{50} of 0.42 mg of protein/mL)	Roy et al. (2000)
Bovine full-fat milk proteins	*S. cerevisiae* K7	30 °C for 72 h	ACE-inhibitory activity of 25%	Rasika et al. (2015)
Sarcoplasmic proteins from canned sardine by-products	*S. pastorianus* (brewer's spent yeast)	Proteolytic activity of 0.725 U/mL, E/S of 0.27:1 (U/mg), 50 °C for 7 h	Antioxidant activity (FRAP of 293 μM TE/mL) and ACE inhibitory activity (IC_{50} of 164 μg protein/mL)	Vieira et al. (2017b)
Sarcoplasmic proteins from canned sardine by-products	*S. pastorianus* (brewer's spent yeast)	Proteolytic activity of 0.725 U/mL, E/S of 0.27:1 (U/mg), 50 °C for 7 h, followed by 10 kDa-UF membrane	Anti-inflammatory activity (increase levels of NO, MCP-1, VEGF, IL-8, ICAM-1, and ROS upon TNF-α treatment)	Vieira et al. (2017c)
Muscle and viscera proteins from canned sardine by-products	*S. pastorianus* (brewer's spent yeast)	Proteolytic activity of 1 U/mL, E/S of 20% (v/v), 50 °C for 7 h	Muscle protein hydrolysates presented EAI, ESI, FE, FS, WBC, and OBC of 54.5 m^2/g, 27.3 min, 57.9%, 39.7%, 4.3 g/g, and 5.4 g/g, respectively. Viscera protein hydrolysates presented EAI, ESI, FE, FS, WBC, and OBC of 80.1 m^2/g, 45.6 min, 79.2%, 41.5%, 4.8 g/g, and 5.8 g/g, respectively.	Vieira et al. (2017b)
Brewer's spent grain proteins	*S. pastorianus* (brewer's spent yeast)	E/S of 0.29:1 (U/mg), 50 °C for 6 h	Antioxidant activity (TPC of 1.65 mg GAE/mL and FRAP value of 1.88 mg Trolox equivalent/mL)	Vieira et al. (2016a)
Whey protein	*S. cerevisiae* (baker's yeast)	28 °C for 48 h, followed by in vitro GI digestion	ACE inhibitory activity (IC_{50} of 0.08 mg/mL)	Vermeirssen et al. (2003)

(continued)

Table 2.2 (continued)

Substrate	Enzyme	Conditions	Biological activity	Reference
Pea protein	*S. cerevisiae* (commercial baking yeast)	28 °C for 48 h, followed by in vitro GI digestion	ACE inhibitory activity (IC_{50} of 0.18 mg/mL)	Vermeirssen et al. (2003)
Yam peel proteins	*S. cerevisiae* (BY4743) proteases	27 °C for 96 h	High protein content with good essential amino acid profile (leucine, valine, lysine, and threonine)	Aruna et al. (2017)
S. cerevisiae (PTCC 5269) inner content	*S. cerevisiae* (PTCC 5269)	Autolysis: 2.5% dry yeast cell/water, 52 °C for 96 h	Antioxidant activity (DPPH of 52.23 μM TE/mg protein and ABTS of 2211.59 μM TE/mg protein) and ACE inhibitory activity (IC_{50} of 2.18 mg/mL)	Mirzaei et al. (2015)
Brewer's spent yeast inner content (autolysis)	*S. pastorianus* (brewer's spent yeast)	Autolysis: Yeast extract containing 10 mg of proteins/mL, protease activity of 0.220 U/mL, 36.0 °C for 6 h	Antioxidant activity (TPC of 385 mM GAE/mL and FRAP of 374 mM TE/mL) and ACE inhibitory activity (IC_{50} value of 379 mg/mL)	Vieira et al. (2017d)

GI gastrointestinal, *IC_{50}* 50% inhibitory concentration, *UF* ultrafiltration, *TNF-α* tumor necrosis factor-α, *NO* nitric oxide, *ROS* reactive oxygen species, *ICAM-1* intercellular adhesion molecule-1, *IL-8* interleukin-8, *MCP-1* monocyte chemoattractant protein 1, *VEGF* vascular endothelial growth factor, *E/S* enzyme/substrate ratio, *TPC* total phenolic content, *FRAP* ferric ion-reducing antioxidant power, *DPPH* 2,2-diphenyl-1-picryl-hydrazyl, *ABTS* 2,2′-azinobis (3-ethyl-benzothiazoline-6-sulfonate), *EAI (m^2/g)* emulsifying activity index, *ESI (min)* emulsifying stability index, *FE (%)* foaming expansion, *FS (%)* foaming stability, *WBC (g/g)* water-binding capacity, *OBC (g/g)* oil-binding capacity

(Vieira et al. 2017d). Further data from in vitro GI digestion, cellular antioxidant assays, and cell monolayer permeation studies confirmed the potential of BSY autolysate obtained from inner yeast cell content as a functional food ingredient (Vieira et al. 2016b).

2.5.6 Cosmeceuticals

Glycolic acid, a useful industrial compound with applications in cosmetics and chemical processes and as a biopolymer precursor, has been successfully produced from xylose in engineered *S. cerevisiae*. Also, the fatty alcohol 1-hexadecanol has applications in detergents, emulsifiers, and cosmetics (Lane et al. 2018).

2.5.7 Biofuel Industry

S. cerevisiae is the most frequent yeast strain for bioethanol production due to its high fermentation rate, ethanol tolerance, and GRAS status. However, this species is unable to hydrolyze cellulose and hemicellulose available in lignocellulosic biomass (van Zyl et al. 2015). Thus, several technologies have been employed to engineer *S. cerevisiae* strains capable to synthesize and secrete amylolytic enzymes for the direct conversion of starch-rich materials to ethanol (Kim et al. 2010). For instance, Nakatani et al. (2013) reported the development of a cellulase- and expansin-co-expressing strain of *S. cerevisiae* with direct ethanol fermentation performance. The low price and ready availability makes starch the most common substrate for biofuel production (Mobini-Dehkordi and Afzal Javan 2012). Basically, the process of making ethanol from starch involves saccharification, where starch is converted into sugar using enzymes such as gluco-amylase and α-amylase, followed by fermentation of glucose to ethanol by *S. cerevisiae*, and finally recovery (of ethanol) (Mobini-Dehkordi and Afzal Javan 2012; Sundarram and Murthy 2014; Liu and Kokare 2017). Recent advances have included other starch feedstocks such as sugarcane juice, cassava starch, and other carbon sources (Wangpor et al. 2017). For instance, Grohmann et al. (1994) studied the enzymatic hydrolysis of polysaccharides in orange peel by *S. cerevisiae*. These authors suggested that the fermentation of enzymatic hydrolysates of orange peel to ethanol by *S. cerevisiae* was relatively simple technically, but with low economic viability. More recently, Wangpor et al. (2017) studied the enzymatic hydrolysis of cassava starch by *S. cerevisiae*; the optimum liquefaction conditions for dextrin concentration were 0.9 mg/g of α-amylase, 85 °C, and 180 min, and the saccharification conditions for glucose concentration were 1.5 mg/g of gluco-amylase, 60 °C, and 90 min. Moreover, engineered *S. cerevisiae* has been employed for the bioconversion of lignocellulosic sugars to a large variety of non-ethanol value-added products, such as xylitol, 1-hexadecanol, isobutanol, 2,3-butanediol, lactic acid, poly-3-D-hydroxybutyrate (PHB), 3-hydroxypropionic acid (3HPA), glycolic acid, D-xylonate, isoprenoids, and spermidine (Lane et al. 2018).

2.6 Conclusion

Many studies have demonstrated the potential use of *S. cerevisiae* enzymes in different industrial products and processes. In fact, *S. cerevisiae* provides several enzymes, such as pectinases, chitinases, invertases, catalases, tannases, amylases, proteases, cellulases, lipases, and laccases, among others, with a large application in the food, beverage, bioethanol, nutraceuticals, pharmaceuticals, and cosmeceuticals industries. Additionally, the advances in modern biotechnology, namely the recent developments in genomics, proteomics, and emerging recombinant DNA techniques, have been crucial to discover new areas of applications of *S. cerevisiae* enzymes, namely in the nutraceuticals and pharmaceuticals industries.

Acknowledgments *Conflicts of Interest*: The authors declare no conflict of interest.

References

Abbas CA (2006) Production of antioxidants, aromas, colours, flavours, and vitamins by yeasts. In: Querol A, Fleet G (eds) Yeasts in food and beverages. Springer, Berlin, Heidelberg, pp 285–334

Acourene S, Ammouche A (2012) Optimization of ethanol, citric acid, and α-amylase production from date wastes by strains of *Saccharomyces cerevisiae*, *Aspergillus niger*, and *Candida guilliermondii*. J Ind Microbiol Biotechnol 39:759–766

Adrio J, Demain A (2014) Microbial enzymes: tools for biotechnological processes. Biomol Ther 4:117–139

Agrawal PB, Pandit AB (2003) Isolation of α-glucosidase from *Saccharomyces cerevisiae*: cell disruption and adsorption. Biochem Eng J 15:37–45

Agyepong JK, Barimah J (2017) Evaluation of crude preparations of *Saccharomyces cerevisiae* (ATCC 52712) pectolytic enzymes in cassava starch extraction: effects of variety on yield and starch recovery rates. Afr J Biotechnol 16:2031–2042

Ahmad R, Sardar M (2015) Enzyme immobilization: an overview on nanoparticles as immobilization matrix. Biochem Anal Biochem 4:178

Akaracharanya A, Kesornsit J, Leepipatpiboon N, Srinorakutara T, Kitpreechavanich V, Tolieng V (2011) Evaluation of the waste from cassava starch production as a substrate for ethanol fermentation by *Saccharomyces cerevisiae*. Ann Microbiol 61:431–436

Akardere E, Özer B, Çelem EB, Önal S (2010) Three-phase partitioning of invertase from Baker's yeast. Sep Purif Technol 72:335–339

Alimardani-Theuil P, Gainvors-Claisse A, Duchiron F (2011) Yeasts: an attractive source of pectinases - from gene expression to potential applications: a review. Process Biochem 46:1525–1537

AL-Sa'ady AJR (2014) Optimization of invertase production from *Saccharomyces cerevisiae* by solid state fermentation. Curr Res Microbiol Biotechnol 2:373–377

Ametefe GD, Dzogbefia VP, Apprey C, Kwatia S (2017) Optimal conditions for pectinase production by *Saccharomyces cerevisiae* (ATCC 52712) in solid state fermentation and its efficacy in orange juice extraction. IOSR J Biotechnol. Biochemist 3:78–86

Aruna TE, Aworh OC, Raji AO, Olagunju AI (2017) Protein enrichment of yam peels by fermentation with *Saccharomyces cerevisiae* (BY4743). Ann Agric Sci 62:33–37

Bae S-M, Park Y-C, Lee T-H, Kweon D-H, Choi J-H, Kim S-K, Ryu Y-W, Seo J-H (2004) Production of xylitol by recombinant *Saccharomyces cerevisiae* containing xylose reductase gene in repeated fed-batch and cell-recycle fermentations. Enzym Microb Technol 35:545–549

Bauer FF, NDLOVU T (2015) Method for preventing wine haze formation. WO2015162573A1

Belmares R, Contreras-Esquivel JC, Rodŕiguez-Herrera R, ARr C, Aguilar CN (2004) Microbial production of tannase: an enzyme with potential use in food industry. LWT - Food Sci Technol 37:857–864

Białas W, Szymanowska D, Grajek W (2010) Fuel ethanol production from granular corn starch using *Saccharomyces cerevisiae* in a long term repeated SSF process with full stillage recycling. Bioresour Technol 101:3126–3131

Białecka-Florjańczyk E, Krzyczkowska J, Stolarzewicz I (2010) Catalytic activity of baker's yeast in ester hydrolysis. Biocatal Biotransformation 28:288–291

Blanco P, Sieiro C, Díaz A, Reboredo N, Villa T (1997) Grape juice biodegradation by polygalacturonases from *Saccharomyces cerevisiae*. Int Biodeterior Biodegradation 40:115–118

Blanco P, Sieiro C, Villa TG (1999) Production of pectic enzymes in yeasts. FEMS Microbiol Lett 175:1–9

Bulter T, Alcalde M, Sieber V, Meinhold P, Schlachtbauer C, Arnold FH (2003) Functional expression of a fungal laccase in *Saccharomyces cerevisiae* by directed evolution. Appl Environ Microbiol 69:987–995

Cherry JR, Fidantsef AL (2003) Directed evolution of industrial enzymes: an update. Curr Opin Biotechnol 14:438–443
Cortivo PRD, Hickert LR, Hector R, Ayub MAZ (2018) Fermentation of oat and soybean hull hydrolysates into ethanol and xylitol by recombinant industrial strains of *Saccharomyces cerevisiae* under diverse oxygen environments. Ind Crop Prod 113:10–18
Costenoble R, Adler L, Niklasson C, Lidén G (2003) Engineering of the metabolism of *Saccharomyces cerevisiae* for anaerobic production of mannitol. FEMS Yeast Res 3:17–25
Degrassi G, Uotila L, Klima R, Venturi V (1999) Purification and properties of an esterase from the yeast *Saccharomyces cerevisiae* and identification of the encoding gene. Appl Environ Microbiol 65:3470–3472
Djokoto D, Dzogbefia VP, Oldham JH (2006) Rapid extraction of pawpaw juice with the application of locally produced pectic enzymes from *Saccharomyces cerevisiae* ATCC 51712. Food Biotechnol 20:31–41
dos Santos Aguilar JG, Sato HH (2018) Microbial proteases: production and application in obtaining protein hydrolysates. Food Res Int 103:253–262
Dzogbefia V, Amoke E, Oldham J, Ellis W (2001) Production and use of yeast pectolytic enzymes to aid pineapple juice extraction. Food Biotechnol 15:25–34
Dzogbefia V, Ofosu G, Oldham J (2008) Evaluation of locally produced *Saccharomyces cerevisiae* pectinase enzyme for industrial extraction of starch from cassava in Ghana. Sci Res Essay 3:365–369
Faria-Oliveira F, Puga S, Ferreira C (2013) Yeast: world's finest chef. In Food Industry Innocenzo Muzzalupo, IntechOpen
Fernandes P, Carvalho F (2017) Chapter 19 - microbial enzymes for the food industry. In: Brahmachari G, Demain AL, Adrio JL (eds) Biotechnology of microbial enzymes - production, biocatalysis and industrial applications. Elsevier, Academic Press, San Diego, CA, pp 513–544
Ferreira IM, Pinho O, Vieira E, Tavarela JG (2010) Brewer's *Saccharomyces* yeast biomass: characteristics and potential applications. Trends Food Sci Technol 21:77–84
Gainvors A, Frezier V, Lemaresquier H, Lequart C, Aigle M, Belarbi A (1994) Detection of polygalacturonase, pectin-lyase and pectin-esterase activities in a *Saccharomyces cerevisiae* strain. Yeast 10:1311–1319
Ganeva V, Galutzov B, Teissié J (2003) High yield electroextraction of proteins from yeast by a flow process. Anal Biochem 315:77–84
Grohmann K, Baldwin EA, Buslig BS (1994) Production of ethanol from enzymatically hydrolyzed orange peel by the yeast *Saccharomyces cerevisiae*. Appl Biochem Biotechnol 45:315
Gummadi SN, Panda T (2003) Purification and biochemical properties of microbial pectinases - a review. Process Biochem 38:987–996
Hashem M, Darwish SM (2010) Production of bioethanol and associated by-products from potato starch residue stream by *Saccharomyces cerevisiae*. Biomass Bioenergy 34:953–959
Hecht KA, O'Donnell AF, Brodsky JL (2014) The proteolytic landscape of the yeast vacuole. Cell Logist 4:e28023
Huang G, Chen S, Dai C, Sun L, Sun W, Tang Y, Xiong F, He R, Ma H (2017) Effects of ultrasound on microbial growth and enzyme activity. Ultrason Sonochem 37:144–149
Kanauchi O, Igarashi K, Ogata R, Mitsuyama K, Andoh A (2005) A yeast extract high in bioactive peptides has a blood-pressure lowering effect in hypertensive model. Curr Med Chem 12:3085–3090
Karthik N, Binod P, Pandey A (2017) 15 - Chitinases. In: Larroche C, Sanroman MA, Du G, Pandey A (eds) Current developments in biotechnology and bioengineering. Elsevier, Amsterdam, pp 335–368
Kim B-W, Kim H-W, Nam S-W (1997) Continuous production of fructose-syrups from inulin by immobilized inulinase from recombinant *Saccharomyces cerevisiae*. Biotechnol Bioprocess Eng 2:90–93
Kim J-H, Kim H-R, Lim M-H, Ko H-M, Chin J-E, Lee HB, Kim I-C, Bai S (2010) Construction of a direct starch-fermenting industrial strain of *Saccharomyces cerevisiae* producing glucoamylase, α-amylase and debranching enzyme. Biotechnol Lett 32:713–719

Kumar R, Kesavapillai B (2012) Stimulation of extracellular invertase production from spent yeast when sugarcane pressmud used as substrate through solid state fermentation. Springerplus 1:81
Lane S, Dong J, Jin Y-S (2018) Value-added biotransformation of cellulosic sugars by engineered *Saccharomyces cerevisiae*. Bioresour Technol 260:380–394
Lee W-J, Ryu Y-W, Seo J-H (2000) Characterization of two-substrate fermentation processes for xylitol production using recombinant *Saccharomyces cerevisiae* containing xylose reductase gene. Process Biochem 35:1199–1203
Liu D, Lebovka N, Vorobiev E (2013) Impact of electric pulse treatment on selective extraction of intracellular compounds from *Saccharomyces cerevisiae* yeasts. Food Bioprocess Technol 6:576–584
Liu X, Kokare C (2017) Chapter 11 - microbial enzymes of use in industry. In: Brahmachari G, Demain AL, Adrio JL (eds) Biotechnology of microbial enzymes - production, biocatalysis and industrial applications. Elsevier, Academic Press, London, pp 267–298
Lopes LMM, Costa Batista LH, Gouveia MJ, Leite TCC, de Mello MRF, de Assis SA, de Sena AR (2018) Kinetic and thermodynamic parameters, and partial characterization of the crude extract of tannase produced by S*accharomyces cerevisiae* CCMB 520. Nat Prod Res 32:1068–1075
Lushchak VI, Gospodaryov DV (2005) Catalases protect cellular proteins from oxidative modification in *Saccharomyces cerevisiae*. Cell Biol Int 29:187–192
Madeira JV Jr, Contesini FJ, Calzado F, Rubio MV, Zubieta MP, Lopes DB, de Melo RR (2017) Chapter 18—Agro-industrial residues and microbial enzymes: an overview on the eco-friendly bioconversion into high value-added products. In: Brahmachari G, Demain AL, Adrio JL (eds) Biotechnology of microbial enzymes - production, biocatalysis and industrial applications. Elsevier, Academic Press, London, pp 475–511
Marangon M, Van Sluyter SC, Neilson KA, Chan C, Haynes PA, Waters EJ, Falconer RJ (2011) Roles of grape thaumatin-like protein and chitinase in white wine haze formation. J Agric Food Chem 59:733–740
Marsit S, Dequin S (2015) Diversity and adaptive evolution of *Saccharomyces* wine yeast: a review. FEMS Yeast Res 15:fov067-fov067
Martín C, Galbe M, Wahlbom CF, Hahn-Hägerdal B, Jönsson LJ (2002) Ethanol production from enzymatic hydrolysates of sugarcane bagasse using recombinant xylose-utilising *Saccharomyces cerevisiae*. Enzym Microb Technol 31:274–282
Mirzaei M, Mirdamadi S, Ehsani MR, Aminlari M, Hosseini E (2015) Purification and identification of antioxidant and ACE-inhibitory peptide from S*accharomyces cerevisiae* protein hydrolysate. J Funct Foods 19:259–268
Mobini-Dehkordi M, Afzal Javan F (2012) Application of alpha-amylase in biotechnology. J Biol Today's World 1:15–20
Mogharabi M, Faramarzi MA (2014) Laccase and laccase-mediated systems in the synthesis of organic compounds. Adv Synth Catal 356:897–927
Nadir N, Mel M, Karim A, Ismail M, Mohd Yunus R (2009) Comparison of sweet sorghum and cassava for ethanol production by using *Saccharomyces cerevisiae*. J Appl Sci 9:3068–3073
Nakatani Y, Yamada R, Ogino C, Kondo A (2013) Synergetic effect of yeast cell-surface expression of cellulase and expansin-like protein on direct ethanol production from cellulose. Microb Cell Factories 12:66
Nandy SK, Srivastava RK (2018) A review on sustainable yeast biotechnological processes and applications. Microbiol Res 207:83–90
Nuanpeng S, Thanonkeo S, Klanrit P, Thanonkeo P (2018) Ethanol production from sweet sorghum by *Saccharomyces cerevisiae* DBKKUY-53 immobilized on alginate-loofah matrices. Braz J Microbiol 49:140
Oh EJ, Ha S-J, Kim SR, Lee W-H, Galazka JM, Cate JH, Jin Y-S (2013) Enhanced xylitol production through simultaneous co-utilization of cellobiose and xylose by engineered *Saccharomyces cerevisiae*. Metab Eng 15:226–234
Onilude AA, Fadaunsi IF, Garuba EO (2012) Inulinase production by *Saccharomyces* sp. in solid state fermentation using wheat bran as substrate. Ann Microbiol 62:843–848

Park Y-C, Oh EJ, Jo J-H, Jin Y-S, Seo J-H (2016) Recent advances in biological production of sugar alcohols. Curr Opin Biotechnol 37:105–113

Parrella A, Caterino E, Cangiano M, Criscuolo E, Russo C, Lavorgna M, Isidori M (2012) Antioxidant properties of different milk fermented with lactic acid bacteria and yeast. Int J Food Sci Tech 47:2493–2502

Pedrolli DB, Monteiro AC, Gomes E, Carmona EC (2009) Pectin and pectinases: production, characterization and industrial application of microbial pectinolytic enzymes. Open Biotechnol J 3:9–18

Rasika D, Ueda T, Jayakody L, Suriyagoda L, Silva K, Ando S, Vidanarachchi J (2015) ACE-inhibitory activity of milk fermented with *Saccharomyces cerevisiae* K7 and *Lactococcus lactis* subsp. *lactis* NBRC 12007. J Natl Sci Found 43:141–151

Ribereau-Gayon P, Dubourdieu D, Donèche B, Lonvaud A (2006) Handbook of enology, volume 1: the microbiology of wine and vinification. John Wiley & Sons, Ltd, Chichester, UK. ISBN 0, 471(97362)

Roy MK, Watanabe Y, Tamai Y (1999) Induction of apoptosis in HL-60 cells by skimmed milk digested with a proteolytic enzyme from the yeast *Saccharomyces cerevisiae*. J Biosci Bioeng 88:426–432

Roy MK, Watanabe Y, Tamai Y (2000) Yeast protease B-digested skimmed milk inhibits angiotensin-I-converting-enzyme activity. Biotechnol Appl Biochem 31:95–100

Saranraj P, Sivasakthivelan P, Suganthi K (2017) Baker's yeast: historical development, genetic characteristics, biochemistry, fermentation and downstream processing. J Acad Ind Res (JAIR) 6:111–119

Schousboe I (1976) Properties of triacylglycerol lipase in a mitochondrial fraction from baker's yeast (*Saccharomyces cerevisiae*). Biochim Biophys Acta (BBA)-Lipids and Lipid Metabolism 450:165–174

Seip JE, Di Cosimo R (1992) Optimization of accessible catalase activity in polyacrylamide gel-immobilized *Saccharomyces cerevisiae*. Biotechnol Bioeng 40:638–642

Sharma VK, Kumar N, Prakash T, Taylor TD (2009) MetaBioME: a database to explore commercially useful enzymes in metagenomic datasets. Nucleic Acids Res 38:D468–D472

Shi S, Valle-Rodríguez JO, Khoomrung S, Siewers V, Nielsen J (2012) Functional expression and characterization of five wax ester synthases in *Saccharomyces cerevisiae* and their utility for biodiesel production. Biotechnol Biofuels 5:7

Shynkaryk M, Lebovka N, Lanoisellé J-L, Nonus M, Bedel-Clotour C, Vorobiev E (2009) Electrically-assisted extraction of bio-products using high pressure disruption of yeast cells (*Saccharomyces cerevisiae*). J Food Eng 92:189–195

Singh RS, Singh RP (2017) 18 - Inulinases. In: Pandey A, Negi S, Soccol CR (eds) Current developments in biotechnology and bioengineering - production, isolation and purification of industrial products. Elsevier, Amsterdam, pp 423–446

Singh SK, Sczakas G, Soccol CR, Pandey A (2008) Production of enzymes by solid-state fermentation. In: Pandey A, Negi S, Soccol CR (eds) Current developments in solid-state fermentation. Springer, pp 183–204

Spier M, Nogueira A, Alberti A, Gomes T, Dhillon G (2016) Potential applications of enzymes in brewery and winery. In: Dhillon GS, Kaur S (eds) Agro-industrial wastes as feedstock for enzyme production - apply and exploit the emerging and valuable use options of waste biomass. Elsevier, Academic Press, London, pp 261–278

Sundarram A, Murthy TPK (2014) α-Amylase production and applications: a review. J Appl Environ Microbiol 2:166–175

Taketani S, Nishino T, Katsuki H (1981) Purification and properties of sterol-ester hydrolase from *Saccharomyces cerevisiae*. J Biochem 89:1667–1673

Tavano OL (2013) Protein hydrolysis using proteases: an important tool for food biotechnology. J Mol Catal B Enzym 90:1–11

van Zyl W, den Haan R, Rose S, la Grange D (2015) Expression of fungal hydrolases in Saccharomyces cerevisiae. In: Himmel ME (ed) Direct microbial conversion of biomass to advanced biofuels. Elsevier, Waltham, MA, pp 153–175

Venkateshwaran G, Somashekar D, Prakash MH, Agrawal R, Basappa SC, Joseph R (1999) Production and utilization of catalase using *Saccharomyces cerevisiae*. Process Biochem 34:187–191

Vermeirssen V, Van Camp J, Decroos K, Van Wijmelbeke L, Verstraete W (2003) The impact of fermentation and in vitro digestion on the formation of angiotensin-I-converting enzyme inhibitory activity from pea and whey protein. J Dairy Sci 86:429–438

Vieira E, Brandão T, Ferreira IM (2013) Evaluation of Brewer's spent yeast to produce flavor enhancer nucleotides: influence of serial repitching. J Agric Food Chem 61:8724–8729

Vieira E, Teixeira J, Ferreira IM (2016a) Valorization of brewers' spent grain and spent yeast through protein hydrolysates with antioxidant properties. Eur Food Res Technol 242:1975–1984

Vieira EF, Carvalho J, Pinto E, Cunha S, Almeida AA, Ferreira IM (2016c) Nutritive value, antioxidant activity and phenolic compounds profile of brewer's spent yeast extract. J Food Compost Anal 52:44–51

Vieira EF, da Silva DD, Carmo H, Ferreira IM (2017a) Protective ability against oxidative stress of brewers' spent grain protein hydrolysates. Food Chem 228:602–609

Vieira EF, das Neves J, Vitorino R, Dias da Silva D, Carmo H, Ferreira IM (2016b) Impact of in vitro gastrointestinal digestion and transepithelial transport on antioxidant and ACE-inhibitory activities of Brewer's spent yeast autolysate. J Agric Food Chem 64:7335–7341

Vieira EF, Ferreira IM (2017) Antioxidant and antihypertensive hydrolysates obtained from by-products of cannery sardine and brewing industries. Int J Food Prop 20:662–673

Vieira EF, Melo A, Ferreira IM (2017d) Autolysis of intracellular content of Brewer's spent yeast to maximize ACE-inhibitory and antioxidant activities. LWT—Food Sci Technol 82:255–259

Vieira EF, Pinho O, Ferreira IM (2017b) Bio-functional properties of sardine protein hydrolysates obtained by brewer's spent yeast and commercial proteases. J Sci Food Agric 97:5414–5422

Vieira EF, Van Camp J, Ferreira IM, Grootaert C (2017c) Protein hydrolysate from canned sardine and brewing by-products improves TNF-α-induced inflammation in an intestinal-endothelial co-culture cell model. Eur J Nutr:1–12

Wangpor J, Prayoonyong P, Sakdaronnarong C, Sungpet A, Jonglertjunya W (2017) Bioethanol production from cassava starch by enzymatic hydrolysis, fermentation and ex-situ nanofiltration. Energy Procedia 138:883–888

Wilkins MR, Widmer WW, Grohmann K (2007) Simultaneous saccharification and fermentation of citrus peel waste by *Saccharomyces cerevisiae* to produce ethanol. Process Biochem 42:1614–1619

Younes B, Cilindre C, Jeandet P, Vasserot Y (2013) Enzymatic hydrolysis of thermo-sensitive grape proteins by a yeast protease as revealed by a proteomic approach. Food Res Int 54:1298–1301

Younes B, Cilindre C, Villaume S, Parmentier M, Jeandet P, Vasserot Y (2011) Evidence for an extracellular acid proteolytic activity secreted by living cells of *Saccharomyces cerevisiae* PlR1: impact on grape proteins. J Agric Food Chem 59:6239–6246

Zhang C, Kim S-K (2012) Chapter 28—Application of marine microbial enzymes in the food and pharmaceutical industries. In: Kim S-K (ed) Advances in food and nutrition research. Academic Press, Cambridge, MA, pp 423–435

Laccase: Recombinant Expression, Engineering and Its Promising Applications

3

P. Sai Preethi, Ajitha Gomathi, Ramachandran Srinivasan, J. G. S. Pavan Kumar, K. Murugesan, and Gothandam Kodiveri Muthukailannan

Abstract

Laccases are cuproproteins belonging to the oxidoreductase family that catalyse the oxidation of various aromatic and non-aromatic compounds. With wide occurrence in bacteria, fungi, higher plants and insects, laccases have been majorly exploited due to their pivotal role in bioremediation and industrial uses. Low productivity and high cost limit the use of native laccases. Besides these drawbacks, native sources yield a mixture of isozymes, which is not desirable for commercial scopes. Thus, in conjunction with this, recombinant laccases have gained importance for sustainable processing due to their improved catalysis, stabilities, and solvent and anionic tolerance. Henceforth, this review mainly focuses to accentuate the recombinant laccases from fungal, mushroom, bacterial and actinobacterial sources, laccase engineering and their proficient industrial applications. The characteristic traits of the recombinant forms of laccase from distinct sources emphasise the requirement to study them more for their upcoming, hidden environmental and biotechnological applications.

3.1 Introduction

Laccases are oxidoreductases or *p*-diphenol oxidases (EC 1.10.3.2). They belong to the enzyme superfamily of multicopper oxidases, which is a widely distributed protein family among prokaryotes and eukaryotes (Hoegger et al. 2006). Laccases can

P. S. Preethi · A. Gomathi · R. Srinivasan · J. G. S. Pavan Kumar ·
G. Kodiveri Muthukailannan (✉)
School of Bio Sciences and Technology, Vellore Institute of Technology, Vellore, India

K. Murugesan
Department of Environmental Sciences, Periyar University, Salem, India

N. K. Arora et al. (eds.), *Microbial Enzymes: Roles and Applications in Industries*, Microorganisms for Sustainability 11,
https://doi.org/10.1007/978-981-15-1710-5_3

be produced from fungi, bacteria, actinobacteria, insects, mushrooms and plants (Murugesan et al. 2007; Devasia and Nair 2016; Alves et al. 2014; Arakane et al. 2005; Zhang et al. 2010; Jaiswal et al. 2014). The catalytic activity of laccase is driven by the presence of four copper atoms in their redox sites and they do not require H_2O_2 for substrate oxidation. Laccase production can be enhanced by using dry biomass of *Anacystis nidulans* (a cyanobacteria) as a supplement in solid-state fermentation (Mishra and Kumar 2007).

Laccases oxidise a wide range of phenolic compounds and non-phenolic compounds either directly or through redox mediator systems. Laccases are able to oxidise various aromatic pollutants such as bisphenol A (Zeng et al. 2017), anthracene (Wu et al. 2010) and triclosan (Murugesan et al. 2010). Laccases potentially degrade anthroquinone dyes such as acid green (AG) 25 and diazo dye acid red (AR) 18 (Wang et al. 2012), remazol brilliant blue R, reactive blue 4 (Afreen et al. 2016) and azo dyes such as reactive black 5 and reactive orange 16 (Lade et al. 2015).

Laccases widely catalyze 2–2′-azine *bis* (3-ethylbenzothiazoline-6-sulphonic acid) (ABTS) as the main source of synthetic substrate. Several studies for characterisation of microbial laccases (Lu et al. 2013; Murugesan et al. 2009) and plant laccases (Jaiswal et al. 2014) have used ABTS as the indispensable substrate for assessing laccase activity. 1-Hydroxybenzotriazole is the phenolic substrate as well as a mediator compound used for distinct applications such as degradation of endocrine-disrupting chemicals (Zeng et al. 2017) and environmental pollutants (Murugesan et al. 2010). Guaiacol, *p*-cresol, *p*-aminophenol, *p*-phenylenediamine, hydroquinone and tropolone are the other phenolic substrates that are widely exploited to study laccase activity (Devasia and Nair 2016). Laccases are highly exploited in commercial industries like textile (Luciana et al. 2009), food (Renzetti et al. 2010), pulp and bleaching (Arias et al. 2003; Shankar et al. 2018), cosmetics (Jeon et al. 2010), nanobiotechnology (Franzoi et al. 2009b), green chemistry (Jeon et al. 2012) and bioremediation (Balcázar-lópez et al. 2016; Naraian et al. 2018).

Recombinant DNA molecules are new artificial DNA strands that are produced by combining two unrelated (non-homologous) genes (Glick et al. 2010), for example hybrid of *Escherichia coli* plasmid with *Kurthia huakuii* LAM0618 laccase gene (Lam et al. 2016). Recombinant laccases have successfully accomplished promising roles for several applications. For instance, the recombinant laccase, Lac15 expressed in *E. coli*, proved its wide potential for decolourising reactive azo dyes under alkaline environment (Ge et al. 2011). Laccase expressed from *Trametes versicolor* in maize seeds is used as an excellent polymerisation agent (Bailey et al. 2004). Figure 3.1 illustrates the steps involved in recombinant DNA technology using bacterial system as host.

Laccases also exuberate multifarious medical properties. For instance, *Fusarium* laccase exhibited cytotoxic effect on Hep2 cell line, revealing its anticancer activity (Arul and Shanmugam 2012). Additionally, laccase inhibit the hepatitis C virus entry into peripheral blood cells and hepatoma cells (El-Fakharany et al. 2010). Contradicting with these, laccase from *Albatrellus dispansus* failed to inhibit HIV-1 reverse transcriptase enzyme, even at 1 mg/mL concentration (Wang and Ng 2004). These applications of laccases prove for their enhanced production as recombinant forms for discrete applications.

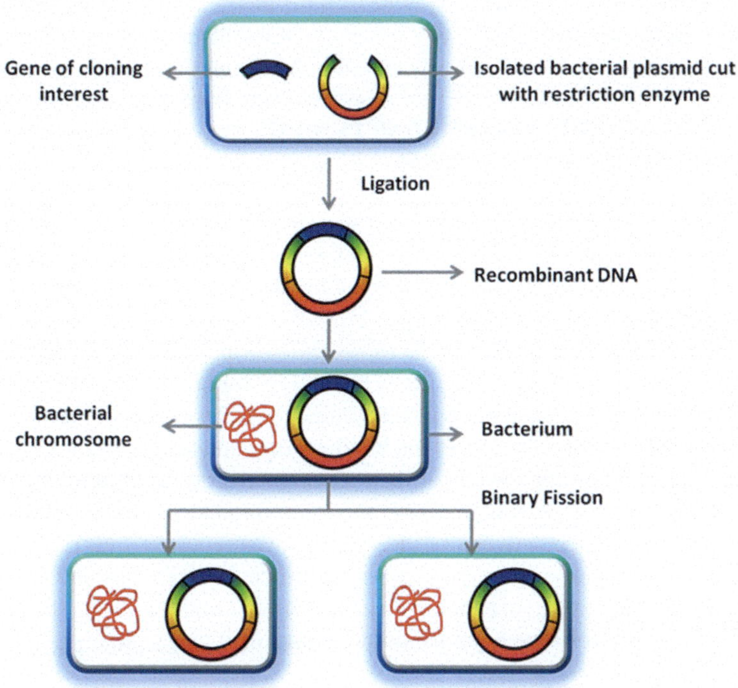

Fig. 3.1 A schematic diagram of basic steps involved in the construction of recombinant or chimeric DNA using a bacterial host for enhancing the production of required product

So far, around 50 cloning and expression studies have been effectively carried out in various sources of laccases such as plants, bacteria, fungi, actinobacteria and mushroom by using the expression systems such as *E. coli* BL21, *E. coli* DH5 α, *E. coli* XL1, *Kluyveromyces lactis, Saccharomyces cerevisiae, Yarrowia lipolytica, Pichia pastoris, Aspergillus oryzae, Aspergillus sojae* strain 1860, *Pichia methanolica* and transgenic maize. Laccase produced from the native microbial sources are not applicable at industrial scale due to their high cost of preparation, purification processes and low production yields. Furthermore, the laccases synthesised from native sources cannot meet the exceeding market demand due to the incompatibility of standard industrial fermentations with optimal growth conditions of several microbes (Piscitelli et al. 2010). Also, the scaling up properties exhibited by the recombinant laccases have opened up attractive commercial opportunities; adding to these advantages, the process of expressing the laccase genes recombinantly becomes a necessity to meet large-scale enzymatic needs in shorter time with desired properties such as high stability, substrate specificity and purity (Ferrer-Miralles et al. 2009). Recombinant laccases can be produced according to either the need of the user or the process (Demain and Vaishnav 2009).

Thus, considering all these desirable traits, this comprehensive review mainly aims to focus on recombinant laccases and their engineering prospective, attesting the cloning reports of laccases available in the market from a wide category of microbial sources and their biotechnological applications.

3.2 Cloning and Expression of Fungal and Mushroom Laccases

Laccases are quite commonly produced by basidiomycetes. Laccase-encoding genes from a wide range of fungi have been cloned in suitable vectors and expressed in *P. pastoris* host system (Colao et al. 2006; Chao and Bao 2003; Soden et al. 2002; Chiara et al. 2009). The white-rot fungus *Trametes trogii* is highly capable of producing five isoforms of laccases. In the three-dimensional structure of recombinant laccase (Lcc2) of *T. trogii,* the copper ions and ABTS-binding pocket are found. The basic architecture of Lcc2 copper and substrate sites is very identical to that of previously reported laccase (Lcc1) and the residues coordinating the four copper ions are conserved. The binding cavity for the reducing substrate is close to the T1 copper site, and is rather wide, which allows the accommodation of molecules of various sizes. The difference between Lcc1 and Lcc2 was reported with respect to two spatially close residues surrounding the substrate cavity. Thr 164 and Ser 264 of Lcc1 are replaced by two hydrophobic residues Phe 163 and Ile 265, respectively, in Lcc2. This variation restricts Lcc2 from interacting with bulky polar groups in substrates (Chiara et al. 2009).

Laccase-encoding gene from *Trametes versicolor* was expressed in different expression systems such as *Yarrowia lipolytica* (Jolivalt et al. 2005), *Saccharomyces cerevisiae* (Necochea et al. 2005), *Pichia methanolica* (Guo et al. 2006) transgenic maize (Hood et al. 2003) and *P. pastoris* (Jönsson et al. 1997; Brown et al. 2002). Supplementary to the above investigations, laccase expression has also been achieved in other *Trametes* strains. *Trametes* sp. AH28–2 laccase was heterologously overexpressed in the yeast, *P. pastoris* strain GS115 (Li et al. 2007). Expressing the enzyme-coding genes heterologously in suitable host systems could enhance the yields without impeding the subsequent downstream processing and enzyme applications.

The laccase genes from *Pleurotus ostreatus* expressed recombinantly in *K. lactis and S. cerevisiae* resulted in higher yield of the recombinant enzyme only in *K. lactis*, which can be highly recommended for biotechnological applications such as dye decolourisation and effluent treatment (Piscitelli et al. 2005). Cloning and expression of laccase gene from *Pycnoporus cinnabarinus* produced higher yield in *Aspergillus niger* compared to *P. pastoris* due to the usage of laccase signal peptide, glucoamylase preprosequence (Otterbein et al. 2000; Record et al. 2002). The expression of laccase-coding gene from *P. eryngii* in immobilised cells of *S. cerevisiae* led to 1.6-fold higher yield than that of free cells, which contributes to the potential of immobilised cells to get explored widely (Bleve et al. 2008). Hence, it follows that immobilisation of microorganisms capable of degrading specific contaminants significantly supports bioremediation processes, reduces their costs and allows for the multifarious use of biocatalysts. Ultimately, the expression of the enzyme-coding genes in entrapped microbial cells could confer higher protein outturn than the free microbial cells. Figure 3.2 presents a list of vital immobilisation carriers, applicable in bioremediation.

Mushroom laccases have been assigned several physiological functions similar to those of other fungal laccases. Several studies have been examined using mushroom laccases. For instance, a laccase from *Hypsizygus ulmarius* and *Volvariella volvacea* was recombinantly expressed in *E. coli*. In the former case, copper was found to enhance the laccase regulation and production but, in the latter, no recombinant laccase was induced by copper (Ravikumar et al. 2013; Chen et al. 2004). Thus, supplementing copper into growth medium might increase the laccase yield (due to the presence of copper atoms in active sites).

Another study reported on the laccase genes that are related to the stipe elongation, examined in stipe tissue of *V. volvacea* (an edible mushroom). In this investigation, the transcription patterns of 11 laccase genes revealed that the expression of Vvlcc3 was highest among other genes. Hence, only this gene was expressed in *P. pastoris* and the corresponding Vvlcc3 laccases were found to be highly identical with basidiomycete laccase. The Vvlcc3 product comprised of 515 amino acids, which included 496 signal peptides having 19 amino acids. The expression of Vvlcc3 was reported to exceed, peak and decrease in egg stage, elongation stage and mature stage, respectively (Lu et al. 2015). Thus, additionally gene knockout can be performed for laccase-encoding genes for enlightening the other characteristic roles of laccase in the development of *V. volvacea*, whereby an explicit idea is to apply them further for application in the food and commercial industries.

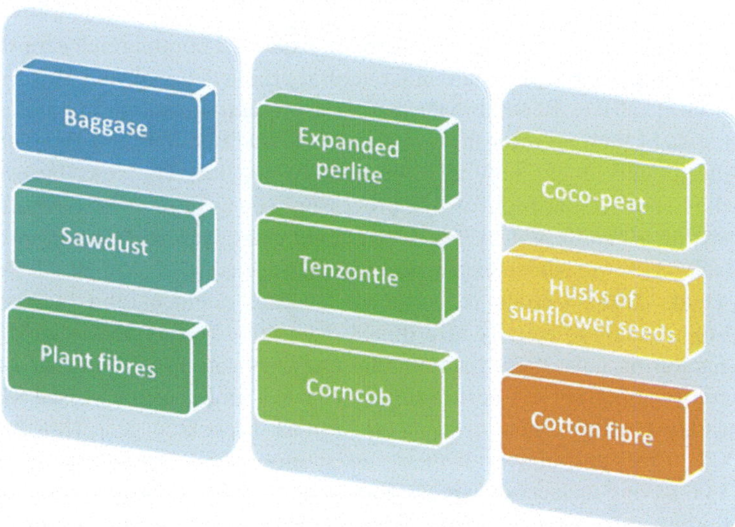

Fig. 3.2 A pictorial view of currently used, natural, inexpensive, promising and biocompatible carriers for immobilisation of microbial cells to accomplish bioremediation

Lentinula edodes, the white-rot fungus which is one of the most widely cultivated mushrooms in Asia, has laccase and manganese peroxidase activity with little lignin peroxidase activity. *L. edodes* laccase, expressed in tobacco BY-2 cells, had smaller molecular weight and exhibited a different pattern of post-translational modification to that of native *L. edodes* laccase and both the enzymes were able to decolourise the same set of dyes (Sakamoto et al. 2008). As laccases, manganese peroxidases and lignin peroxidases are ligninolytic enzymes representing the single family of oxidoreductase, it could be feasible for the same organism to accomplish the coproduction of these variegated enzymes at a time.

Rigidoporus lignosus is a basidiomycete causing white rot of roots in about 100 species of trees. *R. lignosus* laccase was expressed in *E. coli*, and its redox reaction monitored with potentiometry indicated that current increased in a solution containing recombinant laccase with increase in syringaldazine concentration. Furthermore, it was suggested that this enzyme can be used as a potential biosensor (Nicolini et al. 2013). A novel expression study from the laccase genes of the white-rot fungus, *Auricularia auricula-judae* Au916, was reported previously, in which seven laccase genes containing signature sequences were isolated from *A. auricula-judae* Au916 based on the mycelium-derived transcriptome and expressed in *E. coli* DH5α (Fan et al. 2014). Besides these studies, the laccase genes from a number of other fungal and mushroom sources have been expressed heterologously in various host systems (Table 3.1).

3.3 Cloning and Expression of Actinobacterial Laccases

Actinomycetes comprise a large and diverse group of filamentous bacteria, many of which have important ecological roles and are exploited commercially for the production of antibiotics, enzymes and other natural products (Cook and Meyers 2003; Waksman et al. 2010). Recombinant laccase from actinobacteria exhibits special characteristics of thermal robustness, resistance to metallic ions and chemical inhibition. Additionally, they show alkaline activity profiles. They can be used and employed as an effectual tool for dye processing (Feng et al. 2015). An identical feature was also manifested by Sil A, a recombinant laccase from *Streptomyces ipomoea* CECT 3341, whose expression was performed in *E. coli* BL21 (Molina-guijarro et al. 2009). Certain recombinant laccases could exhibit high stability retaining their catalytic activity as a dimer, even after boiling and treatment with SDS in denaturing gels; these recombinant enzymes can also exhibit an unprecedented activity at high pH. A best example is SLAC, an *E. coli*-expressed recombinant biocatalyst, with *Streptomyces coelicolor* as the source gene (Machczynski et al. 2004). Over and above all these, an actinobacteria, *Streptomyces cyaneus* CECT 3335, having its laccase expression host as *E. coli*, was able to benchmark its family of other actinomycetes, by producing a recombinant enzyme with out-put of 104 mg/L, which is a challenging for production by other microbes (Ece et al. 2017). Along these lines, it is certain that actinomycetes could be one of the leading producers of the commercial enzyme apart from executing their sterling role in the production of antibiotics and other commercial products.

Table 3.1 List of various recombinantly expressed fungal and mushroom laccases in different host expression systems

Gene	Source	Host	Comments	Reference
clac1, clac3	*Trametes* sp. strain C30	*Saccharomyces cerevisiae*	2 mg/L of recombinant LAC3 was produced	Klonowska et al. (2005)
poxc and poxa1b	*Pleurotus ostreatus*	*Kluyveromyces lactis, Saccharomyces cerevisiae*	1.1 and 1.4 mg of rPOXC and rPOXA1b were produced, respectively, by *Kluyveromyces lactis,* which showed higher production than *Saccharomyces cerevisiae*	Piscitelli et al. (2005)
Laccase IIIb	*Trametes versicolor*	*Yarrowia lipolytica*	2.5 mg/L of recombinant laccase was produced by yeast (host)	Jolivalt et al. (2005)
lac4	*Pleurotus sajor-caju*	*Pichia pastoris*	4.85 mg/L of recombinant laccase was produced	Soden et al. (2002)
mrl2	*Moniliophthora roreri*	*Pichia pastoris*	The recombinant enzyme was able to degrade endocrine-disrupting chemicals and non-steroidal anti-inflammatory drugs efficiently than faster than more potent laccase from *Trametes versicolor*	Bronikowski et al. (2017)
ery3	*Pleurotus eryngii*	*Saccharomyces cerevisiae*	139 mU/mL of laccase was produced in the immobilised cells of the yeast, which was increased 1.6-fold times higher than that of free cells	Bleve et al. (2008)
poxc and *poxa1b*	*P. ostreatus*	*Saccharomyces cerevisiae*	The clone 1M9B selected out of 1100 showed a single mutation (L112F), enhancing the enzyme activity but making it less stable than the wild-type enzyme (POXA1b)	Festa et al. (2008)
Laccase	*Schizophyllum commune*	*Aspergillus sojae* strain 1860	The maximum laccase activity was found to be 774 U/mL	Hatamoto et al. (1999)
Laccase I	*Trametes* sp.	Transgenic maize	The molecular parameters that induce high expression were the maize embryo-preferred globulin 1 promoter and along with protein targeting to the cell wall	Hood et al. (2003)

(continued)

Table 3.1 (continued)

Gene	Source	Host	Comments	Reference
lac3, *lac4*	*Coprinus comatus*	*Pichia pastoris*	Expressed enzymes were able to resist sodium dodecyl sulphate (SDS) and were capable of degrading triarylmethane dyes but were not suitable for azo and anthraquinone dyes	Gu et al. (2014)
lacB	*Trametes* sp. AH28–2	*Pichia pastoris GS115*	Maximum yield of recombinant enzyme was (32,000 U/L), higher than the native enzyme LacB, (30,000 U/L)	Li et al. (2007)
fvlac7	*Flammulina velutipes*	*Escherichia coli* DH10B	The fvlac7 laccase amino acid sequence was similar to lac2 (45.5% identity) and lacFv (48.1% identity) from *F. velutipes*. Thus, fvlac7 laccase is a novel laccase	Kim et al. (2013)
lcc1	*Trametes versicolor*	*Pichia pastoris*	Using, *P. pastoris pep4* mutant strain SMD1168 was reported to result in twofold higher level of recombinant enzyme activity than *P. pastoris* GS115	Jonsson et al. (1997)
lccIV	*Trametes versicolor*	*P. pastoris* strain GS115	The usage of native signal peptide led to the correct proteolytic protein processing of LccIV to mature form, whereas *Saccharomyces α*-mating factor signal peptide resulted in additional tetra-peptide in the N-terminal end of the enzyme and 25% lower specific activity	Brown et al. (2002)
lcc1	*Trametes versicolor*	*Pichia methanolica* (PMAD11 and PMAD16)	PMAD11 strain of *Pichia methanolica* had higher laccase extracellular activity compared to PMAD16. Laccase activity in case of PMAD11 recombinant was about 12.6 U/mL	Guo et al. (2006)

(continued)

Table 3.1 (continued)

Gene	Source	Host	Comments	Reference
Lcc	*Fome lignosus*	*Pichia pastoris* GS115	Mutagenesis enhanced expression 3.7-fold to 144 mg/L of recombinant laccase in yeast, together with a 1.4-fold increase in kcat	Hu et al. (2007)
Lccα	*Trametes versicolor*	*Saccharomyces cerevisiae*	200 μg/mL of recombinant laccase was produced in *Saccharomyces cerevisiae*	Necochea et al. (2005)
lac1	*Pycnoporus cinnabarinus I-937*	*Pichia pastoris*	8 mg/L of the recombinant protein was produced in expression system	Otterbein et al. (2000)
lac1	*Pycnoporus cinnabarinus*	*Aspergillus niger*	When compared to wild-type laccase (45 mg/L), the recombinant laccase obtained was 1 mg/L, from the dikaryotic strain I-937 of *P. cinnabarinus*, and about 145 mg/L of recombinant enzyme was produced by the monokaryotic strain ss3 of *P. cinnabarinus*	Record et al. (2002)
Lac-T16	*Polyporus grammocephalus* TR16	*Pichia pastoris*	The activity of the laccase expressed in *P. pastoris was* 893.3 U/mL with α-factor secretion signal peptide	Huang et al. (2011a)
MtL	*Myceliophthora thermophila*	*Saccharomyces cerevisiae*	18 mg/L of laccase produced as the highest yield in the host by its heterologous expression	Bulter et al. (2003)
poxa3	*Pleurotus ostreatus*	*Kluyveromyces lactis, E. coli*	20 mU/mL of laccase activity was obtained between 14th and 17th days	Faraco et al. (2008)
lcc1-lcc7	*Auricularia auricula-judae* Au916	*Escherichia coli, strain DH5*	The lcc3 gene was highly expressed in both free-living and substrate mycelium; the lcc5 gene was mostly expressed during the fruiting-body formation and maturation revealing that lcc5gene is vital during the sexual reproduction stage	Fan et al. (2014)

(continued)

Table 3.1 (continued)

Gene	Source	Host	Comments	Reference
lacD	*Trametes* sp. 420	*Pichia pastoris* strain GS115	Laccase (8.3×10^4 U/L) activity was obtained for the recombinant laccase in the host	Hong et al. (2007)
Laccase	*Hypsizygus ulmarius*	*E. coli* (BL 21)	The recombinant laccase was characterised to be highly purified (23.23-fold) with 52% recovery	Ravikumar et al. (2013)
lac4	*Volvariella volvacea*	*E. coli* DH5 α	Lac3 amino acid sequence was a highly identical (77–80%) with sequences of basidiomycetes such as *Polyporus ciliatus*, *L. edodes* and *Pleurotus sajor-caju*	Chen et al. (2004)
Vvlcc3	*Volvariella volvacea*	*Pichia pastoris*	293.86 U/L of recombinant enzyme was obtained after 21 days of cultivation	Lu et al. (2015)

3.4 Cloning and Expression of Bacterial Laccases

In comparison to fungal laccases, bacterial laccases are tackling the drawbacks of instability and industrial operational applications. Bacterial laccases are very active and are more stable at high pH and temperature. Recently, more attention is paid to bacterial laccases to produce novel biocatalysts with respect to industrial prospects (Capalash et al. 2007). Novel green biocatalysts of *Bacillus licheniformis* from putative sources were identified as multicopper oxidase, whose gene was cloned and expressed in *E. coli*, and its catalytic properties were analysed specifically on oxidation of phenolic and non-phenolic acids (Koschorreck et al. 2008b). However, for future investigations, similar oxidative studies on nitroaromatic acids (NAA) are highly encouraged as NAA are military explosives, tending to accumulate widely in the environment resulting in undesirable human health hazards.

Bacillus subtilis, an aerobic Gram-positive and rod-shaped bacterium, can produce laccase by using fructose and peptone as the indispensable and vital sources. Laccase from *B. subtilis* is an important ligninolytic enzyme which can be produced in liquid culture in bulk volumes by submerged fermentation. A laccase gene from *B. subtilis* expressed recombinantly in *E. coli* DH5α potentially mineralised chlorophenol, which is a vital raw material in the chemical industry for producing preservatives, dyes and pesticides but on the other hand are also hazardous compounds which could affect the environment and humans (Menaka et al. 2015). Apart from pesticides, herbicides are also major polluting agents, as their overusage results in negative effects on environment. Hence, executing highly accountable studies on herbicide mineralisation cannot be averted.

A gene, namely cotA-encoding laccase from *Bacillus* sp. HR03, has been cloned and expressed in *E. coli* BL2 where the recombinant enzyme demonstrated an onliest L-Dopa oxidation, which is uncommon amidst its *Bacillus* family (Fathi-roudsari et al. 2010). An *E. coli*-expressed laccase from *Bacillus halodurans* C-125 has been recorded to be a dexterous biocatalyst for applications for which classical laccases are not suitable; such as kraft-pulp bio-bleaching for paper production due to its alkaline pH optimum of 7.5–8. The anions such as fluoride, chloride and cyanide inhibited the laccase activity by binding to the various copper sites of the enzyme, as a consequence of interrupting the internal electron transfer among the different copper active sites. Similar kind of properties were also recounted in recombinant laccase produced by *Ochrobactrum* sp. 531 (Ruijssenaars and Hartmans 2004; Li et al. 2012). Being very unique and advanced than *B. halodurans* C-125 and *Ochrobactrum* sp. laccases, a bacterial laccase, lac15 expressed recombinantly in *E. coli*, revealed both its alkalescence and anion tolerance activity (Fang et al. 2011). Laccase of *Lactobacillus plantarum* J16 (CECT 8944), expressed in *E. coli* BL21, showed active role in the degradation of biogenic amines, thereby acquiring an aptitude to eliminate toxic compounds in fermented foods and beverages (Callejón et al. 2016). The splendid quality of *Bacillus* strains to produce and secrete the extracellular enzymes in bulk (20–25 g/L) has made them among the most important industrial enzyme producers. Table 3.2 represents recombinant expression of bacterial laccases in *E.coli* expression host.

3.5 Laccase Engineering

Enzyme engineering is a trending and fast-growing research platform, mainly as a consequence of recent developments in rational mutagenesis, directed evolution and in silico techniques (Nanda and Koder 2010; Tracewell and Arnold 2009). To offer an insight into the plasticity of laccases and their functional evolutionary path within the broad copper-binding domain family of proteins is a challenging and inspirational task. It is also a laborious task to engineer laccases into the novel and robust artificial enzymes with fascinating properties. The commencement of enzymes to multiple cupredoxin domain (laccase, ceruloplasmin or ferroxidases, nitrite reductase) from single cupredoxin domain due to evolutionary factors has involved several methodologies like gene duplications, loss or acquisition of domain recruitment and copper-binding sites (Nakamura and Go 2005; Nersissian and Shipp 2002). With the advanced technological improvements several investigations were made to shed light on the enzyme engineering and scopes. Hence, a recent novel report has addressed the computer-aided laccase engineering and the improvement in substrate oxidation with directed evolution. Besides the evolutionary studies, with respect to the bioremediatory prospects, oxidation of aniline compounds has also been accomplished with laccase engineering strategies (Monza et al. 2015; Santiago et al. 2016). Thus, the computer-aided laccase engineering could be highly recommended for improved degradation of toxic dye effluents, polyaromatic hydrocarbons (PAHs), plastics and pharmaceuticals.

In the panel of the goals raised out for laccase engineering, heterologous functional expression, thermostability, improved activity, tolerance to organic solvents, physiological fluids, ionic liquids and resistance to distinct inhibitors are the major challenges for the clear comprehending of the laccase structural and functional relationships. Figure 3.3 illustrates the various laccase engineering techniques and their advantages.

Table 3.2 List of various recombinantly expressed bacterial laccases in suitable host expression systems

Gene	Source	Host	Comments	Reference
Cot A	*Bacillus subtilis* PTCC1023	*E. coli* BL21 (DE3)	The recombinant laccase produced in the host exhibited 99% sequence identity with *Bacillus subtilis*	Ghasemi et al. (2014)
cotA	*Bacillus licheniformis*	*Escherichia coli*	CotA protein was not able to oxidise coumaric acid, vanillic acid and cinnamic acid, except syringic acid. Dimerisation of caffeic acid, sinapic acid and ferulic acid by cotA protein was investigated. Among all the non-phenolic acids, cotA exhibited the highest activity towards sinapic acid	Koschorreck et al. (2008b)
sufI	*Lactobacillus plantarum* J16 (CECT 8944)	*Escherichia coli* BL21	The heterogeneously expressed recombinant laccase had wide potentiality to oxidise biological amines like tyramine, which will be useful to eliminate toxic compounds from beverages and fermented food	Callejón et al. (2016)
Laccase	*Bacillus subtilis*	*E. coli* DH5 α	123 mg/mL of recombinant product was produced with the highest enzyme activity of 200 U/mL	Menaka et al. (2015)
CotA	*Bacillus* sp. HR03	*E. coli* BL21	CotA from *Bacillus* sp. HR03 showed 98.2% identity and 98.8% similarity with CotA from *B. subtilis*	Fathi-roudsari et al. (2010)
lbh1	*Bacillus halodurans* C-125	*Escherichia coli* BL21	Activity with dimethoxyphenol was much higher (54 U/mL) than ABTS (0.4 U/mL) and syringaldazine (2.6 U/mL)	Ruijssenaars and Hartmans (2004)
Laccase	*Ochrobactrum*. Sp. 531	*E. coli* BL21 (DE3)	The expressed MCO protein oxidised syringaldazine and dimethoxyphenol under alkaline conditions unlike the classical laccases	Li et al. (2012)
lac15	Marine microbial metagenome	*E. coli* BL21	High chloride tolerance and alkalescence-dependent activity reveal that the recombinantly expressed laccase can be used for industrial applications	Fang et al. (2011)

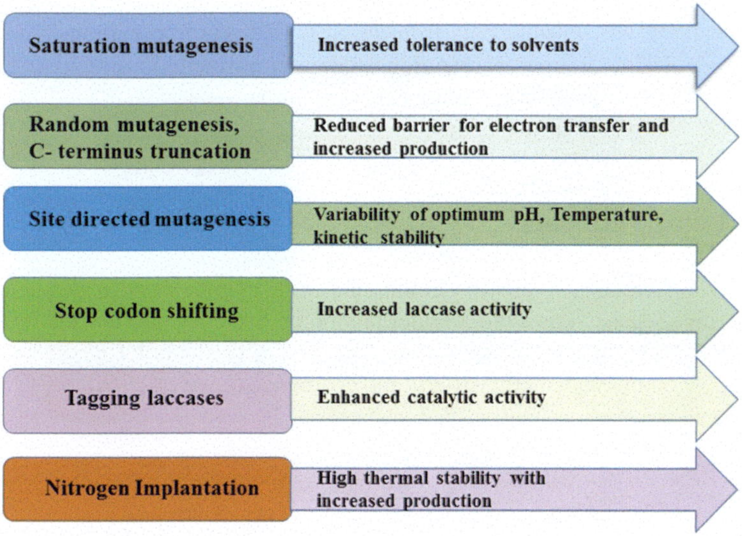

Fig. 3.3 An overview of various promising techniques for laccase engineering with their corresponding advantages and effects on the engineered biocatalyst

3.6 Multifaceted Applications of Recombinant Laccases

Recombinant laccases are one of the unbeatable options, for various biotechnological applications. For example, the overexpression of a potato laccase (PPO) in tomato conferred an enhanced resistance to the transgenic plant, against a bacterial pathogen (Li and Steffens 2002). Recombinant laccases could also carry out effective degradation of recalcitrant synthetic dyes such as congo red, remazol brilliant blue R (RBBR), aniline blue and even the biopolymer lignin (Shi et al. 2015). Apart from these biocatalysts, a recombinant laccase from *P. pastoris* was efficacious to biotransform various types of dyes such as crystal violet (azo), congo red (triphenylmethane), indigo carmine (indigo) and RBBR without any redox mediators (Lu et al. 2009). A potential use of recombinant *P. ostreatus* laccase for degradation of complex triazo and polyazo dyes, such as acid yellow 49, acid red 266, acid blue 62, direct blue 71, direct red 80 and direct yellow 106, was also suggested (Miele et al. 2010). Out of ordinary to other recombinant laccases, the enzyme from *B. subtilis* can potentially produce oligomers as a result of degradation of an azo dye, Sudan Orange G, through radical coupling reaction (Pereira et al. 2009).

Extending their application to wider prospects, laccases also become promising candidates for degradation of micropollutants and xenobiotic compounds. Laccase from *Trametes versicolor*, expressed in *P. pastoris*, favoured the deterioration of acenaphthene, fluorene, anthracene and acenaphthylene (Koschorreck et al. 2008a). Some potential laccases are capable of degrading a panel of phenolic compounds. Particularly, a recombinant laccase of *Trametes sanguineus*, heterologously expressed *in Trichoderma atroviride*, revealed to degrade PAHs, like

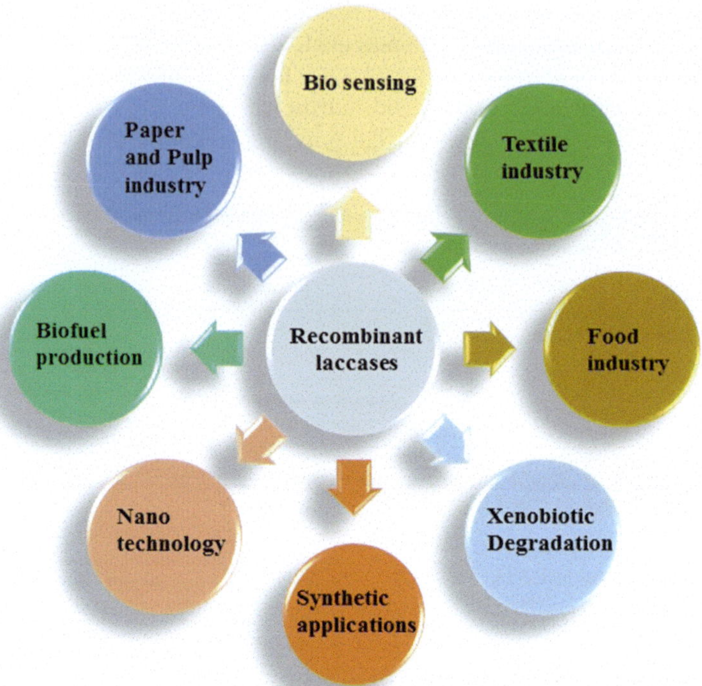

Fig. 3.4 A schematic illustration of major applications of recombinant laccases in bioremediation and industries

benzo[α]pyrene and phenanthrene with endocrine disruptors like bisphenol A (Balcázar-lópez et al. 2016). Figure 3.4 enlists the distinct applications of recombinant laccases.

Mineralisation of noxious micropollutants such as polychlorinated biphenyls (PCBs) can also be exacerbated by recombinant laccases (Fujihiro et al. 2009). The catalytic activity of a novel laccase from *L. edodes*, expressed in *P. pastoris*, has been analysed on benchmark substrates, PAHs mineralisation and dye biotransformation (Wong et al. 2013).

Laccase systems are also well-recruited green biocatalysts for lignocellulose degradation processes which aids in biofuel production. Recently *Myceliophthora thermophila* recombinant laccase was suggested to be an excellent enzyme for degradation of lignin present in ground eucalypt wood (Rico et al. 2014).

Interestingly, the same mechanism of lignin degradation process aided by laccase is used for paper manufacturing in the bio-bleaching and biopulping industries. To accomplish this objective, several studies have been carried out using laccases. For instance, Ravalson and colleagues generated a chimeric laccase by performing the fusion of carbohydrate-binding module (CBM) of *A. niger* cellobiohydrolase to *P. cinnabarinus* laccase. The resultant recombinant laccase has been investigated to have a high potential in softwood pulp delignification compared to the native enzyme due to the additional CBM moiety (Ravalason et al. 2009). Expression of *P. cinnabarinus* laccase in two different Aspergilli hosts resulted in the host-specific recombinant product (Sigoillot et al. 2004). A similar kind of approach of linking certain peptide sequences with *Stachybotrys chartarum* laccase at the C-terminal has been studied to remove carotenoid pigmental stains on fabrics. The resultant laccase moiety was suggested to be a robust biocatalyst for usage in the bleaching industry (Janssen et al. 2004).

Laccases are also involved in synthesis apart from degradative processes. Several natural products formed by laccases are coloured, which are synthesised by oxidative coupling reactions. This versatile enzyme also produces pigments due to oxidation reactions. The interesting fact is that most of the coloured products formed by laccases are soluble in water, and thus these compounds can be used as dyes in the textile and paper industries. For instance, an engineered laccase isozyme, Lac3 from *Trametes* sp. expressed in the yeast *S. cerevisiae*, was employed to form a trimer 2, 2-*bis*(3′-indolyl)indoxyl (yellow compound) from indole dye using TEMPO and dioxygen as substrate partners under pressure through oxidative coupling reaction (Ganachaud et al. 2008).

Stability of laccases to higher concentration of organic solvents is always an additional advantage with respect to the organic chemistry. Due to the higher solvent concentration, the enzyme undergoes unfolding, resulting in reduced catalytic activity, and also highly concentrated solvents are usually preferred to carry out various transformations. In this context, recombinant laccase from *M. thermophila* have been addressed by Zumárraga and colleagues to exhibit tolerance to co-solvents (Zumárraga et al. 2007). Overexpressed laccase from *Kurthia huakuii* LAM0618 in *E. coli* has also been addressed to exhibit increased tolerance to various organic solvents (Lam et al. 2016). A similar expression study of laccase genes from *Ganoderma fornicatum* (Huang et al. 2011b) and *Cyathus bulleri* (Garg et al. 2012) in *P. pastoris* host system has been ascribed to be highly tolerant to organic solvents.

Recombinant laccases also represent an attractive route as biosensors for detection of various aromatic compounds. Ionic liquid-based biosensors, obtained from recombinant laccase preparation, have been used to detect rosmarinic acid in luteolin (Franzoi et al. 2009b), plant extracts (Franzoi et al. 2009a) and rutin (Cristina et al. 2009). Covalently immobilised recombinant laccases from *Polyporus pinsitus* and *M. thermophila* are also regarded as promising agents for the detection of phenolic compounds (Kulys and Vidziunaite 2003). Laccases as commercial biosensors have been shown to remarkably exhibit long-term stability towards catechol, catecholamine (Quan and Shin 2004) and *p*-phenylenediamine (Quan et al. 2002) with micromolar detection limits.

Apart from the above-mentioned applications, recombinant laccases have also been employed as biocontrol tools to improve the acceptability and safety of various food and beverages. For instance, an engineered laccase from *M. thermophila* expressed in *Aspergillus oryzae* has been investigated for anti-toxicological applications against a wide range of oral care products such as mouthwash, mints and toothpastes which are regarded as chromosomal mutagens. As a result, this recombinant laccase added to various oral care products was addressed to result in negligible mutagenic effect on cultured human lymphocytes (Brinch and Pedersen 2002). In another interesting report, a recombinant *A. niger* laccase was recounted to eliminate the toxicological and mutagenic activity of aflatoxin B in food industry sectors (Alberts et al. 2009). Recombinant laccase from *P. pinsitus* and *M. thermophila*, expressed in *A. oryzae*, was used for examining the antioxidant activity in the food industry (Kulys and Bratkovskaja 2007). These versatile scopes encourage the researchers to engineer and design laccases for establishing a sustained, cheap, vogue and fascinating technologies.

3.7 Conclusion

The application of laccases as dexterous green biocatalysts is highly multifaceted with promising scope and benefits. Apart from these prospectives, the first and foremost quality required is designing and comprehending of bio-based processes around what happens in nature and mechanisms involved in it. For instance, understanding the synergy between all the enzymes involved in the breakdown of intoxicating aromatic pollutants, and utilisation of this knowledge for designing the new bio-based processes, will effectively contribute towards more efficient processes. With the consistent view, mediators are also playing an important role in determining the enzymatic catalysis. Even though the contribution of synthetic mediators' systems tends to enhance the laccase activity and their production, some other synthetic mediators like 1-hydroxybenzotriazole (HBT) might result in inactivation of laccase in case of mixed-enzyme systems, due to the generation of by-products during the reaction. However, this is not consistent with the other synthetic mediators like N-hydroxyacetanilide (NHA), as it is addressed to exhibit low toxicity as a mediator.

On the other hand, natural mediators such as vanillin and acetosyringone could also equally contribute their role. The application of engineering in mining enzymes from microbial communities has become more feasible due to advances in sequencer technology, tagging, oligonucleotide-directed mutagenesis, artificial gene synthesis and CRISPER/cas9-mediated genome editing. These techniques pose a directed channel for the search of novel laccases. The establishment of new bioinformatics tools such as the Laccase Engineering Database (LccED) has been a cumulative support for researchers to design better laccases. Such technologies can be adapted in the search of better commercial enzyme systems.

Laccases have been exploited widely in the textile industry for bleaching, in the food industry for baking, for wine oxidation prevention, in the pulp and paper industry for bleaching and for delignification of lignocelluloses. They also play a conversant role in bioremediation, effluent treatment, biosensing, cosmetic industry and nanobiotechnology. Despite all the adroit applications of laccases, there is a need for their exploration in the biofuel production, immunomodulation and brain dissemination of mice in case of medical aspects and herbicidal and pharmaceutical compound-removing systems. Phrmaceuticals and several other chemicals are considered as the emerging pollutants exhibiting toxic impacts on marine and aquatic creatures such as fishes. The pharmaceutical drugs such as propranolol exhibited feminisation in male fishes; likewise, they synthetically produced estrogens resulting in structural variation of the endocrine glands of fishes, molluscs, frogs, etc. Thus, for abating these adverse effects of drugs on the environment, it is necessary to attest clone and produce abundant recombinant and engineered laccases in order to use them as a robust and environment-friendly tool.

References

Afreen S, Anwer R, Singh RK, Fatma T (2016) Extracellular laccase production and its optimization from Arthrospira maxima catalyzed decolorization of synthetic dyes. SAUDI J Biol Sci. https://doi.org/10.1016/j.sjbs.2016.01.015

Alberts JF, Gelderblom WCA, Botha A, Van ZWH (2009) Degradation of aflatoxin B 1 by fungal laccase enzymes. Int J Food Microbiol 135:47–52. https://doi.org/10.1016/j.ijfoodmicro.2009.07.022

Alves T, Fernandes R, Batista Da Silveira W et al (2014) Laccases from actinobacteria—what we have and what to expect. Adv Microbiol 4:285–296. https://doi.org/10.4236/aim.2014.46035

Arakane Y, Muthukrishnan S, Beeman RW et al (2005) Laccase 2 is the phenoloxidase gene required for beetle cuticle tanning. Proc Natl Acad Sci 102:11337–11342. https://doi.org/10.1073/pnas.0504982102

Arias ME, Arenas M, Rodriguez J et al (2003) Kraft pulp biobleaching and mediated oxidation of a nonphenolic substrate by laccase from Streptomyces cyaneus CECT 3335. Appl Environ Microbiol 69:1953–1958. https://doi.org/10.1128/AEM.69.4.1953-1958.2003

Arul DCS, Shanmugam S (2012) Analysis of antiproliferative activity of Laccase from Fusarium Species. Res J Pharm Biol Chem Sci 3:912–920

Bailey MR, Woodard SL, Callaway E et al (2004) Improved recovery of active recombinant laccase from maize seed. Appl Microbiol Biotechnol 63:390–397. https://doi.org/10.1007/s00253-003-1362-z

Balcázar-lópez E, Méndez-lorenzo LH, Batista- RA et al (2016) Xenobiotic compounds degradation by heterologous expression of a Trametes sanguineus Laccase in Trichoderma atroviride. PLoS One 11(2):e0147997. https://doi.org/10.1371/journal.pone.0147997

Bleve G, Lezzi C, Mita G et al (2008) Molecular cloning and heterologous expression of a laccase gene from Pleurotus eryngii in free and immobilized Saccharomyces cerevisiae cells. Appl Microbiol Biotechnol 79:731–741. https://doi.org/10.1007/s00253-008-1479-1

Brinch DS, Pedersen PB (2002) Toxicological studies on Laccase from Myceliophthora thermophila expressed in Aspergillus oryzae. Regul Toxicol Pharmacol 35:296–307. https://doi.org/10.1006/rtph.2002.1538

Bronikowski A, Peter LH, Katja K, Vlada BU (2017) Expression of a new laccase from Moniliophthora roreri at high levels in Pichia pastoris and its potential application in micropollutant degradation. Appl Microbiol Biotechnol 7:73. https://doi.org/10.1186/s13568-017-0368-3

Brown MA, Zhao Z, Grant Mauk A (2002) Expression and characterization of a recombinant multi-copper oxidase: Laccase IV from Trametes versicolor. Inorg Chim Acta 331:232–238. https://doi.org/10.1016/S0020-1693(01)00814-3

Bulter T, Alcalde M, Sieber V et al (2003) Functional expression of a fungal laccase Saccharomyces cerevisiae by directed evolution. Appl Environ Microbiol 69:987–995. https://doi.org/10.1128/AEM.69.2.987-995.2003

Callejón S, Sendra R, Ferrer S, Pardo I (2016) Cloning and characterization of a new laccase from Lactobacillus plantarum J16 CECT 8944 catalyzing biogenic amines degradation. Appl Microbiol Biotechnol 100(7):3113–3124. https://doi.org/10.1007/s00253-015-7158-0

Capalash N, Sharma P, Goel R (2007) Bacterial laccases. World J Microbiol Biotechnol 23(6):823–832. https://doi.org/10.1007/s11274-006-9305-3

Chao WLY, Bao SLH (2003) Molecular cloning and characterization of a laccase gene from the basidiomycete Fome lignosus and expression in Pichia pastoris. Appl Microbiol Biotechnol 63(2):174–181. https://doi.org/10.1007/s00253-003-1398-0

Chen S, Ge W, Buswell JA (2004) Molecular cloning of a new laccase from the edible straw mushroom Volvariella volvacea: possible involvement in fruit body development. FEMS Microbiol Lett 230:171–176. https://doi.org/10.1016/S0378-1097(03)00878-4

Chiara M, Carlo C, Silvestri F, Ruzzi M (2009) Modeling the 3-D structure of a recombinant laccase from Trametes trogii Active at a pH close to neutrality. Protein J 28(9–10):375–383. https://doi.org/10.1007/s10930-009-9204-1

Colao MC, Lupino S, Garzillo AM et al (2006) Heterologous expression of *lcc1* gene from *Trametes trogii* in *Pichia pastoris* and characterization of the recombinant enzyme. Microb Cell Fact 5:31. https://doi.org/10.1186/1475-2859-5-31

Cook AE, Meyers PR (2003) Rapid identification of filamentous actinomycetes to the genus level using genus-specific 16S rRNA gene restriction fragment patterns. Int J Syst Evol Microbiol 53:1907–1915. https://doi.org/10.1099/ijs.0.02680-0

Cristina A, Migowski P, Dupont J, Cruz I (2009) Development of biosensors containing laccase and imidazolium bis(trifluoromethylsulfonyl) imide ionic liquid for the determination of rutin. Anal Chim Acta 639(1–2):90–95. https://doi.org/10.1016/j.aca.2009.03.012

Demain AL, Vaishnav P (2009) Production of recombinant proteins by microbes and higher organisms. Biotechnol Adv 27(3):297–306. https://doi.org/10.1016/j.biotechadv.2009.01.008

Devasia S, Nair AJ (2016) Screening of potent laccase producing organisms based on the oxidation pattern of different phenolic substrates. Int J Curr Microbiol App Sci 5(5):127–137. https://doi.org/10.20546/ijcmas.2016.505.014

Ece S, Lambertz C, Fischer R, Commandeur U (2017) Heterologous expression of a Streptomyces cyaneus laccase for biomass modification applications. Appl Microbiol Biotechnol 7(1):86. https://doi.org/10.1186/s13568-017-0387-0

El-Fakharany EM, Haroun BM, Ng TB, Redwan ER (2010) Oyster mushroom laccase inhibits hepatitis C virus entry into peripheral blood cells and hepatoma cells. Protein Pept Lett 17(8):1031–1039. https://doi.org/10.2174/092986610791498948

Fan X, Zhou Y, Xiao Y et al (2014) Cloning, expression and phylogenetic analysis of a divergent laccase multigene family in Auricularia auricula-judae. Microbiol Res 169:453–462. https://doi.org/10.1016/j.micres.2013.08.004

Fang Z, Li T, Wang Q (2011) A bacterial laccase from marine microbial metagenome exhibiting chloride tolerance and dye decolorization ability. Appl Microbiol Biotechnol 89(4):1103–1110. https://doi.org/10.1007/s00253-010-2934-3

Faraco V, Ercole C, Festa G et al (2008) Heterologous expression of heterodimeric laccase from Pleurotus ostreatus in Kluyveromyces lactis. Appl Microbiol Biotechnol 77(6):1329–1335. https://doi.org/10.1007/s00253-007-1265-5

Fathi-roudsari M, Mohammadian M, Molina N et al (2010) Enhanced expression of a recombinant bacterial laccase at low temperature and microaerobic conditions: purification and biochemical characterization. J Ind Microbiol Biotechnol 37(8):863–869. https://doi.org/10.1007/s10295-010-0734-5

Feng H, Zhang D, Sun Y (2015) Expression and characterization of a recombinant Laccase with alkalistable and thermostable properties from Streptomyces griseorubens JSD-1. Appl Biochem Biotechnol 176(2):547–562. https://doi.org/10.1007/s12010-015-1594-2

Ferrer-Miralles N, Domingo-Espín J, Corchero J et al (2009) Microbial factories for recombinant pharmaceuticals. Microb Cell Fact 8:17. https://doi.org/10.1186/1475-2859-8-17

Festa G, Autore F, Fraternali F et al (2008) Development of new laccases by directed evolution: Functional and computational analyses. Proteins 72(1):25–34. https://doi.org/10.1002/prot.21889

Franzoi AC, Dupont J, Spinelli A, Vieira IC (2009a) Biosensor based on laccase and an ionic liquid for determination of rosmarinic acid in plant extracts. Talanta 77:1322–1327. https://doi.org/10.1016/j.talanta.2008.09.010

Franzoi AC, Vieira C, Dupont J, Scheeren W (2009b) Biosensor for luteolin based on silver or gold nanoparticles in ionic liquid and laccase immobilized in chitosan modified with cyanuric chloride. Analyst 134:2320–2328. https://doi.org/10.1039/b911952c

Fujihiro S, Higuchi R, Hisamatsu S (2009) Metabolism of hydroxylated PCB congeners by cloned laccase isoforms. Appl Biochem Biotechnol 82(5):853–860. https://doi.org/10.1007/s00253-008-1798-2

Ganachaud C, Garfagnoli V, Tron T, Iacazio G (2008) Trimerisation of indole through laccase catalysis. Tetrahedron Lett 49:2476–2478. https://doi.org/10.1016/j.tetlet.2008.02.021

Garg N, Bieler N, Kenzom T et al (2012) Cloning, sequence analysis, expression of Cyathus bulleri laccase in Pichia pastoris and characterization of recombinant laccase. BMC Biotechnol 12:75. https://doi.org/10.1186/1472-6750-12-75

Ge H, Xu P, Xu Y et al (2011) Purification, crystallization and preliminary crystallographic analysis of recombinant Lac15 from a marine microbial metagenome. Acta Crystallogr Sect F Struct Biol Cryst Commun 67:956–958. https://doi.org/10.1107/S1744309111024912

Ghasemi Y, Yarahmadi E, Ghoshoon MB et al (2014) Cloning, expression and purification of laccase enzyme gene from Bacillus subtilis in Escherichia coli. Minerva Biotecnol 26(3):295–300

Glick BR, Pasternak JJ, Patten CL (2010) Molecular Biotechnology: Principles and Applications of Recombinant DNA, 4th edn. America Society for Microbiology, United States

Gu C, Zheng F, Long L, Wang J, Ding S (2014) Engineering the expression and characterization of two novel Laccase isoenzymes from *Coprinus comatus* in *Pichia pastoris* by fusing an additional ten amino acids Tag at N-terminus. PLoS One 9(4):e93912. https://doi.org/10.1371/journal.pone.0093912

Guo M, Lu F, Du L et al (2006) Optimization of the expression of a laccase gene from Trametes versicolor in Pichia methanolica. Appl Microbiol Biotechnol 71(6):848–852. https://doi.org/10.1007/s00253-005-0210-8

Hatamoto O, Sekine H, Nakano E, Abe K (1999) Cloning and expression of a cDNA encoding the laccase from Schizophyllum commune. Biosci Biotechnol Biochem 63(1):58–64. https://doi.org/10.1271/bbb.63.58

Hoegger PJ, Kilaru S, James TY et al (2006) Phylogenetic comparison and classification of laccase and related multicopper oxidase protein sequences. FEBS J 273:2308–2326. https://doi.org/10.1111/j.1742-4658.2006.05247.x

Hong YZ, Zhou HM, Tu XM et al (2007) Cloning of a laccase gene from a novel basidiomycete Trametes sp. 420 and its heterologous expression in Pichia pastoris. Curr Microbiol 54(4):260–265. https://doi.org/10.1007/s00284-006-0068-8

Hood EE, Bailey MR, Beifuss K et al (2003) Criteria for high-level expression of a fungal laccase gene in transgenic maize. Plant Biotechnol J 1(2):129–140. https://doi.org/10.1046/j.1467-7652.2003.00014.x

Hu MR, Chao YP, Zhang GQ et al (2007) Molecular evolution of Fome lignosus laccase by ethyl methane sulfonate-based random mutagenesis in vitro. Biomol Eng 24:619–624. https://doi.org/10.1016/j.bioeng.2007.08.020

Huang SJ, Liu ZM, Huang XL et al (2011a) Molecular cloning and characterization of a novel laccase gene from a white-rot fungus Polyporus grammocephalus TR16 and expression in Pichia pastoris. Lett Appl Microbiol 52:290–297. https://doi.org/10.1111/j.1472-765X.2010.02997.x

Huang W, Tai R, Hseu R, Huang C (2011b) Overexpression and characterization of a thermostable, pH-stable and organic solvent-tolerant Ganoderma fornicatum laccase in Pichia pastoris. Process Biochem 46:1469–1474. https://doi.org/10.1016/j.procbio.2011.03.020

Jaiswal N, Pandey VP, Dwivedi UN (2014) Purification of a thermostable laccase from Leucaena leucocephala using a copper alginate entrapment approach and the application of the laccase in dye decolorization. Process Biochem 49:1196–1204. https://doi.org/10.1016/j.procbio.2014.04.002

Janssen GG, Baldwin TM, Winetzky DS et al (2004) Selective targeting of a laccase from Stachybotrys chartarum covalently linked to a carotenoid-binding peptide. J Pept Res 64:10–24. https://doi.org/10.1111/j.1399-3011.2004.00150.x

Jeon JR, Kim EJ, Murugesan K et al (2010) Laccase-catalysed polymeric dye synthesis from plant-derived phenols for potential application in hair dyeing: Enzymatic colourations driven by homo- or hetero-polymer synthesis. J Microbial Biotechnol 3:324–335. https://doi.org/10.1111/j.1751-7915.2009.00153.x

Jeon JR, Baldrian P, Murugesan K, Chang YS (2012) Laccase-catalysed oxidations of naturally occurring phenols: from in vivo biosynthetic pathways to green synthetic applications. J Microbial Biotechnol 5:318–332. https://doi.org/10.1111/j.1751-7915.2011.00273.x

Jolivalt C, Madzak C, Brault A et al (2005) Expression of laccase IIIb from the white-rot fungus Trametes versicolor in the yeast Yarrowia lipolytica for environmental applications. Appl Microbiol Biotechnol 66:450–456. https://doi.org/10.1007/s00253-004-1717-0

Jönsson LJ, Saloheimo M, Penttilä M (1997) Laccase from the white-rot fungus Trametes versicolor: cDNA cloning of lcc1 and expression in Pichia pastoris. Curr Genet 32:425–430. https://doi.org/10.1007/s002940050298

Kim JK, Lim SH, Kang HW (2013) Cloning and characterization of a novel laccase gene, fvlac7, based on the genomic sequence of Flammulina velutipes. Mycobiology 41:37–41. https://doi.org/10.5941/MYCO.2013.41.1.37

Klonowska A, Gaudin C, Asso M et al (2005) LAC3, a new low redox potential laccase from Trametes sp. strain C30 obtained as a recombinant protein in yeast. Enzyme Microb Technol 36:34–41. https://doi.org/10.1016/j.enzmictec.2004.03.022

Koschorreck K, Richter SM, Swierczek A et al (2008a) Comparative characterization of four laccases from Trametes versicolor concerning phenolic C–C coupling and oxidation of PAHs. Arch Biochem Biophys 474:213–219. https://doi.org/10.1016/j.abb.2008.03.009

Koschorreck K, Richter SM, Urlacher VB (2008b) Cloning and characterization of a new laccase from Bacillus licheniformis catalyzing dimerization of phenolic acids. Appl Microbiol Biotechnol 79:217–224. https://doi.org/10.1007/s00253-008-1417-2

Kulys J, Bratkovskaja I (2007) Antioxidants determination with laccase. Talanta 72:526–531. https://doi.org/10.1016/j.talanta.2006.11.011

Kulys J, Vidziunaite R (2003) Amperometric biosensors based on recombinant laccases for phenols determination. Biosens Bioelectron 18:319–325. https://doi.org/10.1016/S0956-5663(02)00172-0

Lade H, Kadam A, Paul D, Govindwar S (2015) Biodegradation and detoxification of textile azo dyes by bacterial consortium under sequential microaerophilic/aerobic processes. EXCLI J 14:158–174. https://doi.org/10.17179/excli2014-642

Lam T, Guo X, Zhou S et al (2016) Characterization of a highly thermostable and organic solvent-tolerant copper- containing polyphenol oxidase with dye- decolorizing ability from Kurthia huakuii. PLoS One 11(10):e0164810. https://doi.org/10.1371/journal.pone.0164810

Li L, Steffens JC (2002) Overexpression of polyphenol oxidase in transgenic tomato plants results in enhanced bacterial disease resistance. Planta 215:239–247. https://doi.org/10.1007/s00425-002-0750-4

Li JF, Hong YZ, Xiao YZ et al (2007) High production of laccase B from Trametes sp. in Pichia pastoris. World J Microbiol Biotechnol 23:741–745. https://doi.org/10.1007/s11274-006-9286-2

Li Y, Zuo W, Li Y, Wang X (2012) Cloning of multicopper oxidase gene from Ochrobactrum sp. 531 and characterization of its alkaline laccase activity towards phenolic substrates. Adv Biol Chem 2:248–255. https://doi.org/10.4236/abc.2012.23031

Lu L, Zhao M, Liang S et al (2009) Production and synthetic dyes decolourization capacity of a recombinant laccase from Pichia pastoris. J Appl Microbiol 107:1149–1156. https://doi.org/10.1111/j.1365-2672.2009.04291.x

Lu L, Zeng G, Fan C et al (2013) Characterization of a laccase-like multicopper oxidase from newly isolated Streptomyces sp. C1 in agricultural waste compost and enzymatic decolorization of azo dyes. Biochem Eng J 72:70–76. https://doi.org/10.1016/j.bej.2013.01.004

Lu Y, Wu G, Lian L et al (2015) Cloning and expression analysis of Vvlcc3, a novel and functional laccase gene possibly involved in stipe elongation. Int J Mol Sci 16:28498–28509. https://doi.org/10.3390/ijms161226111

Luciana P, Ana VC, Viegas CA (2009) Enzymatic biotransformation of the azo dye Sudan Orange G with bacterial CotA-laccase. J Biotechnol 139:68–77. https://doi.org/10.1016/j.jbiotec.2008.09.001

Machczynski MC, Vijgenboom E, Samyn B, Canters GW (2004) Characterization of SLAC: a small laccase from Streptomyces coelicolor with unprecedented activity. Protein Sci 13:2388–2397. https://doi.org/10.1110/ps.04759104

Menaka S, Lone TA, Lone RA (2015) Cloning of Laccase Gene from a newly isolated 2, 4-dichlorophenol degrading Bacillus subtilis from dyeing industry sites. Am Euras J Agric Environ Sci 15:1602–1608. https://doi.org/10.5829/idosi.aejaes.2015.15.8.94202

Miele A, Giardina P, Sannia G, Faraco V (2010) Random mutants of a Pleurotus ostreatus laccase as new biocatalysts for industrial effluents bioremediation. J Appl Microbiol 108:998–1006. https://doi.org/10.1111/j.1365-2672.2009.04505.x

Mishra A, Kumar S (2007) Cyanobacterial biomass as N-supplement to agro-waste for hyper-production of laccase from Pleurotus ostreatus in solid state fermentation. Process Biochem 42:681–685. https://doi.org/10.1016/j.procbio.2006.09.022

Molina-guijarro JM, Pérez J, Muñoz-dorado J et al (2009) Detoxification of azo dyes by novel pH-versatile, salt-resistant laccase from Streptomyces ipomoea. Int Microbiol 12:13–21. https://doi.org/10.2436/20.1501.01.77

Monza E, Lucas MF, Camarero S et al (2015) Insights into laccase engineering from molecular simulations: toward a binding-focused strategy. J Phys Chem Lett 6:1447–1453. https://doi.org/10.1021/acs.jpclett.5b00225

Murugesan K, Nam IH, Kim YM, Chang YS (2007) Decolorization of reactive dyes by a thermostable laccase produced by Ganoderma lucidum in solid state culture. Enzyme Microb Technol 40:1662–1672. https://doi.org/10.1016/j.enzmictec.2006.08.028

Murugesan K, Yang IH, Kim YM et al (2009) Enhanced transformation of malachite green by laccase of Ganoderma lucidum in the presence of natural phenolic compounds. Appl Microbiol Biotechnol 82:341–350. https://doi.org/10.1007/s00253-008-1819-1

Murugesan K, Chang YY, Kim YM et al (2010) Enhanced transformation of triclosan by laccase in the presence of redox mediators. Water Res 44:298–308. https://doi.org/10.1016/j.watres.2009.09.058

Nakamura K, Go N (2005) Function and molecular evolution of multicopper blue proteins. Cell Mol Life Sci 62:2050–2066. https://doi.org/10.1007/s00018-004-5076-5

Nanda V, Koder RL (2010) Designing artificial enzymes by intuition and computation. Nat Chem 2:15–24. https://doi.org/10.1038/nchem.473

Naraian R, Kumari S, Gautam RL (2018) Biodecolorization of brilliant green carpet industry dye using three distinct *Pleurotus* spp. Env Sust 1(2):141–148

Necochea R, Valderrama B, Díaz-Sandoval S et al (2005) Phylogenetic and biochemical characterisation of a recombinant laccase from Trametes versicolor. FEMS Microbiol Lett 244:235–241. https://doi.org/10.1016/j.femsle.2005.01.054

Nersissian AM, Shipp EL (2002) Blue copper-binding domains. Adv Protein Chem 60:271–340. https://doi.org/10.1016/S0065-3233(02)60056-7

Nicolini C, Bruzzese D, Cambria MT et al (2013) Recombinant laccase: I. Enzyme cloning and characterization. J Cell Biochem 114:599–605. https://doi.org/10.1002/jcb.24397

Otterbein L, Record E, Longhi S et al (2000) Molecular cloning of the cDNA encoding laccase from Pycnoporus cinnabarinus I-937 and expression in Pichia pastoris. Eur J Biochem 267:1619–1625. https://doi.org/10.1046/j.1432-1327.2000.01166.x

Pereira L, Coelho AV, Viegas CA et al (2009) Enzymatic biotransformation of the azo dye Sudan Orange G with bacterial CotA-laccase. J Biotechnol. 139:68–77. https://doi.org/10.1016/j.jbiotec.2008.09.001

Piscitelli A, Giardina P, Mazzoni C, Sannia G (2005) Recombinant expression of Pleurotus ostreatus laccases in Kluyveromyces lactis and Saccharomyces cerevisiae. Appl Microbiol Biotechnol 69:428–439. https://doi.org/10.1007/s00253-005-0004-z

Piscitelli A, Pezzella C, Giardina P et al (2010) Heterologous laccase production and its role in industrial applications. Bioeng Bugs 1:252–262. https://doi.org/10.4161/bbug.1.4.11438

Quan D, Shin W (2004) Amperometric detection of catechol and catecholamines by immobilized Laccase from DeniLite. Electroanalysis 16:1576–1582. https://doi.org/10.1002/elan.200302988

Quan D, Kim Y, Yoon KB, Shin W (2002) Assembly of Laccase over platinum oxide surface and application as an amperometric biosensor. Bullet Kor Chem Soc 23:385–390. https://doi.org/10.5012/bkcs.2002.23.3.385

Ravalason H, Herpoël-gimbert I, Record E et al (2009) Fusion of a family 1 carbohydrate binding module of Aspergillus niger to the Pycnoporus cinnabarinus laccase for efficient softwood kraft pulp biobleaching. J Biotechnol 142:220–226. https://doi.org/10.1016/j.jbiotec.2009.04.013

Ravikumar G, Kalaiselvi M, Gomathi D et al (2013) Cloning, expression in *E. coli* and the enzymatic properties of laccase from *Hypsizygus ulmarius*. World J Pharm Pharm Sci 3:493–505

Record E, Punt PJ, Chamkha M et al (2002) Expression of the Pycnoporus cinnabarinus laccase gene in Aspergillus niger and characterization of the recombinant enzyme. Eur J Biochem 269:602–609. https://doi.org/10.1111/jfpe.12411

Renzetti S, Courtin CM, Delcour JA, Arendt EK (2010) Oxidative and proteolytic enzyme preparations as promising improvers for oat bread formulations: Rheological, biochemical and microstructural background. Food Chem 119:1465–1473. https://doi.org/10.1016/j.foodchem.2009.09.028

Rico A, Rencoret J, Río JC et al (2014) Pretreatment with laccase and a phenolic mediator degrades lignin and enhances saccharification of Eucalyptus feedstock. Biotechnol Biofuels 7:6. https://doi.org/10.1186/1754-6834-7-6

Ruijssenaars HJ, Hartmans S (2004) A cloned Bacillus halodurans multicopper oxidase exhibiting alkaline laccase activity. Appl Microbiol Biotechnol 65(2):177–182. https://doi.org/10.1007/s00253-004-1571-0

Sakamoto Y, Nakade K, Yano A et al (2008) Heterologous expression of lcc1 from Lentinula edodes in tobacco BY-2 cells results in the production an active, secreted form of fungal laccase. Appl Microbiol Biotechnol 79:971–980. https://doi.org/10.1007/s00253-008-1507-1

Santiago G, De Salas F, Lucas MF et al (2016) Computer-aided laccase engineering toward biological oxidation of arylamines. ACS Catal 6:5415–5423. https://doi.org/10.1021/acscatal.6b01460

Shankar S, Shikha BC, Chandra R, Tyagi S (2018) Laccase based de-inking of mixed office waste and evaluation of its impact on physico-optical properties of recycled fiber. Env Sust 1(3):223–244

Shi X, Liu Q, Ma J, Liao H (2015) An acid-stable bacterial laccase identified from the endophyte Pantoea ananatis Sd-1 genome exhibiting lignin degradation and dye decolorization abilities. Biotechnol Lett 37:2279–2288. https://doi.org/10.1007/s10529-015-1914-1

Sigoillot C, Record E, Belle V et al (2004) Natural and recombinant fungal laccases for paper pulp bleaching. Appl Microbiol Biotechnol 64:346–352. https://doi.org/10.1007/s00253-003-1468-3

Soden DM, O'Callaghan J, Dobson ADW (2002) Molecular cloning of a laccase isozyme gene from Pleurotus sajor-caju and expression in the heterologous Pichia pastoris host. Microbiology 148:4003–4014. https://doi.org/10.1099/00221287-148-12-4003

Tracewell CA, Arnold FH (2009) Directed enzyme evolution: climbing fitness peaks one amino acid at a time. Curr Opin Chem Biol 13:3–9. https://doi.org/10.1016/j.cbpa.2009.01.017

Waksman AS, Albert S, Donald MR (2010) Production of antibiotic substances by actinomycetes. Ann NY Acad Sci 1213:112–124. https://doi.org/10.1111/j.1749-6632.2010.05861.x

Wang HX, Ng TB (2004) A novel laccase with fair thermostability from the edible wild mushroom (Albatrellus dispansus). Biochem Biophys Res Commun 319:381–385. https://doi.org/10.1016/j.bbrc.2004.05.011

Wang W, Zhang Z, Ni H et al (2012) Decolorization of industrial synthetic dyes using engineered Pseudomonas putida cells with surface-immobilized bacterial laccase. Microb Cell Fact 11:75. https://doi.org/10.1186/1475-2859-11-75

Wong K, Cheung M, Au C, Kwan H (2013) A novel lentinula edodes laccase and its comparative enzymology suggest guaiacol-based laccase engineering for bioremediation. PLoS One 8(6):e66426. https://doi.org/10.1371/journal.pone.0066426

Wu YR, Luo ZH, Kwok-Kei Chow R, Vrijmoed LLP (2010) Purification and characterization of an extracellular laccase from the anthracene-degrading fungus Fusarium solani MAS2. Bioresour Technol 101:9772–9777. https://doi.org/10.1016/j.biortech.2010.07.091

Zeng S, Zhao J, Xia L (2017) Simultaneous production of laccase and degradation of bisphenol A with Trametes versicolor cultivated on agricultural wastes. Bioprocess Biosyst Eng 40:1237–1245. https://doi.org/10.1007/s00449-017-1783-1

Zhang GQ, Wang YF, Zhang XQ et al (2010) Purification and characterization of a novel laccase from the edible mushroom Clitocybe maxima. Process Biochem 45:627–633. https://doi.org/10.1016/j.procbio.2009.12.010

Zumárraga M, Bulter T, Shleev S et al (2007) In vitro evolution of a fungal laccase in high concentrations of organic cosolvents. Chem Biol 14:1052–1064. https://doi.org/10.1016/j.chembiol.2007.08.010

Advantages and Progress Innovations of Solid-State Fermentation to Produce Industrial Enzymes

4

Liliana Londoño-Hernandez, Héctor A. Ruiz, Cristina Ramírez Toro, Alberto Ascacio-Valdes, Raúl Rodriguez-Herrera, Antonio Aguilera-Carbo, Gisela Tubio, Guillermo Pico, Arely Prado-Barragan, Gerardo Gutierrez-Sanchez, and Cristóbal N. Aguilar

Abstract

Industrial enzymes are biocatalysts that are commercially used in a variety of commercial sectors such as pharmaceuticals, chemical production, biofuels,

L. Londoño-Hernandez · A. Ascacio-Valdes · R. Rodriguez-Herrera · C. N. Aguilar (✉)
Bioprocesses and Bioproducts Group, Food Research Department, School of Chemistry, Universidad Autónoma de Coahuila, Saltillo, Coahuila, Mexico
e-mail: cristobal.aguilar@uadec.edu.mx

H. A. Ruiz
Biorefinery Group, Food Research Department, Faculty of Chemistry Sciences, Autonomous University of Coahuila, Saltillo, Coahuila, Mexico

Cluster of Bioalcoholes, Mexican Centre for Innovation in Bioenergy (Cemie-Bio), Mexico City, Mexico

C. R. Toro
School of Food Engineering, Universidad del Valle, Cali, Valle del Cauca, Colombia

A. Aguilera-Carbo
Department of Food and Nutrition, Universidad Autónoma Agraria Antonio Narro, Buenavista, Saltillo, Coahuila, Mexico

G. Tubio · G. Pico
Institute of Biotechnological and Chemical Processes—Rosario-School of Biochemical and Pharmaceutical Sciences, Universidad Nacional de Rosario, Rosario, Argentina

A. Prado-Barragan
Department of Biotechnology, Universidad Autonoma Metropolitana, Unidad Iztapalapa, Mexico City, Mexico

G. Gutierrez-Sanchez
Complex Carbohydrate Research Center, The University of Georgia, Athens, GA, USA

N. K. Arora et al. (eds.), *Microbial Enzymes: Roles and Applications in Industries*, Microorganisms for Sustainability 11,
https://doi.org/10.1007/978-981-15-1710-5_4

food and beverages, and consumer products. Due to advancements in recent years, biocatalysts are considered more economical than use of whole cells and can be used as a unit operation within a process to generate the desired product of interest. Industrial biological catalysis through enzymes has experienced rapid growth in recent years due to their ability to operate at mild conditions, high specificity, and high productivity. Industrial enzymes can be produced by both submerged fermentation (SmF) and solid-state fermentation (SSF). In contrast to the first process, the second bioprocess (SSF) is the cultivation of microorganisms under controlled conditions in the absence of free water. Examples of bioproducts of SSF include industrial enzymes, fuels, and nutrient-enriched animal feeds. Most industrial enzymes are manufactured using the traditional bioprocess of SmF, where microbial cells are suspended in a large volume of water that is stirred and aerated using mechanical devices; such culture conditions dictate the overall physiological behavior of microorganisms provoking biochemical and structural changes affecting the quantity and activity of biocatalysts produced. Among the main advantages of SSF over SmF is a higher volumetric productivity, secretion facilities to get extracellular bioproducts with higher stability, being usually simpler with lower energy requirements, resembling of the natural habitat of some microorganisms, and easier downstream processing. In this chapter we summarize, compare, analyze, and discuss the technological, biochemical, and microbiological advantages of SSF to produce industrial enzymes. Furthermore, culture conditions, aggregation and diffusional phenomena, bioreactors, genetic expression, and protein regulation are covered.

4.1 Introduction

The enzymes are biomolecules of protein that accelerate the speed of reaction until reaching a balance. Enzymes constitute the type of proteins which are most numerous and specialized and act as catalysts of chemical reactions specific in living beings or biological systems. Enzymes are available from variable sources such as from extreme environment or from microorganisms present in nature. Enzymes are important to build and maintain the life of all organisms. These biological entities participate in the buildup of new chemical compounds, or in the hydrolysis and modification of these. Many of the enzymes do not work alone, but organize themselves in sequences or metabolic pathways, and many of them can regulate their enzymatic activity. Enzymes make it possible for these reactions to occur faster. Due to the multiple industrial applications that enzymes have, the global market of these biomolecules has grown in recent years. In 2011, its value was about \$4 billion and it is supposed to be \$6 billion in 2016 (Oliveira et al. 2017). Likewise, research around the processes for obtaining and purifying enzymes has increased. One of the processes that have presented good productivity for obtaining enzymes is fermentation.

Fermentation is the process by which the microorganisms and their activities convert nutrients to produce metabolites. Fermentation can be divided into two methods, submerged fermentation/liquid-state fermentation (SmF) and solid-state fermentation (SSF) (Sukma et al. 2018). SSF can be defined as the fermentation processes carried out by the microorganisms in solid substrates in nearly no free water, but with the amount of water sufficient to allow the growth of the microorganisms (Chen 2013). In recent years, many researchers have shown that SSF has a major impact on productivity, higher yields, and improved product characteristics compared to SmF.

SSF has received recognition as a potential biotechnological process to produce different metabolites of industrial interest such as enzymes, pigments, bioethanol, aroma compounds, antibiotics, and mycotoxins, among others. One of the main advantages of the SSF is the use of agro-industrial by-products as a substrate or source of nutrients. This makes it an attractive technology to add value to these by-products and reduce environmental pollution due to their accumulation (Ghoshal et al. 2012). One of the metabolites of greatest industrial interest is enzyme. Therefore, research and production of enzymes using biotechnological processes have increased in recent years. The SSF has advantages in terms of the production of enzymes over SmF which include the use of agro-industrial wastes, higher volumetric productivity, secretion facilities to get extracellular bioproducts with higher stability, being usually simpler with lower energy requirements, resembling of the natural habitat of some microorganisms, and easier downstream processing.

To produce enzymes by SSF, different microorganisms are used, including bacteria, fungi, and yeasts. However, due to the biochemical characteristics and its adaptation to the substrate, filamentous fungi are the most used. Among the most used fungi are those belonging to the genera *Aspergillus sp.* and *Rhizopus sp.* Different enzymes have been produced from these microorganisms, among which are proteases, amylases, cellulases, pectinases, etc. (Schuster et al. 2002; Londoño-Hernández et al. 2017).

In recent years, research has focused on increasing the productivity of the process of obtaining enzymes. In this way, the design of new bioreactors and optimization of the process conditions have been carried out (Biz et al. 2016; Ashok et al. 2017; Pitol et al. 2017), and new genetic engineering tools have been used to improve the yields in the production of enzymes. In this chapter, advances in SSF to produce enzymes of industrial interest will be reviewed. There is a brief discussion of the microorganisms and enzymes produced, and a focus is presented on the design aspects of bioreactors and the factors that affect the production of enzymes.

4.2 Microorganisms Used in SSF

One of the key factors in the development of SSF processes is the selection of the microorganism to be used according to the metabolite or the enzyme of interest. There are a wide range of microorganisms capable of producing enzymes by SSF such as aerobic and anaerobic bacteria, yeast, anaerobic fungi, soft-rot fungi, white-rot fungi, and brown-rot fungi (Yoon et al. 2014). However, the most commonly used are filamentous fungi, since SSF simulates the natural habitat of these microorganisms. Due to their ability to grow in low-humidity conditions, yeasts have also been used to obtain various enzymes by SSF, finding good results. Likewise, by the use of bacteria and Actinomycetes such as *Streptomyces sp*, production of a wide range of degradative enzymes and high resistance to extreme conditions have been reported (Soccol et al. 2017).

Due to the high tolerance to low water activities, the high potential to excrete hydrolytic enzymes, and the morphology, the filamentous fungi are the microorganisms more used to produce enzymes. Filamentous fungi are featured as modular organisms, which grow by the repeated iteration of modules usually to yield a branching pattern. The tubular hypha that emerges from the spore elongates at the tip and at the same time along the hypha new branches are formed. The branches continue to form a porous three-dimensional network of hyphae or mycelium. Their morphology allows filamentous fungi to colonize and penetrate the solid substrate in search for nutrients, producing various enzymes for it (Rahardjo 2005).

With the increase in industrialization processes, the production of agricultural by-products has increased, which has caused an increase in environmental pollution due to inadequate management of solid waste. At a global level, alternatives are necessary for reutilizing these wastes, considering environmental and economic aspects. One of the alternatives is the use of these residues in biotechnological processes to obtain enzymes. The use of these agro-industrial residues in SSF processes is of interest due to the high availability, its low cost, and the composition that allows obtaining different types of enzymes. In addition, the SSF processes are considered as an eco-friendly technology. The filamentous fungi can effectively hydrolyze such type of substrates, and this is an added advantage for the use of these microorganisms in SSF (Ozcirak Ergun and Ozturk Urek 2017).

Among the microorganisms used for enzyme production in recent years are *Paecilomyces variotii*, *Aspergillus oryzae*, *Rhizopus oryzae*, and *Aspergillus terreus* to produce amylases (Chen et al. 2014; Kaur et al. 2015; Sahnoun et al. 2015); *Aspergillus fumigatus*, *Rhizopus oryzae*, *Lactobacillus plantarum*, and *Aspergillus niger* to produce cellulases (Djoulde Darman et al. 2011; Liu et al. 2011; Kupski et al. 2014); *Penicillium simplicissimum* to produce lipases (Godoy et al. 2011); *A. oryzae* and *A. niger* to produce pectinases (Baladhandayutham and Thangavelu 2011; Biz et al. 2016); *Aspergillus versicolor*, *A. oryzae*, *Aspergillus terreus*, and *A. niger* to produce proteases (Veerabhadrappa et al. 2014; de Castro et al. 2015; Homaei et al. 2016); and *Aspergillus tubingensis*, *R. oryzae, Mucor indicus, Mucor hiemalis*, *Peniophora*, and *Trichoderma viride* to produce xylanases and laccases (Irfan et al. 2014; Adhyaru et al. 2016; Behnam et al. 2016; Shankar et al. 2018).

4.3 Enzymes Produced by SSF

During the last few years, the use of biotechnological tools has improved the quality of the products obtained while simplifying industrial processes through the use of biological techniques such as SSF. Not only these tools made possible the reduction in production energy costs and wastewater generation, but also they allowed the use of by-products and agricultural wastes as substrate for microorganism's growth (agricultural waste valorization). The reutilization of agricultural residues or by-products in bioprocesses reduces the costs of production and solves the problem of their disposal into the environment, a major cause for environmental pollution and an important loss of biomass that could be used to produce different metabolites with added commercial value.

Agricultural residuals such as straw from wheat, corn, rice, soy, and cotton; sugarcane bagasse and orange bagasse; and by-products like bakery wastes are complex networks of cellulose, hemicelluloses, and lignin and are considered lignocellulosic wastes, some of which may contain significant concentrations of soluble carbohydrates and inducers of enzyme synthesis (Lizardi-Jiménez and Hernández-Martínez 2017). Cellulose and hemicellulose represent the primary structural polysaccharides of the plant cell wall that can be digested by hydrolytic extracellular enzymes produced by for example amylases (EC 3.2.1.1), glucanases (EC 3.2.1.4), xylanases (EC 3.2.1.8), and polygalacturonases (EC 3.2.1.67) among others; these can be produced by filamentous fungi. On the other hand, lignin is a highly irregular and insoluble polymer, chemically bonded by covalent linkages with hemicellulose. White-rot fungi are highly effective microbes to degrade lignin. Their lignin-degrading ability depends on several ligninolytic extracellular enzymes including laccases (EC 3.10.3.2), peroxidases (EC 1.11.1.7), and oxidases (EC 1.6.3.1); low-molecular-weight metabolites; and activated oxygen species (Ozcirak Ergun and Ozturk Urek 2017).

Fungi and yeast are the most suitable microorganisms for SSF, while bacteria have been considered unsuitable, according to the theoretical concept of water activity (Singh et al. 2012). According to Polizeli et al. (2005) many advantages can be obtained with SSF systems when filamentous fungi are used as enzyme producers. In recent decades, the use of fungi in bioprocesses has grown in importance because of the successful production of numerous enzymes with different biochemical properties and excellent potential for biotechnological applications (Knob et al. 2014).

In the last decade, there has been an increasing trend towards the utilization of fungal extracellular enzymes for food and nonfood applications (Gutiérrez-Correa et al. 2012; Buenrostro Figueroa et al. 2014; Poggi-Varaldo et al. 2014). These applications range from straightforward industrial processes to pharmaceutical discovery and development. Thus, industrial enzymes represent the center of biotechnology, in terms of research and applications. Table 4.1 summarizes different enzymes used in the industries (Thomas et al. 2013).

Today, consumer demand requires higher levels of quality in food in terms of natural flavor, taste, digestibility, and nutritional value. This trend triggered the need for development of enzymes like amylase, xylanase, and glucanase, with

Table 4.1 Enzyme production with industrial applications using solid-state fermentation (SSF)

Enzymes	Microorganism	Industrial use	References
Protease	*Engyodontium album, Aspergillus versicolor, and Aspergillus niger*	Food	Abraham et al. (2013)
Elagitanase	*Aspergillus niger*	Food	Buenrostro Figueroa et al. (2014)
Polygalacturonase	*Aspergillus sojae*	Textile and food	Demir and Tarı (2014)
Xylanase	*Aspergillus niger and Aspergillus fumigatus*	Food and paper	Loureiro et al. (2016)
Peroxidase	*Phanerochaete chrysosporium*	Biosensors	Li et al. (2015)
Chitinase	*Penicillium ochrochloron MTCC 517*	Biopesticides	Thomas et al. (2013)
Lipase	*Aspergillus versicolor*	Food	Veerabhadrappa et al. (2014)
Glucoamylase	*Aspergillus niger*	Starch	Pandey (1991a)
Fibrinolytic enzyme	*Fusarium oxysporum*	Food	Tao et al. (1997)
Aminoacylase	*Trichoderma*	Pharmaceutical	Wakayama et al. (2004)
Pectinase	*Aspergillus*	Drinks	de Oliveira et al. (2018)
Lactase	*Aspergillus*	Dairy	Neri et al. (2009)
Laccase	*Pleurotus ostreatusIE-8*	Food	Membrillo et al. (2008)
Amylase	*Aspergillus oryzae*	Baking	Porfiri et al. (2012)

applications in food processing like clarification of juices, preparation of dextrans for use as food thickeners, and production of fluids and juices from plant materials, among others (Polizeli et al. 2005).

In the textile industry, the enzymes amylase, pectinase, lipase, catalase, peroxidase, glucanase, and laccase, all obtained by SSF from various fungi species, are applied for the treatment of natural protein fibers (wool and silk), cellulose fibers (cotton, linen, and hemp), and synthetic fibers with a high impact on both the production of threads, yarns, and fabrics and the manufacture of various products.

In the pulp and paper industry, the incorporation of xylanase leads to the reduction of the use of chemical pollutants in the pre-bleaching stage, pulp bleaching, improved drainage speed of recycled fibers, increase in leaf density of paper and reduction in its rusticity, and decrease of waste and contaminants in the recycling process.

Furthermore, the use of hydrolytic enzymes in animal feed may improve cellulose accessibility to ruminal digestion and thus improve the nutritional value of the feed. The inclusion of enzymes as phytase (EC 3.1.3.8), in the formulation of feeding diets for monogastric animals, has increased in recent times (Rodríguez-Fernández et al. 2010).

One area of great importance is the production of clean energy based on sustainable development, such as renewable resources. Currently, biomasses of lignocellulosic origin are pretreated enzymatically with lipase, glucanases, xylanase, and amylases for the generation of fermentable sugars used in the production of fuels of biological origin: biodiesel generation or second-generation ethanol production (Anwar et al. 2014).

Optimization in the process to produce enzymes by SSF requires the study of various farming parameters, which depend on the enzyme that will be produced, the substrate used, and the microorganism that produces it.

4.4 Biotechnological Engineering in the Production of Enzymes

Commercial demand for enzymes of industrial interest has increased in recent years, and research focused on attending this demand has been developed. Some biotechnological processes or bioprocesses have been used to produce enzymes of industrial interest, which are produced by different microorganisms. One of the most suitable bioprocesses for this purpose is SSF or solid-state culture using fungi (Lopez-Trujillo et al. 2017). SSF has been distinguished because it is a bioprocess that allows the growth of the microorganism (mainly fungi) in the fermentable material that in turn acts as a support, or allows the growth of the microorganism on an inert support that is enriched with a culture media and humidity (Krishna 2005). Once the bioprocess begins, the microorganism excretes diverse enzymes to the environment to carry out the catabolism of the nutrients and ensure their growth; in this way it is possible to take advantage of this bioprocess to obtain various metabolites (Medina-Morales et al. 2017; Soccol et al. 2017). From the 1980s to the present, several applications of SSF have been found, such as the production of secondary metabolites, including antioxidants, alkaloids, or antibiotics; biodegradation of certain agro-industrial residues or biotransformation of crop residues; and one of the most important is production of a wide variety of enzymes, such as cellulases, lipases, and proteases (Pandey 2001).

For a microorganism to carry out the SSF, it is necessary to control several factors. As already mentioned, the most commonly used microorganisms to carry out a fermentation in a solid medium are filamentous fungi, because they can grow more easily on the substrate and tolerate a lower water activity; in addition they produce greater amount of enzymes that help to obtain the products of interest (Raimbault 1998). Second are the substrates. The most commonly used substrates are agricultural and forest residues; this is because they are very abundant and do not have a specific use; besides they represent a very high source of nuisance value. About agricultural waste, some of the most used are bagasse (of certain plants mainly cereals). The husks or pulps of numerous fruits and other residues of different plants have also been used. All these substrates, coming from plants, have a high content of molecules, such as cellulose, hemicellulose, starch, lignin, pectins, and polyphenols which are used by microorganisms for their growth (Soccol et al. 2017).

Moisture is required for an SSF to be carried out in the best way. Humidity plays a very important role in the SSF. Fungi are microorganisms that grow in environments with high humidity. A low moisture value can cause an inadequate absorption of nutrients and affects the stability of the enzymes (Moo-Young et al. 1983). In general, in bioprocesses like this, humidity values range between 30 and 85%, but for fungi it should normally be between 20 and 70% (Pandey 2001).

The pH value is variable due to the production of organic acids during fermentation, which will cause the pH to decrease. There may also be an increase in pH caused by the assimilation of certain organic acids present and by the alkalinization of the urea present (Raimbault 1998). Filamentous fungi tolerate a pH range of 2–9, and their optimum pH is in the range of 3.8–6. Finally, temperature is considered the most important physical factor for the development of SSF, because the enzymes and certain metabolites produced are highly sensitive to temperature. Fungi have the ability to withstand temperature ranges between 20 and 55 °C; therefore they are the most efficient microorganisms for this bioprocess (Yadav 1988). If the abovementioned factors are controlled in an appropriate way, SSF occurs in a better way, and if this bioprocess is regulated it can be used to produce different metabolites of industrial interest, for example, enzymes.

Some applications of the SSF to produce enzymes of industrial interest are known. In some reported works it has been demonstrated that filamentous fungus is able to produce a variety of enzymes that have application in the industry (Ramírez-Coronel et al. 2003). It has been reported that the strain *A. niger* Aa-20 is capable to produce hydrolase enzymes, such as tannase (Aguilar et al. 2001). In addition, other strains of filamentous fungi, such as *A. niger* PSH and *Penicillium commune* EH2, are capable to produce enzymes of industrial importance (Hernández et al. 2005), and all the above by SSF. One of the most interesting and important species that has been used for SSF is *A. niger* GH1 (isolated from creosote bush (*Larrea tridentata*)) (Belmares et al. 2009). This strain is capable to produce tannase enzyme and, as has already been mentioned, tannase is a hydrolase enzyme that has application in the food industry, for example, in the production of some beverages, mainly for its clarification (Belmares Cerda et al. 2003).

Fungi of the genus *Aspergillus* used in SSF in addition to tannase produce other enzymes of industrial interest. The production of the polyphenol oxidase enzyme by *A. niger* has been reported. This enzyme is used as a catalyst for the release of phenolic compounds present in fruit drinks to increase their antioxidant potential (Shi et al. 2005). *Aspergillus* also produces by SSF another enzyme, β-glucosidase, which hydrolyzes glycosidic bonds to release bioactive compounds from their sugar core, and once the bioactive compounds are free they can be applied to the formulation of functional foods (Ascacio-Valdés et al. 2011). The same function has been reported for enzymes such as xylanase produced by *Aspergillus* by SSF (Huang et al. 2007).

A. niger GH1 is also capable to produce polyphenol oxidase, β-glucosidase, and xylanase by SSF (Ascacio-Valdés et al. 2014), and not only that, it has also been reported that this fungus produces a novel enzyme capable of phytochemical compound biotransformation into biologically active molecules, the novel ellagitannase enzyme. The conditions of SSF to produce this enzyme have already been optimized to guarantee adequate levels of production. Due to the results obtained in the mentioned study it was possible to demonstrate the production of ellagitannase by SSF and also the main reaction mechanisms of the enzyme were described by the identification of the main intermediary compounds for the generation of bioactive compounds (Ascacio-Valdés et al. 2016).

In this section, the potential of SSF as a bioprocess to produce enzymes was described. It is true that it is a bioprocess that needs a rigorous control and that its use on an industrial scale is still scarce. However, there are reports that expose the full potential of application for the enzyme production and it can be concluded that the main challenge is to develop bioprocesses and technologies that allow a better control of fermentation conditions to guarantee the best levels of production.

4.4.1 Culture Conditions

Different culture conditions affect the synthesis of enzymes in an SSF process. Some of these are the type of substrate and the pretreatment made to it, the particle size of the substrate, the water activity (a_w) of the support, the relative humidity, the temperature of the process, the pH of the substrate, the fermentation time, the inoculum size, and the available oxygen. All these aspects have been studied for the improvement in the productivity of the process of obtaining enzymes by SSF. However, the oxygen supply, heat and carbon dioxide removal, temperature and moisture control, and mixing during the cultivation are still constraints for the industrialization of the SSF. Thus, an adequate control of these parameters is very important for an efficient SSF process, and these must be optimized for each bioreactor and process (Poletto et al. 2017).

One of the important parameters is the a_w, which must be adjusted depending on the physical requirements of the microorganisms. Generally, bacteria need an a_w between 0.9 and 0.99, yeast an a_w between 0.8 and 0.9, and filamentous fungi an a_w between 0.6 and 0.7. The increase in water content can impede growth due to the decrease of oxygen in the pores of the substrate (Chen 2013).

In general, for SSF aerobic microorganisms are used. Thus, the oxygen content is a fundamental factor in its development. For this reason, there must be aeration to supply the necessary oxygen and to eliminate both the carbon dioxide and the metabolic heat generated. In this way, the optimum flow of air must be taken into consideration in respect to the nature of the microorganism, the speed of generation of metabolic heat, the concentration of carbon dioxide and other metabolites, and the thickness of the substrate, among others. A variable related to the availability of oxygen is the particle size. The size of the particles determines the degree of porosity; an optimal value of this varies depending on the substrate, the microorganism to be used, and the amount of oxygen present in the medium. Small particles have the advantage of having a greater surface-to-volume ratio, which has a great influence on mass transfer (Pandey 1991a; de Castro et al. 2016).

In the case of temperature also it is necessary to adjust according to the microorganism. Temperature control is important because due to the high concentration of substrate per unit volume and the low thermal conductivity of the system, there may be accumulation of metabolic heat and an increase in the temperature of the crop causing deceleration or arrest of microbial activity, dehydration of the solid medium, and deviation of the metabolism as a defense mechanism against heat or dehydration (Dalsenter et al. 2005).

Another variable to consider during an SSF process is pH. In SSF processes it is very difficult to control pH; for this reason, microorganisms that are not very strict in this aspect and have a wide optimum pH range are desirable. A decrease in pH contributes to the growth of yeasts that negatively influence the process (Torrado et al. 1998; Singhania et al. 2009).

Some studies carried out to obtain enzymes by SSF have been directed to the optimization and control of the process parameters to obtain higher yields. Francis et al. (2003) evaluated the effect of incubation temperature, initial substrate moisture, and inoculum size on the production of α-amylase by *A. oryzae*. They found that temperature is one of the most influential parameters of the process, with 30 °C being at which the microorganism growth is better and there was a greater enzymatic activity. They also found the right composition to supplement the nutrients in the culture medium, increasing the yield by approximately 20%.

In other studies, Yang et al. (2015) optimized culture conditions for the production of glucanase by *Rhizomucor miehei*. They studied the effect of the carbon and nitrogen source, the initial pH of the culture medium, and the incubation temperature. It was found that the source of nitrogen and temperature are some of the parameters that most affect the production of glucanase. High temperatures possibly inhibit ribosomes, altering the entire process of enzymatic production.

Sahnoun et al. (2015) optimized the culture conditions to produce α-amylase by *A. oryzae*. Among the parameters discussed are the initial humidity of the substrate, the carbon/nitrogen ratio, and the size of the inoculum. An increase in yield by 30% was achieved when done in optimal conditions. These studies show the importance of adapting and controlling the conditions of an SSF process.

4.4.2 Transport Phenomena in SSF

SSF is a complex system that involves multiple phases and multiple components, where processes of heat and mass transfer are developed, and multiple reactions are carried out. SSF is accompanied by the development of concentration gradients, which transport substrates and products such as enzymes (Bück et al. 2015). During SSF, the microorganism can grow on the surface of the substrate or within the solid matrix; therefore, the microbial biomass interacts with the solid substrate. In this process, the microbial biomass consumes substrates and secretes metabolites and enzymes. Due to the conditions of the environment, all the transport process of materials and nutrients is carried out by conduction. Therefore, during the process there are different concentration gradients of substrates and products, which can cause local differences in metabolic activity. These concentration gradients are also applicable to the presence of inducers or repressors, affecting the enzyme production. These concentration gradients occur in a general way in the process regardless of the microorganism used. However, in most studies on enzyme production, these gradients are explained for filamentous fungi, since they are the most commonly used microorganisms in SSF and for their potential to excrete hydrolytic enzymes (Rahardjo 2005; Rahardjo et al. 2006).

In order to understand the phenomena occurring within an SSF bioreactor, it is necessary to understand the physical arrangement of the various phases within the system, since the various phenomena occur within and between these phases (Berovic 2006). Transport phenomena in SSF can be divided into micro-phenomena and macro-phenomena, depending on the particle size and scale of the bioreactor (Jou and Lo 2011).

At macroscale, the bioreactor contains three phases: the bioreactor wall, a headspace full of gas, and a substrate bed, composed of particles and air within the interparticle spaces. In general, at macroscale, the bed is considered as a single pseudo-homogeneous phase. Different phenomena of heat and mass transfer occur in the substrate, which can be defined using different mathematical models (Berovic 2006; Chen 2013).

At microscale, the particle air and the interparticle air are treated as different subsystems. Many of the transport processes shown are largely unaffected by the bioreactor and the way it is operated; that is, they are intrinsic to SSF systems due to the presence of the solid phase. These processes include mass transfer processes such as the following (Berovic 2006):

- The diffusion of O_2, CO_2, and water vapor within static regions of the gas phase and their convective movement in regions of airflow, with the extent of static and flowing regions depending on whether the bed is forcefully aerated or not.
- The diffusion of O_2, CO_2, water, nutrients, protons, products, and enzymes within the biofilm phase and the substrate particle.
- Exchanges of O_2, CO_2, and water vapor between the various phases.
- Also, within the particle there will be the reaction of enzymes with their substrates.

Microorganisms most often used in SSF are filamentous fungi. Therefore, the description of the fermentation process and the transport phenomena that occur are generally explained using the growth model of the filamentous fungi on the substrate. The process starts with mixing of a spore inoculum with substrate particles. Each particle initially has a number of spores attached to it, which germinate at different times depending on environmental conditions. Once each spore germinates, a germ tube extends away from the spore and branches to give daughter hyphae, which extend and then branch again, to give an expanding microcolony (Mitchell et al. 2006a; b; Chen 2013).

In order for the process to continue and the nutrients to be used, it is necessary that the enzymes get diffused properly. Enzymes diffuse away from the site of secretion into the particle. The speed of diffusion depends on the size of the enzyme and on the internal structure of the substrate particle. The enzymes begin hydrolyzing the polymer, and the soluble hydrolysis products then diffuse through the substrate. Oxygen consumption causes diffusion of O_2 through the static gas layer to the biomass and any initial O_2 within the substrate also diffuses to the biomass. During the process the hyphae get extended above the surface and penetrate the substrate. During these very early stages, there is a sufficiently high O_2 concentration within the substrate to support this penetration: given the low biomass, the rate of O_2 uptake is low and diffusion can replenish it reasonably effectively. Also due to the low biomass, the overall rate of heat production is very low. So, early in an SSF process, growth is essentially biologically limited. Growth occurs at the maximum specific growth rate, at which the organism is capable of growing on a solid surface, at the prevailing temperature, pH, and water activity, although the extent to which this is true depends on how quickly enzymes are produced to liberate hydrolysis products from polymers (Berovic 2006).

While the biomass continues increasing, the rates of growth-associated activities, such as enzyme production and O_2 consumption, also increase. Subsequently, the biomass begins to decrease the consumption of O_2 and nutrients; at this moment the process is limited by mass transfer. The biomass may continue to penetrate into the substrate, although this might be relatively slow due to O_2 limitations. During this phase the rate of heat production soon exceeds the rate at which heat can be removed, such that the temperature of the substrate bed rises. Likewise, there are structural changes in the bed that could affect the main transport mechanisms (Rahardjo et al. 2006).

Following the process, due to the lack of O_2 and nutrients, and excess heat in the environment, the microorganism initiates sporulation processes, the growth process ends, and cell death occurs. As a result, the growth decelerates, and the rate of heat production falls. As the heat production rate falls, the temperature of the substrate bed falls.

The diffusion of the enzymes can be affected by the reaction rate in SSF, so the use of mathematical models could help in the understanding of these processes. This would allow the improvement of the process and the design of new bioreactors, among others (Sermanni and Tiso 2008; Jou and Lo 2011; Mazaheri and Shojaosadati 2013). There are different mathematical models to explain transport phenomena in SSF. However, there are few mathematical models applied to enzymes. One of the models was proposed by Mitchell et al. (1991) to describe the process of release and diffusion of glucoamylase. During the study the growth of *Rhizopus oligosporus* was described in a thin layer of the substrate. In addition, it is assumed that the glucoamylase release is constant over the period from 0 h to time t_E:

$$J_E|\delta = D_E \left.\frac{\partial C_E}{\partial x}\right|_{\delta} = \mathrm{RGH}\left(t_E - t\right)$$

where $J_E|\delta$ is the flux of glucoamylase across the membrane filter at the surface of the substrate; C_E is the glucoamylase concentration; and D_E is the effective diffusivity of glucoamylase in the model substrate. $H(t_E - t)$ is a Heaviside function. When t is smaller than t_E, this function has the value of 1 and therefore the rate of glucoamylase release is equal to RG. When t is larger than t_E, the function has the value of 0 and therefore there is no further glucoamylase release.

The diffusion of glucoamylase in the model substrate is assumed to follow Fick's law of diffusion (Rahardjo et al. 2006). The mass balance equation for glucoamylase in the substrate matrix is

$$\frac{\partial C_E}{\partial t} = D_E \frac{\partial^2 C_E}{\partial x^2}$$

Considering that the release of glucoamylase is constant for a period of 6 h, the boundary conditions are as follows:

For $0 < t < 6$ h, at $x = 0$, $\frac{\partial C_E}{\partial x} = 0.$

at $x = \delta_S$, $D_E \frac{\partial C_E}{\partial x} = RG$

For $t > 6$ h, at $x = 0$ and $x = \delta_S$, $\frac{\partial C_E}{\partial x} = 0$

The bottom of the substrate is at $x = 0$, and the top at $x = \delta_S$; the fungus grows at the top.

The hydrolysis of starch by glucoamylase was described with Michaelis-Menten kinetics and it was assumed that the glucose is released by the action of glucoamylase and then diffuses towards the mycelium at the substrate matrix:

$$\frac{\partial C_G}{\partial t} = D_G \frac{\partial^2 C_G}{\partial x^2} + \frac{k_{cat} C_E C_S}{K_m + C_S}$$

$$\text{at } x = 0, \quad \frac{\partial C_G}{\partial x} = 0$$

$$\text{at } x = \delta, \quad -D_G \frac{\partial C_G}{\partial x} = X \frac{q_m C_G|_{\delta S}}{K_G + C_G|_{\delta S}} \le X_C \frac{q_m C_G|_{\delta S}}{K_G + C_G|_{\delta S}}$$

where C_G and C_S are glucose and starch concentrations at time t and place x, k_{cat} is the maximum specific activity of glucoamylase (dimensionless), C_E is the enzyme concentration (expressed in activity units per volume), K_m is the apparent Michaelis-Menten constant for glucoamylase, D_G is the effective diffusivity of glucose in the substrate layer, q_m is the maximum specific rate of glucose uptake, $C_G|_{\delta s}$ is the glucose concentration at the top surface of the substrate, K_G is the Monod constant for glucose uptake, x is the amount of biomass present on top of the substrate expressed per unit surface area, and x_c is the maximum amount of biomass that actively takes up glucose. The glucoamylase concentration (C_E) is expressed in terms of its activity and therefore k_{cat} is equal to 1.0. For the purposes of modeling, the starch concentration (C_S) is expressed in terms of the equivalent weight of glucose.

Oxygen transfer is one of the transport phenomena that most limits the behavior of SSF processes. Oxygen supply into the mycelia on the surface of or inside the substrate is hampered by diffusion limitation, which is caused by the presence of liquid-filled pores in a densely packed mycelium layer that is formed close to the substrate surface (Rajagopalan and Modak 1995; Oostra et al. 2001; Rahardjo et al. 2005). The mathematical models described can help predict the effect of nutrient diffusion on the enzymatic productivity of different microorganisms. Rahardjo et al. (2005) carried out a series of experiments to show the effect that a low concentration of oxygen has on the production of a-amylase by *A. oryzae* in SSF. In fact, concentrations lower than 0.25% of oxygen cause a substantial depression in the enzymatic production. Likewise, the researchers stated that because the saturation constant for oxygen concentration was very low, the growth kinetics of *A. oryzae* can be simplified to zero-order kinetics in the coupled diffusion/reaction models.

The equations and the proposed theory for enzymatic diffusion have been used by different authors to explain the phenomena that occur during the obtaining of enzymes by SSF. Nahid et al. (2012) studied the production of glucoamylase by *Aspergillus niger* in SSF using different agro-industrial by-products as support. The studies were carried out in flasks and in a designed tray bioreactor. The results showed that the highest activity was obtained using wheat bran with 10% corn flour. In addition, using the tray bioreactor a slightly higher productivity (7%) is achieved than in the flasks, so at industrial level the use of these bioreactors is proposed.

One of the hypotheses proposed in the diffusion models is that by increasing the interparticle spaces, the diffusion of nutrients and enzymes is greater. Research carried out by Baladhandayutham and Thangavelu (2011) optimizing the production of pectinase by *Aspergillus awamori* in SSF found that the addition of 15% of fibrous materials as sugarcane bagasse increases the productivity of pectinase enzyme, possibly due to the improvement in diffusion processes. However, beyond this amount the phenomena of catabolic repression are present, avoiding the fact that the microorganism hydrolyzes some compounds such as cellulose and hemicellulose. In this study it was found that the best conditions to achieve an enzymatic activity of 103.33 U/mL are temperature 35 °C, pH 5, and process time 72 h. Similar results were found by Suresh and Viruthagiri (2010) using *A. niger* for the production of pectinases in SSF. Martin et al. (2004) evaluated the production of pectinase by *Moniliella* and *Penicillium* using mixtures of agro-industrial by-products such as orange bagasse, sugarcane bagasse, and wheat bran as substrates. Like other researchers (Nagel et al. 1999; Silva et al. 2002), they found that the increase in intraparticle spaces improves enzyme production, so the use of fibrous materials can promote diffusion and in this case improve the productivity in obtaining pectinases.

On the other hand, Viniegra-González et al. (2003) used the logistic models and Luedeking-Piret equations to estimate the values of the coefficients: maximal specific growth rate (μM), maximal biomass level (X_M), enzyme/biomass yield ($Y_{P/X}$), and secondary rate of production or breakdown (k), to compare the yields presented in the production of enzymes tannase, pectinase, and invertase in SSF with SmF. At the end of the study the reaction-diffusion model was proposed to explain the behavior during the process. It was proposed that oxygen diffusion is perpendicular to large cellular aggregates and that diffusion of sugar is horizontal along the thin layer. This makes it possible to create sugar concentrates in the cell aggregates, increasing the diffusion and enzymatic productivity. Following these studies, Montalvo et al. (2005) evaluated the phenomena of catabolic repression in the tannase production by SSF, considering the gradients of diffusion as the greatest limitation in this kind of processes. They concluded that in effect, the existence of diffusional gradients of substrate is one of the main factors responsible for the high enzyme activity. The formation of these gradients can be demonstrated if the substrate uptake rate is higher than substrate diffusion rate and if this relation is inverse the catabolic repression phenomenon will be present due to the accumulation of the substrate (glucose as repressor) in its environment.

4.4.3 SSF Bioreactors

The bioreactor is one of the most important equipment in the development of bioprocess, and numerous types of bioreactors have been used for fermentation technologies in the production of different secondary metabolites (i.e., antibiotics, enzymes, bioactive compounds, pigments) or biofertilizers, biopesticides, and biofuels from solid agro-industrial residues (Barrios-Gonzalez et al. 2005) (please see Fig. 4.1). The use and design of bioreactors for SSF have been reviewed previously (Pandey 1991b, 2003; Mitchell et al. 2000; Raghavarao et al. 2003; Robinson and Nigam 2003; Durand 2003; Prabhakar et al. 2005; Krishna 2005; Mitchell et al. 2006a; b; Couto and Sanromán 2006; Ruiz-Leza et al. 2007; Bhargav et al. 2008; Singhania et al. 2009; Ali and Zulkali 2011; Chen 2013; Ashok et al. 2017; Arora et al. 2018; Krishania et al. 2018), taking into account the advantages and disadvantages in terms of operational conditions, scale-up, substrate, microbial strain, etc. Pino et al. (2018) reported that the bioreactor provides optimal conditions for enzymes and microorganisms, improving the yields in the bioprocess.

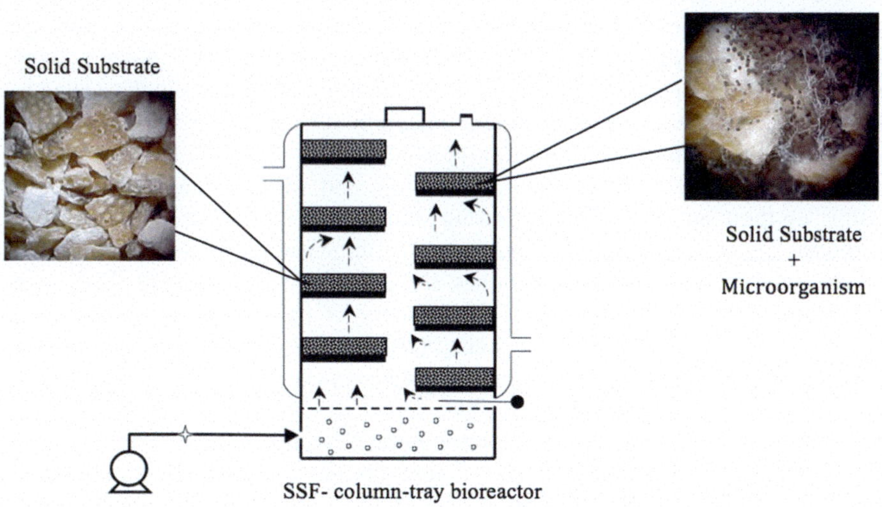

Fig. 4.1 Scheme of an SSF bioreactor (adapted and modified from Ruiz et al. 2012)

The bioreactor for SSF can be operated in different modes, for example:

1. *Batch:* The solid substrate and/or carbon source is added in the reactor; moreover the residence time in the fermentation is long and this mode of operation requires regular reinoculation (Van de Lagemaat and Pyle 2004; Ruiz et al. 2012). Ruiz et al. (2012) used a solid-column-tray bioreactor for pectinase production from lemon peel pomace as support and carbon source; they concluded that the use of this kind of bioreactor allowed the high production of fungal biomass and enzyme production at 96 h.
2. *Fed-batch:* The solid substrate and/or carbon source could be added at determined time intervals (Astolfi et al. 2011). Gonzalez-Figueredo et al. (2010) proposed a mathematical model using a pilot-plant fed-batch SSF reactor for compost of *Agaricus bisporus* mushroom cultivation.
3. *Continuous and/or plug-flow bioreactor operation*: The solid substrate and/or carbon source is passing in one direction; typically a screw system transports the solid substrate in different directions. Van de Lagemaat and Pyle (2004) reported that a continuous screw fermenter could be operated with constant non-inoculated feed. Nigam (2009) mentioned that a plug-flow mode in continuous conditions can be easier in the operation on industrial scale compared to other fermentation systems.

For the design of the bioreactors there are many parameters that affect the process, for example, the substrate/support (particle size), environmental conditions (aeration), sterilization system, and mechanical agitation. In these days, there are some common types of bioreactors that have been developed, for example: static bioreactors (Soccol et al. 2017), drum bioreactors (Wang et al. 2010), tray bioreactors (Figueroa-Montero et al. 2011; Ruiz et al. 2012), packed-bed or fixed-bed bioreactors (Castro et al. 2015; Soccol et al. 2017).

Another important parameter to consider in the design of SSF bioreactors is the possible scaling up; there are different levels: flask level (50–1000 g substrate), laboratory fermenter level (5–20 kg substrate), pilot fermenter level (50–5000 kg), and production fermenter level (25–1000 tonnes of substrate) (Nigam 2009). According to Lonsane et al. (1992), the most important parameters in the design for scaling up are variations in the biomass, inoculum at large scale, medium sterilization, aeration, agitation, heat removal, moisture content in the solids, pH control, contamination control, downstream processing, and waste management. For the scaling up there are different engineering strategies, for example (1) fundamental methods, (2) semi-fundamental methods, (3) dimensional analysis, (4) rule of thumb, and (5) trial-and-error techniques. One of the most common methods is the dimensional analysis, keeping similar geometry of the bioreactor (Votruba and Sobotka 1992; Durand 2003). Soccol et al. (2017) reported that the most important problems in the scaling up are the lack of standardized process and limited reproducibility of the experimental results. Mitchell et al. (2004) mentioned that one of the most important problems is the control in the bed temperature in the large-scale SSF bioreactor. They studied the forced aeration and intermittent agitation in the construction of packed-bed SSF bioreactor of 200 L. Bandelier et al. (1997) used the fed-batch SSF strategy in an aseptic pilot-scale reactor (50 L) in the gibberellic acid production using *Gibberella fujikuroi* reaching levels of up to 3 g per kg of dry matter, and concluded that the bioreactor is a good and efficient system in the production of secondary metabolite production.

In a recent work, Biz et al. (2016) used a pilot-scale packed-bed SSF bioreactor for pectinase production using citrus waste and sugarcane bagasse as substrate; as a result, they reported that the bed bioreactor was controlled and obtained pectinase yields of 33–41 U/g of substrate. Cerda et al. (2017) produced cellulase and xylanase in a pilot-scale SSF bioreactor of 50 L; they used a consortium (bacteria: *Pseudoxanthomonas taiwanensis* and *Sphingobacterium composti* and the yeasts: *Cyberlindnera jadinii* and *Barnettozyma californica*) of microorganisms able to produce cellulases (3.1 FPU/g of substrate), xylanase (48 U/g of substrate), and used coffee husk as substrates.

On the other hand, monitoring, digital system, and automatic process control are of considerable interest in the fermentation industries; the automatic control reduces the production costs, increases the conversion yield, and maintains the quality of the compounds and products (Scheper and Lammers 1994; Rani and Rao 1999). Pérez-Correa et al. (2006) mentioned that the manual control can regulate the operating conditions within the solid substrate bed, but in large-scale SSF systems it is not feasible. In order to properly monitor and control the reactors, it is necessary to have an important series of instruments, in this case primary control elements (sensors) for *temperature in the solid substrate bed*: resistance temperature detectors (RTDs) or thermocouples; *bed water content or water activity*: instruments based on capacitance or conductivity-based devices; *gas flow rate*: pitot tube, Venturi, orifice flow meters, and rotameters; and *pH*: glass electrodes and off-gas analysis: gas chromatography (Fernández and Pérez-Correa 2006). The most common control system is the gain scheduling proportional–integral–derivative (PID) controller; model-based control may be the excellent option in large-scale SSF bioreactors (Rani and Rao 1999; Pérez-Correa et al. 2006). Fernandez et al. (1996) automated a solid substrate cultivation pilot reactor (50 kg); they reported that an excellent performance in the control of temperature and water content in the solid substrate bed was achieved.

4.4.4 Genetic Engineering

Production of industrial enzymes has increased in the last decades, thanks to progress in culture conditions, selection of better microbial strains, and genetic improvement of microorganisms. Regarding genetic engineering or improvement, three different strategies can be highlighted: induced mutation, genetic recombination, and DNA technologies. After the second world war, induction of mutations in different organisms was the focus of extensive research, mainly chemical- and physical-induced mutation. However, application of mutation technologies on plant and animal improvement had low impact and was almost dropped out. Induced mutation technologies found a niche of application on improvement of microorganisms for enzyme production.

Induced chemical mutation relies on application of base analogs, which have similarity to nitrogen bases and can be incorporated into the DNA. Mutagens are not incorporated into DNA, but alter a base, causing in this way base mispairing, and intercalating agents which mimic base pair and are capable to intercalate themselves in between the nitrogen bases. Some examples of chemicals used for induced mutation are ethyl methanesulfonate (EMS), methyl methanesulfonate (MMS), hydroxylamine (HA), nitrosoguanidine (NTG), and 4-nitroquinoline-1-oxide (Adrio and Demain 2003). After application of a mutagenic agent, cell populations must be screened in order to identify those individuals with the desired phenotype. Lipase production increased 2.5-fold in mutants from *A. niger* strains which were treated with nitrous acid in comparison to wild isolates (Mala et al. 2001).

For induced physical mutations in microbial genomes different kinds of radiations (X-ray irradiation, ultraviolet, and gamma rays) have been used. The general effect of irradiation on DNA is production of lesions, which interfere with normal base pairing, insertion, duplication, and deletion of bases that may produce frameshift mutation, producing a very different protein (Lee et al. 2001). An example of the use of induced physical mutation to increase enzyme production is the study performed by Vu et al. (2009), treating spores of *Aspergillus* sp. SU14 using Co60 (gamma rays), γ-rays, ultraviolet irradiation, and *N*-methyl-*N*′-nitro-*N*-nitrosoguanidine, resulting in 2.2-fold increase in cellulase production in comparison to the wild strain. However, use of induced biological mutation has been very rare.

Genetic recombination has been used more frequently after overcoming the main drawback (low frequency of genetic recombination, especially in industrial microorganisms) with the use of strategies such as protoplast fusion. In this approach, the traditional methods of genetic improvement involving crossing and selection in order to generate new genotypes are involved. Three parasexual processes in bacteria (conjugation, transduction, and transformation) are the most common ways of microbial genetic recombination. Conjugation is described as the process where DNA is transferred from one cell to another by contact, while during transduction DNA is transferred from cell to cell via bacteriophages, and transformation is the ability of one competent cell to uptake naked DNA from its surrounding environment and express it (Adrio and Demain 2010). Genetic recombination was used to develop new antibiotics by protoplast fusion. In this case, protoplast of *Streptomyces griseus*, a streptomycin producer, was fused with protoplast of *Streptomyces tenjimariensis*, an istamycin producer, and the hybrid cell produced a new antibiotic (Yamashita et al. 1985). Difficulties to recover the cell membrane are the main drawbacks to this technology dissemination. Transposons or translocatable DNA segments are responsible for internal genetic rearrangements. The insertion sequences have been reported to have broad host specificity in *Streptomyces* (Adrio and Demain 2010). Use of these insertional sequences facilitates physical mapping of insertions, cloning of DNA flanking insertions, efficient construction of mutant libraries, and construction of highly stable mutants. Transposon technology has been employed to increase *tylosin* yield. This compound is an antibiotic and a bacteriostatic feed additive which is utilized in veterinary medicine (Baltz et al. 1997).

DNA technology involves different methodologies, molecular breeding, association analyses, combinatorial biosynthesis, whole-genome shuffling, genome mining, and metabolic engineering. Molecular breeding allows in vitro homologous recombination, being a way of mimicking natural recombination, while association analyses are the study of a DNA sequence variation along the whole genome with the objective to identify its association with an observable trait (Adrio and Demain 2010). On the other hand, combinatorial biosynthesis is defined as a systematic modification and interchange of genes important for biosynthesis of natural products which result in the production of unnatural or hybrid products (Rix et al. 2002). However, whole-genome shuffling is a methodology used for microbial strain improvement which combines the advantages of multiparental crossing by DNA shuffling with recombination of whole genomes. DNA shuffling is used for in vitro homologous recombination of selected mutant genes, which were previously digested at random (Adrio and Demain 2010). Genome mining is defined as the use of computational methodologies to derive information about discovery and characterization of different compounds and biochemical routes based on genome analysis (Ziemert et al. 2016), and metabolic engineering includes modification of specific biochemical ways in order to improve product formation or cell properties; this modification is via recombinant DNA technology (Nandy 2016). Recombinant DNA technology for industrial enzyme production includes the selection of an efficient gene, its introduction into a suitable vector, and transformation in an efficient host to produce a high amount of recombinant protein of interest (Gopinath et al. 2017).

Metabolic engineering is the most used DNA technology for improving yields of industrial enzymes; in order to meet this objective, four different strategies have been followed: incorporation of more than one copy of the gene of interest, use of stronger promoters, use of preferred codons, and alternatives to the signal peptide (Pan et al. 2013). It has been observed that introduction of more than one copy of the gene of interest in microbial cell had increased expression of the industrial enzyme, but the number of copies introduced had a threshold, and after that enzymatic yield decreased. Verdoes et al. (1994) reported an 18-fold increase of glucoamylase yield, after introducing 20 copies of the glucoamylase gene into *A. niger.* Use of stronger promoter, or in some cases introduction of exogenous genes to some host, results in low expression. Ma et al. (2006) overexpressed a lipase gene in *Bacillus subtilis* by use of a strong promoter; in this case lipase yield was 100-fold higher than original strain. Use of preferred codons: low expression of heterologous genes is attributed to codon bias. This can be solved by incorporating host-preferred codons and improving the amount of rare tRNAs (Pan et al. 2013). Veana et al. (2014) using codons preferred by *Pichia pastoris* could express an invertase from *A. niger* GH1 and alternatives to the signal peptide. The leader sequence of 15–30 amino acids found in the N-terminus of the expressed proteins that are destined towards the secretory pathway is called signal peptide. It has been observed that modification of the signal peptide can increase the yield of industrial enzymes (Pan et al. 2013). Liu et al. (2005) improved glucoamylase yield by replacing the signal peptide.

4.5 Perspectives

The biotechnological processes of SSF have a potential application to produce enzymes of industrial interest from agro-industrial residues that are generated in large quantities and whose current disposition is not the most adequate. Through these processes it is possible to obtain enzymes that can work in extreme conditions of temperature or pH, or enzymes that are more resistant during the process.

Currently, there are several studies found in SSF processes. However, the great challenge is to take these on an industrial scale and maintain high productivity. Therefore, it is necessary to continue with studies at different levels of biotechnology, from the identification of new strains of microorganisms that hydrolyze different compounds and produce highly specific enzymes to the design of new larger scale bioreactors and through research in genetic engineering (DasIsha et al. 2019).

Acknowledgments The authors would like to thank the financial support from the Energy Sustainability Fund 2014-05 (CONACYT-SENER), Mexican Centre for Innovation in Bioenergy (Cemie-Bio), and Cluster of Bioalcohols (Ref. 249564). We also gratefully acknowledge support for this research by the Mexican Science and Technology Council (CONACYT, Mexico) for the Infrastructure Project-INFR201601 (Ref. 269461) and Basic Science Project-2015-01 (Ref. 254808).

References

Abraham J, Gea T, Sánchez A (2013) Potential of the solid-state fermentation of soy fibre residues by native microbial populations for bench-scale alkaline protease production. Biochem Eng J 74:15–19

Adhyaru DN, Bhatt NS, Modi HA, Divecha J (2016) Insight on xylanase from Aspergillus tubingensis FDHN1: production, high yielding recovery optimization through statistical approach and application. Biocatal Agric Biotechnol 6:51–57

Adrio JL, Demain AL (2003) Fungal biotechnology. Int Microbiol 6(3):191–199

Adrio J-L, Demain AL (2010) Recombinant organisms for production of industrial products. Bioeng Bugs 1(2):116–131

Aguilar CN, Augur C, Favela-Torres E, Viniegra-González G (2001) Production of tannase by Aspergillus niger Aa-20 in submerged and solid-state fermentation: influence of glucose and tannic acid. J Ind Microbiol Biotechnol 26(5):296–302

Ali HKQ, Zulkali MMD (2011) Design aspects of bioreactors for solid-state fermentation: a review. Chem Biochem Eng 25(2):255–266

Anwar Z, Gulfraz M, Irshad M (2014) Agro-industrial lignocellulosic biomass a key to unlock the future bio-energy: a brief review. J Radiat Res Appl Sci 7(2):163–173

Arora S, Rani R, Ghosh S (2018) Bioreactors in solid state fermentation technology: design, applications and engineering aspects. J Biotechnol 269:16–34

Ascacio-Valdés JA, Buenrostro-Figueroa JJ, Aguilera-Carbo A, Prado-Barragán A, Rodríguez-Herrera R, Aguilar CN (2011) Ellagitannins: biosynthesis, biodegradation and biological properties. J Med Plants Res 5(19):4696–4703

Ascacio-Valdés JA, Buenrostro JJ, De la Cruz R, Sepúlveda L, Aguilera AF, Prado A, Contreras JC, Rodríguez R, Aguilar CN (2014) Fungal biodegradation of pomegranate ellagitannins. J Basic Microbiol 54(1):28–34

Ascacio-Valdés JA, Aguilera-Carbó AF, Buenrostro JJ, Prado-Barragán A, Rodríguez-Herrera R, Aguilar CN (2016) The complete biodegradation pathway of ellagitannins by Aspergillus niger in solid-state fermentation. J Basic Microbiol 56(4):329–336

Ashok A, Doriya K, Rao DRM, Kumar DS (2017) Design of solid state bioreactor for industrial applications: an overview to conventional bioreactors. Biocatal Agric Biotechnol 9:11–18

Astolfi V, Joris J, Verlindo R, Oliveira JV, Maugeri F, Mazutti MA, de Oliveira D, Treichel H (2011) Operation of a fixed-bed bioreactor in batch and fed-batch modes for production of inulinase by solid-state fermentation. Biochem Eng J 58–59(1):39–49

Baladhandayutham S, Thangavelu V (2011) Optimization and kinetics of solid-state fermentative production of pectinase by Aspergillus awamori. Int J ChemTech Res 3(4):1758–1764

Baltz RH, McHenney MA, Cantwell CA, Queener SW, Solenberg PJ (1997) Applications of transposition mutagenesis in antibiotic producing streptomycetes. Antonie Van Leeuwenhoek 71(1–2):179–187

Bandelier S, Renaud R, Durand A (1997) Production of gibberellic acid by fed-batch solid state fermentation in an aseptic pilot-scale reactor. Process Biochem 32(2):141–145

Barrios-Gonzalez J, Fernandez FJ, Tomasini A, Mejia A (2005) Secondary metabolites production by solid-state fermentation. Malaysian J Microbiol 1(1):1–6

Behnam S, Karimi K, Khanahmadi M, Agriculture I, Resources N, Salimian Z (2016) Optimization of xylanase production by Mucor indicus, Mucor hiemalis, and Rhizopus oryzae through solid state fermentation. Biol J Microorg 4(16):1–10

Belmares Cerda R, Reyes Vega ML, Contreras Esquivel JC, Rodríguez Herrera R, Aguilar CN (2003) Efecto de la fuente de carbono sobre la producción de tanasa en dos cepas de Aspergillus niger. Revista Mexicana de Ingeniería Química 2:95–100

Belmares R, Garza Y, Rodríguez R, Contreras-Esquivel JC, Aguilar CN (2009) Composition and fungal degradation of tannins present in semiarid plants. Elec J Env Agricult Food Chem 8:312–318

Berovic M (ed) (2006) Biochemical engineering principles. Doctoral/Post-Doctoral Level Course [Supetar, Island Brac, Croatia, 17 to 23 September 2006]. Faculty of Chemistry and Chemical Technology, University of Ljubljana, Ljubljana

Bhargav S, Panda BP, Ali M, Javed S (2008) Solid-state fermentation: an overview. Chem Biochem Eng 22(1):49–70

Biz A, Finkler ATJ, Pitol LO, Medina BS, Krieger N, Mitchell DA (2016) Production of pectinases by solid-state fermentation of a mixture of citrus waste and sugarcane bagasse in a pilot-scale packed-bed bioreactor. Biochem Eng J 111:54–62

Bück A, Casciatori FP, Thoméo JC, Tsotsas E (2015) Model-based control of enzyme yield in solid-state fermentation. Procedia Eng 102:362–371

Buenrostro Figueroa J, Ascacio-Valdés J, Sepúlveda L, la Cruz R, Prado LA, Aguilar M, Rodriguez R, Aguilar C (2014) Potential use of different agroindustrial by-products as supports of fungal ellagitannase production under solid state fermentation. Food Bioprod Process 92:376–382

Castro AM, Castilho LR, Freire DMG (2015) Performance of a fixed-bed solid-state fermentation bioreactor with forced aeration for the production of hydrolases by Aspergillus awamori. Biochem Eng J 93:303–308

Cerda A, Gea T, Vargas-García MC, Sánchez A (2017) Towards a competitive solid state fermentation: cellulases production from coffee husk by sequential batch operation and role of microbial diversity. Sci Total Environ 589:56–65

Chen, H., 2013. Modern solid state fermentation.

Chen B, Wu Q, Xu Y (2014) Filamentous fungal diversity and community structure associated with the solid state fermentation of Chinese Maotai-flavor liquor. Int J Food Microbiol 179:80–84

Couto SR, Sanromán M a (2006) Application of solid-state fermentation to food industry—a review. J Food Eng 76(3):291–302

Dalsenter FDH, Viccini G, Barga MC, Mitchell D a, Krieger N (2005) A mathematical model describing the effect of temperature variations on the kinetics of microbial growth in solid-state culture. Process Biochem 40(2):801–807

DasIsha D, Mani K, Salgaonkar B, Braganca JM (2019) Characterization of extremely halophilic archaeal isolates from Indian salt pans and their screening for production of hydrolytic enzymes. Environ Sustain 2(3):227–239

de Castro RJS, Ohara A, Nishide TG, Albernaz JRM, Soares MH, Sato HH (2015) A new approach for proteases production by Aspergillus niger based on the kinetic and thermodynamic parameters of the enzymes obtained. Biocatal Agric Biotechnol 4(2):199–207

de Castro AM, Castilho L d R, Freire DMG (2016) Characterization of babassu, canola, castor seed and sunflower residual cakes for use as raw materials for fermentation processes. Ind Crop Prod 83:140–148

de Oliveira RL, Dias JL, da Silva OS, Porto TS (2018) Immobilization of pectinase from Aspergillus aculeatus in alginate beads and clarification of apple and umbu juices in a packed bed reactor. Food Bioprod Process 109:9–18

Demir H, Tarı C (2014) Valorization of wheat bran for the production of polygalacturonase in SSF of Aspergillus sojae. Ind Crop Prod 54:302–309

Djoulde Darman R, Essia Ngand JJ, Etoa FX (2011) Cassava solid-state fermentation with a starter culture of Lactobacillus plantarum and Rhizopus oryzae for cellulase production. Afr J Microbiol Res 5(27):4866–4872

Durand A (2003) Bioreactor designs for solid state fermentation. Biochem Eng J 13(2–3):113–125

Fernández M, Pérez-Correa JR (2006) Instrumentation for monitoring SSF bioreactors. In: Solid-state fermentation bioreactors. Springer, Berlin, Heidelberg, pp 363–374

Fernandez M, Perez-Correa JR, Solar I, Agosin E (1996) Automation of a solid substrate cultivation pilot reactor. Bioprocess Eng 16(1):1–4

Figueroa-Montero A, Esparza-Isunza T, Saucedo-Castañeda G, Huerta-Ochoa S, Gutiérrez-Rojas M, Favela-Torres E (2011) Improvement of heat removal in solid-state fermentation tray bioreactors by forced air convection. J Chem Technol Biotechnol 86(10):1321–1331

Francis F, Sabu A, Nampoothiri KM, Ramachandran S, Ghosh S, Szakacs G, Pandey A (2003) Use of response surface methodology for optimizing process parameters for the production of α-amylase by Aspergillus oryzae. Biochem Eng J 15(2):107–115

Ghoshal G, Basu S, Shivhare U (2012) Solid state fermentation in food processing. Int J Food Eng 8(3):25

Godoy MG, Gutarra MLE, Castro AM, Machado OLT, Freire DMG (2011) Adding value to a toxic residue from the biodiesel industry: production of two distinct pool of lipases from Penicillium simplicissimum in castor bean waste. J Ind Microbiol Biotechnol 38(8):945–953

Gonzalez-Figueredo, C., De La Torre, L.M., and Sanchez, A., 2010. Dynamic modelling and experimental validation of a solid state fermentation reactor. IFAC Proceedings Volumes (IFAC-PapersOnline). IFAC.

Gopinath SCB, Anbu P, Arshad MK, Lakshmipriya T, Voon CH, Hashim U, Chinni SV (2017) Biotechnological processes in microbial amylase production. Biomed Res Int 2017

Gutiérrez-Correa M, Ludeña Y, Ramage G, Villena GK (2012) Recent advances on filamentous fungal biofilms for industrial uses. Appl Biochem Biotechnol 167(5):1235–1253

Hernández MC, Esquivel JCC, Lara F, Rodríguez R, Aguilar CN (2005) Isolation and evaluation of tannin-degrading fungal strains from the Mexican desert. Zeitschrift für Naturforschung C 60(11–12):844–848

Homaei A, Lavajoo F, Sariri R (2016) Development of marine biotechnology as a resource for novel proteases and their role in modern biotechnology. Int J Biol Macromol 88:542–552

Huang W, Niu H, Li Z, Lin W, Gong G, Wang W (2007) Effect of ellagitannin acyl hydrolase, xylanase and cellulase on ellagic acid production from cups extract of valonia acorns. Process Biochem 42(9):1291–1295

Irfan M, Nadeem M, Syed Q (2014) One-factor-at-a-time (OFAT) optimization of xylanase production from Trichoderma viride-IR05 in solid-state fermentation. J Radiat Res Appl Sci 7(3):317–326

Jou RY, Lo CT (2011) Heat and mass transfer measurements for tray-fermented fungal products. Int J Thermophys 32(1–2):523–536

Kaur H, Arora M, Bhatia S, Alam MS (2015) Optimization of α -amylase and glucoamylase production in solid state fermentation of deoiled rice bran (DRB) by Rhizopus oryzae. Int J Pure Appl Biosc 3(6):249–256

Knob A, Fortkamp D, Prolo T, Izidoro SC, Almeida JM (2014) Agro-residues as Alternative for Xylanase Production by Filamentous Fungi. Bioresources

Krishania M, Sindhu R, Binod P, Ahluwalia V, Kumar V, Sangwan RS, Pandey A (2018) Design of bioreactors in solid-state fermentation. In: Current developments in biotechnology and bioengineering. Elsevier, pp 83–96

Krishna C (2005) Solid-state fermentation systems—an overview. Crit Rev Biotechnol 25:1–2), 1–30

Kupski L, Pagnussatt FA, Buffon JG, Furlong EB (2014) Endoglucanase and total cellulase from newly isolated Rhizopus oryzae and Trichoderma reesei: production, characterization, and thermal stability. Appl Biochem Biotechnol 172:458–468

Lee C, Clarkson J, Shu J, Campbell ID, Yudkin MD (2001) Bacillus subtilis mutations that alter the pathway of phosphorylation of the anti-anti-σF factor SpoIIAA lead to a Spo− phenotype. Mol Microbiol 40(1):9–19

Li H, Zhang R, Tang L, Zhang J, Mao Z (2015) Manganese peroxidase production from cassava residue by Phanerochaete chrysosporium in solid state fermentation and its decolorization of indigo carmine. Chin J Chem Eng 23(1):227–233

Liu S-H, Chou W-I, Sheu C-C, Chang MD-T (2005) Improved secretory production of glucoamylase in Pichia pastoris by combination of genetic manipulations. Biochem Biophys Res Commun 326(4):817–824

Liu D, Zhang R, Yang X, Wu H, Xu D, Tang Z, Shen Q (2011) Thermostable cellulase production of Aspergillus fumigatus Z5 under solid-state fermentation and its application in degradation of agricultural wastes. Int Biodeterior Biodegradation 65(5):717–725

Lizardi-Jiménez MA, Hernández-Martínez R (2017) Solid state fermentation (SSF): diversity of applications to valorize waste and biomass. 3 Biotech 7(1):44

Londoño-Hernández L, Ramírez-Toro C, Ruiz HA, Ascacio-Valdés JA, Aguilar-Gonzalez MA, Rodríguez-Herrera R, Aguilar CN (2017) Rhizopus oryzae–Ancient microbial resource with importance in modern food industry. Int J Food Microbiol 257:110–127

Lonsane BK, Saucedo-Castaneda G, Raimbault M, Roussos S, Viniegra-Gonzalez G, Ghildyal NP, Ramakrishna M, Krishnaiah MM (1992) Scale-up strategies for solid state fermentation systems. Process Biochem 27(5):259–273

Lopez-Trujillo J, Medina-Morales MA, Sanchez-Flores A, Arevalo C, Ascacio-Valdes JA, Mellado M, Aguilar CN, Aguilera-Carbo AF (2017) Solid bioprocess of tarbush (Flourensia cernua) leaves for β-glucosidase production by Aspergillus niger: initial approach to fiber–glycoside interaction for enzyme induction. 3 Biotech 7(4):271

Loureiro DB, Romanini D, Tubio G (2016) Structural and functional analysis of Aspergillus niger xylanase to be employed in polyethylene glycol/salt aqueous two-phase extraction. Biocatal Agric Biotechnol 5:204–210

Ma J, Zhang Z, Wang B, Kong X, Wang Y, Cao S, Feng Y (2006) Overexpression and characterization of a lipase from Bacillus subtilis. Protein Expr Purif 45(1):22–29

Mala JGS, Kamini NR, Puvanakrishnan R (2001) Strain improvement of Aspergillus niger for enhanced lipase production. J Gen Appl Microbiol 47(4):181–186

Martin N, De Souza SR, Da Silva R, Gomes E (2004) Pectinase production by fungal strains in solid-state fermentation using agro-industrial bioproduct. Braz Arch Biol Technol 47(5):813–819

Mazaheri D, Shojaosadati SA (2013) Mathematical models for microbial kinetics in solid-state fermentation: a review. Iran J Biotechnol 11(3):156–167

Medina-Morales MA, López-Trujillo J, Gómez-Narváez L, Mellado M, García-Martínez E, Ascacio-Valdés JA, Aguilar CN, Aguilera-Carbó A (2017) Effect of growth conditions on β-glucosidase production using Flourensia cernua leaves in a solid-state fungal bioprocess. 3 Biotech 7(5):355

Membrillo I, Sánchez C, Meneses M, Favela E, Loera O (2008) Effect of substrate particle size and additional nitrogen source on production of lignocellulolytic enzymes by Pleurotus ostreatus strains. Bioresour Technol 99(16):7842–7847

Mitchell DA, Do DD, Greenfield PF, Doelle HW (1991) A semimechanistic mathematical model for growth of Rhizopus oligosporus in a model solid-state fermentation system. Biotechnol Bioeng 38(4):353–362

Mitchell D a, Krieger N, Stuart DM, Pandey A (2000) New developments in solid-state fermentation II. Rational approaches to the design, operation and scale-up of bioreactors. Process Biochem 35(10):1211–1225

Mitchell, D.A., Von Meien, O.F., Junior, L.F.L.L., and Krieger, N., 2004. Construction of a pilot-scale solid-state fermentation bioreactor with forced aeration and intermittent agitation. Process System Eng, 1–8.

Mitchell DA, Berovič M, Nopharatana M, Krieger N (2006a) The bioreactor step of SSF: a complex interaction of phenomena. In: Solid-state fermentation bioreactors. Springer, Berlin, Heidelberg, pp 13–32

Mitchell DA, Kriger N, Berovic M (2006b) Solid-state fermentation bioreactors-fundamentals of design and operation. Springer, Heidelberg. Mitchell DA

Montalvo MLC, Herrera RR, Esquivel JCC, Aguilar CN (2005) Glucose diffusion on support for solid state fermentation and its influence on tannase production profiles. Int J Chem React Eng 3:1–14

Moo-Young, M., Moreira, A.R., and Tengerdy, R.P., 1983. Principles of solid-substrate fermentation. The filamentous fungi, 4 (S E).

Nagel FJ, Oostra J, Tramper J, Rinzema A (1999) Improved model system for solid-substrate fermentation: effects of pH, nutrients and buffer on fungal growth rate. Process Biochem 35(1–2):69–75

Nahid P, Vossoughi M, Roostaazad R, Ahmadi M, Zarrabi A, Hosseini SM (2012) Production of glucoamylase by Aspergillus niger under solid state fermentation. Int J Eng Trans B: Appl 25(1):1–7

Nandy SK (2016) Bioprocess technology governs enzyme use and production in industrial biotechnology: an overview. Enz Eng 5:1–5

Neri DFM, Balcão VM, Costa RS, Rocha ICAP, Ferreira EMFC, Torres DPM, Rodrigues LRM, Carvalho LB, Teixeira JA (2009) Galacto-oligosaccharides production during lactose hydrolysis by free Aspergillus oryzae β-galactosidase and immobilized on magnetic polysiloxane-polyvinyl alcohol. Food Chem 115(1):92–99

Nigam PS (2009) Production of bioactive secondary metabolites. In: Biotechnology for agro-industrial residues utilisation. Springer, Netherlands, pp 129–145

Oliveira F, Salgado JM, Abrunhosa L, Pérez-Rodríguez N, Domínguez JM, Venâncio A, Belo I (2017) Optimization of lipase production by solid-state fermentation of olive pomace: from flask to laboratory-scale packed-bed bioreactor. Bioprocess Biosyst Eng 40(7):1123–1132

Oostra J, Le Comte EP, Van Den Heuvel JC, Tramper J, Rinzema A (2001) Intra-particle oxygen diffusion limitation in solid-state fermentation. Biotechnol Bioeng 75(1):13–24

Ozcirak Ergun S, Ozturk Urek R (2017) Production of ligninolytic enzymes by solid state fermentation using Pleurotus ostreatus. Ann Agrarian Sci 15(2):273–277

Pan H, Chen Y, Yu P (2013) Advanced strategies for improving the production of industrial enzymes in heterologous host systems. Enz Eng 2(114):2

Pandey A (1991a) Effect of particle size of substrate of enzyme production in solid-state fermentation. Bioresour Technol 37(2):169–172

Pandey A (1991b) Aspects of fermenter design for solid-state fermentations. Process Biochem 26(6):355–361

Pandey A (2001) Solid-state fermentation in biotechnology: fundamentals and applications. Asiatech Publishers

Pandey A (2003) Solid-state fermentation. Biochem Eng J 13(2):81–84

Pérez-Correa JR, Fernández M, von Meien OF, Luz LFL Jr, Mitchell DA (2006) Application of automatic control strategies to SSF bioreactors. In: Solid-state fermentation bioreactors. Springer, Berlin-Heidelberg, pp 387–402

Pino MS, Rodríguez-Jasso RM, Michelin M, Flores-Gallegos AC, Morales-Rodriguez R, Teixeira JA, Ruiz HA (2018) Bioreactor design for enzymatic hydrolysis of biomass under the biorefinery concept. Chem Eng J 347:119–136

Pitol LO, Finkler ATJ, Dias GS, Machado AS, Zanin GM, Mitchell DA, Krieger N (2017) Optimization studies to develop a low-cost medium for production of the lipases of Rhizopus microsporus by solid-state fermentation and scale-up of the process to a pilot packed-bed bioreactor. Process Biochem 62(July):37–47

Poggi-Varaldo HM, Munoz-Paez KM, Escamilla-Alvarado C, Robledo-Narváez PN, Ponce-Noyola MT, Calva-Calva G, Ríos-Leal E, Galíndez-Mayer J, Estrada-Vázquez C, Ortega-Clemente A, Rinderknecht-Seijas NF (2014) Biohydrogen, biomethane and bioelectricity as crucial components of biorefinery of organic wastes: a review. Waste Manag Res 32(5):353–365

Poletto P, Polidoro TA, Zeni M, da Silveira MM (2017) Evaluation of the operating conditions for the solid-state production of pectinases by Aspergillus niger in a bench-scale, intermittently agitated rotating drum bioreactor. LWT Food Sci Technol 79:92–101

Polizeli MLTM, Rizzatti ACS, Monti R, Terenzi HF, Jorge JA, Amorim DS (2005) Xylanases from fungi: properties and industrial applications. Appl Microbiol Biotechnol 67(5):577–591

Porfiri MC, Farruggia BM, Romanini D (2012) Bioseparation of alpha-amylase by forming insoluble complexes with polyacrylate from a culture of Aspergillus oryzae grown in agricultural wastes. Sep Purif Technol 92:11–16

Prabhakar A, Krishnaiah K, Janaun J, Bono A (2005) An overview of engineering aspects of solid state fermentation. Malaysian J Microbiol 1(2):10–16

Raghavarao KSMS, Ranganathan TV, Karanth NG (2003) Some engineering aspects of solid-state fermentation. Biochem Eng J 13:127–135

Rahardjo, Y.S.P. (2005) Fungal mats in solid-state fermentation. Rahardjo, YSP, p. 177

Rahardjo YSP, Sie S, Weber FJ, Tramper J, Rinzema A (2005) Effect of low oxygen concentrations on growth and α-amylase production of Aspergillus oryzae in model solid-state fermentation systems. Biomol Eng 21(6):163–172

Rahardjo YSP, Tramper J, Rinzema A (2006) Modeling conversion and transport phenomena in solid-state fermentation: a review and perspectives. Biotechnol Adv 24(2):161–179

Raimbault M (1998) General and microbiological aspects of solid substrate fermentation. Electron J Biotechnol 1(3):26–27

Rajagopalan S, Modak JM (1995) Evaluation of relative growth limitation due to depletion of glucose and oxygen during fungal growth on a spherical solid particle. Chem Eng Sci 50(5):803–811

Ramírez-Coronel MA, Viniegra-Gonzalez G, Darvill A, Augur C (2003) A novel tannase from Aspergillus niger with β-glucosidase activity. Microbiology 149(10):2941–2946

Rani KY, Rao VSR (1999) Control of fermenters–a review. Bioprocess Eng 21(1):77–88

Rix U, Fischer C, Remsing LL, Rohr J (2002) Modification of post-PKS tailoring steps through combinatorial biosynthesis. Nat Prod Rep 19(5):542–580

Robinson T, Nigam P (2003) Bioreactor design for protein enrichment of agricultural residues by solid state fermentation. Biochem Eng J 13((2–3)):197–203

Rodríguez-Fernández DE, Rodríguez-León JA, de Carvalho JC, Thomaz-Soccol V, Parada JL, Soccol CR (2010) Recovery of phytase produced by solid-state fermentation on citrus peel. Braz Arch Biol Technol 53(6):1487–1496

Ruiz HA, Rodríguez-Jasso RM, Rodríguez R, Contreras-Esquivel JC, Aguilar CN (2012) Pectinase production from lemon peel pomace as support and carbon source in solid-state fermentation column-tray bioreactor. Biochem Eng J 65:90–95

Ruiz-Leza H a, Rodriguez-Jasso RM, Rodriguez-Herrera R, Contreras-Esquivel JC, Aguilar CN (2007) Bioreactor design for solid state fermentation. Revista Mexicana De Ingenieria Quimica 6(1):33–40

Sahnoun M, Kriaa M, Elgharbi F, Ayadi DZ, Bejar S, Kammoun R (2015) Aspergillus oryzae S2 alpha-amylase production under solid state fermentation: optimization of culture conditions. Int J Biol Macromol 75:73–80

Scheper T-H, Lammers F (1994) Fermentation monitoring and process control. Curr Opin Biotechnol 5(2):187–191

Schuster E, Dunn-Coleman N, Frisvad J, Van Dijck P (2002) On the safety of Aspergillus niger—a review. Appl Microbiol Biotechnol 59((4–5)):426–435

Sermanni GG, Tiso N (2008) Aspects of design of bioreactors in SSF. In: Current developments in solid-state fermentation, pp 117–144

Shankar S, Bhan C, Chandra R, Tyagi S (2018) Laccase based de-inking of mixed office waste and evaluation of its impact on physico-optical properties of recycled. Environ Sustain 1(3):233–244

Shi B, He Q, Yao K, Huang W, Li Q (2005) Production of ellagic acid from degradation of valonea tannins by Aspergillus niger and Candida utilis. J Chem Technol Biotechnol 80(10):1154–1159

Silva D, Da Silva Martins E, Da Silva R, Gomes E (2002) Pectinase production by Penicillium viridicatum Rfc3 by solid state fermentation using agricultural wastes and agro-industrial by-products. Braz J Microbiol 33(4):318–324

Singh R, Kapoor V, Kumar V (2012) Utilization of agro-industrial wastes for the simultaneous production of amylase and xylanase by thermophilic actinomycetes. Braz J Microbiol 43(4):1545–1552

Singhania RR, Patel AK, Soccol CR, Pandey A (2009) Recent advances in solid-state fermentation. Biochem Eng J 44(1):13–18

Soccol CR, da Costa ESF, Letti LAJ, Karp SG, Woiciechowski AL, Vandenberghe LP d S (2017) Recent developments and innovations in solid state fermentation. Biotechnol Res Innov 1(1):52–71

Sukma A, Jos B, Sumardiono S (2018) Kinetic of biomass growth and protein formation on rice bran fermentation using Rhizopus oryzae. MATEC Web of Conf 156(01023):1–6

Suresh B, Viruthagiri T (2010) Optimization and kinetics of pectinase enzyme using Aspergillus niger by solid-state fermentation. Indian J Sci Technol 3(8):867–870

Tao S, Peng L, Beihui L, Deming L, Zuohu L (1997) Solid state fermentation of rice chaff for fibrinolytic enzyme production by Fusarium oxysporum. Biotechnol Lett 19(5):465–467

Thomas L, Larroche C, Pandey A (2013) Current developments in solid-state fermentation. Biochem Eng J 81:146–161

Torrado A, González MP, Murado MA (1998) pH regulation in solid state culture through the initial ratio between oxidized and reduced sources of nitrogen. A model applicable to the amylase production by Aspergillus oryzae. Biotechnol Tech 12(5):411–415

Van de Lagemaat J, Pyle DL (2004) Solid-state fermentation: a continuous process for fungal tannase production. Biotechnol Bioeng 87(7):924–929

Veana F, Fuentes-Garibay JA, Aguilar CN, Rodríguez-Herrera R, Guerrero-Olazarán M, Viader-Salvadó JM (2014) Gene encoding a novel invertase from a xerophilic Aspergillus niger strain and production of the enzyme in Pichia pastoris. Enzym Microb Technol 63:28–33

Veerabhadrappa MB, Shivakumar SB, Devappa S (2014) Solid-state fermentation of Jatropha seed cake for optimization of lipase, protease and detoxification of anti-nutrients in Jatropha seed cake using Aspergillus versicolor CJS-98. J Biosci Bioeng 117(2):208–214

Verdoes JC, van Diepeningen AD, Punt PJ, Debets AJM, Stouthamer AH, van den Hondel CA (1994) Evaluation of molecular and genetic approaches to generate glucoamylase overproducing strains of Aspergillus niger. J Biotechnol 36(2):165–175

Viniegra-González G, Favela-Torres E, Aguilar CN, Rómero-Gomez S d J, Díaz-Godínez G, Augur C (2003) Advantages of fungal enzyme production in solid state over liquid fermentation systems. Biochem Eng J 13(2–3):157–167

Votruba J, Sobotka M (1992) Physiological similarity and bioreactor scale-up. Folia Microbiol 37(5):331–345

Vu VH, Pham TA, Kim K (2009) Fungal strain improvement for cellulase production using repeated and sequential mutagenesis. Mycobiology 37(4):267–271

Wakayama M, Kitahata S, Manoch L, Tachiki T, Yoshimune K, Moriguchi M (2004) Production, purification and properties of d-aminoacylase from a newly isolated Trichoderma sp. SKW-36. Process Biochem 39(9):1119–1124

Wang E-Q, Li S-Z, Tao L, Geng X, Li T-C (2010) Modeling of rotating drum bioreactor for anaerobic solid-state fermentation. Appl Energy 87(9):2839–2845

Yadav JS (1988) SSF of wheat straw with alcaliphilic Coprinus. Biotechnol Bioeng 31(5):414–417

Yamashita F, Hotta K, Kurasawa S, Okami Y, Umezawa H (1985) New antibiotic-production generated by protoplast fusion treatment between S. griseus and S. tenjimareinsis. J Antibiot 38:58–63

Yang SQ, Xiong H, Yang HY, Yan QJ, Jiang ZQ (2015) High-level production of β-1,3-1,4-glucanase by Rhizomucor miehei under solid-state fermentation and its potential application in the brewing industry. J Appl Microbiol 118(1):84–91

Yoon LW, Ang TN, Ngoh GC, Chua ASM (2014) Fungal solid-state fermentation and various methods of enhancement in cellulase production. Biomass Bioenergy 67:319–338

Ziemert N, Alanjary M, Weber T (2016) The evolution of genome mining in microbes–a review. Nat Prod Rep 33(8):988–1005

Industrial Production and Optimization of Microbial Enzymes

5

Francois Niyongabo Niyonzima, S. M. Veena, and Sunil S. More

Abstract

Microbial enzymes are the biological catalysts due to their ability to favor reactions more quickly and more efficiently. Various enzymes are produced by microorganisms for industrial uses. They must possess the desired properties like diversified functionality, and stability over pH and temperature ranges. Microorganisms have to produce extracellular enzymes in higher amounts and the produced enzymes have to be safe, stable, and more active. Microbial enzymes with the desired properties can be produced by optimizing fermentation conditions. To make the fermentation cost effective, the utilization of low-cost substrates such as agricultural and spent residues for microbial enzyme production is necessary. Some industrial enzymes used together for the same purpose (like amylase, lipase, and protease used in detergent formulation) are co-produced in a single fermentation to reduce the cost and to maintain the enzyme stability. In addition, for some microorganisms, recombinant DNA technology is used as an alternative strategy for overproducing huge amounts of microbial enzymes with improved substrate specificity and stability. Furthermore, novel techniques like genetic fusion of coding open reading frames or connection of proteins in a posttranslational process are used to manufacture the fused industrial enzymes having combined properties of their parental molecules. The public and private companies have thus to work together with academicians and researchers in

F. N. Niyonzima
Department of Biotechnologies, Faculty of Applied Fundamental Sciences, INES-Ruhengeri, Ruhengeri, Rwanda

S. M. Veena
Department of Biotechnology, Sapthagiri College of Engineering, Bangalore, India

S. S. More (✉)
School of Basic and Applied Sciences, Dayananda Sagara University, Bangalore, India
e-mail: drsunil@dsu.edu.in

N. K. Arora et al. (eds.), *Microbial Enzymes: Roles and Applications in Industries*, Microorganisms for Sustainability 11,
https://doi.org/10.1007/978-981-15-1710-5_5

order to increase production of microbial enzymes needed by the industries. This chapter reviews the production of industrial enzymes and optimization of culture and fermentation conditions in order to increase production of microbial enzymes in huge amounts.

5.1 Introduction

Enzymes are considered as wonderfully specific and efficient biocatalysts for a large number of biochemical reactions. They are not toxic and generate few by-products as compared to chemical catalysts. Industrial enzymes with specific characteristics can be obtained from microorganisms by optimizing process parameters and by enzyme engineering. Various enzymes such as proteases, amylases, cellulases, and lipases were engineered in order to work under industrial conditions. This is necessary as wild microbial strains produce a lesser amount of enzyme, compared to engineered microorganisms. This can be achieved using special methods like use of mutagens (Nelson and Cox 2008; Willey et al. 2008).

The industrial enzymes are derived from plants, animals, and microorganisms. However, the microorganisms are mostly in use to produce these enzymes owing to better yields obtained from them, and reduction of cost and labor. Most of the industrial enzymes are produced using *Bacillus* (Beena et al. 2012; Roy et al. 2012; Asha et al. 2013; Niyonzima et al. 2013; Roohi et al. 2013; Saracoglu et al. 2013; Mathur et al. 2014; More et al. 2015; Waleed et al. 2015; Sinha and Nigam 2016; Sriariyanun et al. 2016; More et al. 2016; Hasan et al. 2017; Shwetha et al. 2017) and *Aspergillus* species (Choudhary 2012; Mini et al. 2012; Pundir et al. 2012; Dhital et al. 2013; Niyonzima and More 2013a; Sandhya et al. 2015; Souza et al. 2015; Xiao et al. 2015; Cavalcanti et al. 2017; Lincoln and More 2018; Pachauri et al. 2018). Some industrial enzymes having same applications can be co-produced in a single fermentation medium. In this case, the production process becomes cost effective and the enzyme stability is assured.

The role of medium composition optimization is to maintain the balance between different ingredients, thereby preventing the number of unused components at the end of fermentation process (Ire et al. 2011). Kumar and Takagi (1999) reported that there is no specific growth medium for the optimum industrial enzyme production by bacteria or fungi. Each bacterial or fungal species has its own growth conditions to produce industrial enzymes in a significant amount. According to Hajji et al. (2008), the growth medium for industrial enzyme production by bacteria and fungi is mainly optimized with one parameter at a time method. Statistical methods are also employed to produce the industrial enzymes in adequate amounts.

Industries are still searching for new microbial strains with desired aspects in order to produce various industrial enzymes to fulfil the current enzyme demand. The proper selection of different industrial microorganisms and the optimization of fermentation conditions are thus necessary to produce inexpensive industrial enzymes. The production of microbial industrial enzymes under optimized conditions to get enzymes with desirable properties is a continuous exercise. The concomitant production of industrial enzymes in a single-economic production medium from microorganisms is also a new challenge. Although some important studies have been reported for the production and optimization of enzymes from microorganisms, there is no specific report for the industrial production and optimization of culture conditions for maximum production. This chapter therefore reports the production and optimization of culture and fermentation conditions to produce industrial enzymes from microorganisms in optimum amounts (Roy and Mukherjee 2013).

5.2 Production of Microbial Industrial Enzymes by Microorganisms

Industrial enzymes are used by various industries for commercial purpose. Production of industrial enzymes by microorganisms is a necessary step in industrial sectors. The production and product delivery of industrial enzymes from microorganisms involve various steps. These are isolation, screening, and identification of enzyme-producing microorganisms; optimization of process parameters and fermentation for industrial enzyme production; purification and characterization of purified enzymes; and industrial enzyme formulation for sale, customer liaison, and working with the regulatory bodies (Fig. 5.1). Most of the bacteria and fungi used to produce industrial enzymes are genetically modified to overproduce them in significant amount (Sarrouh et al. 2012; Sandhya et al. 2015). Solid-state and submerged fermentation are often used to produce industrial enzymes. However, the submerged fermentation is repeatedly reported to be the method of choice for industrial enzyme secretion from microorganisms owing to the extracellular nature of the industrial enzymes that get liberated into the production medium. Sarrouh et al. (2012) reported that pH and temperature stability, specificity, influence of activators and inhibitors, and reaction velocity are some of the criteria used in the selection of the industrial enzymes to be produced by microorganisms.

Industrial enzymes are generally produced under carefully controlled conditions by fermentation, using microorganisms, especially bacteria or fungi. *Bacillus* and *Aspergillus* species are known to be the main producers of industrial enzymes (Tables 5.1 and 5.2). Indeed, most of the species of these genera are safe and do not produce any toxin, grow on inexpensive substrates, and secrete extracellularly adequate amount of enzymes in a reasonable time period. They can also be genetically manipulated easily to give novel industrial enzymes with desirable characteristics (Rao et al. 1998). The microorganisms belonging to the genera *Acinetobacter*, *Pseudomonas*, *Staphylococcus*, *Streptomyces*, *Fusarium*, *Mucor*, *Penicillium*, and *Trichoderma* are also used for industrial enzyme production (Tables 5.1 and 5.2).

For the industrial production of microbial enzymes, submerged fermentation (SMF) and/or solid-state fermentation (SSF) are employed. Each fermentation has its advantages and disadvantages. Indeed, submerged fermentation is often utilized to produce industrial microbial enzymes (Sinha and Khare 2012; Beena et al. 2012; Pathak and Deshmukh 2012; Niyonzima and More 2013a, b; Rodrigues et al. 2013;

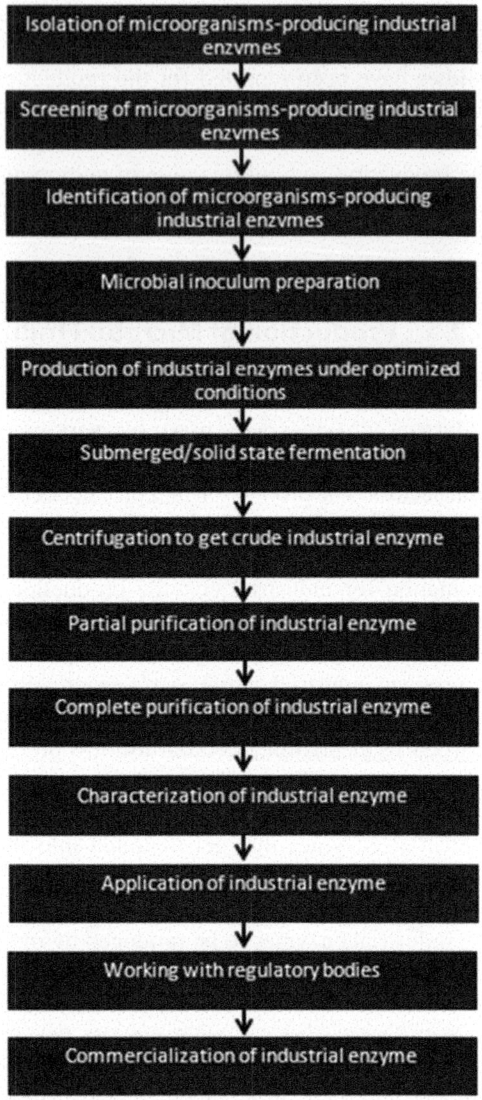

Fig. 5.1 Schematic representation of sequential steps used in the production of industrial enzymes by microorganisms

Mathur et al. 2014; More et al. 2015; Sandhya et al. 2015; Souza et al. 2015; Waleed et al. 2015; Xiao et al. 2015; Sinha and Nigam 2016; Sriariyanun et al. 2016; More et al. 2016; Cavalcanti et al. 2017; Hasan et al. 2017; Shwetha et al. 2017; Pachauri et al. 2018). It allows extracellular industrial enzyme secretion in important amounts in the production medium and thus industrial enzyme recovery is high. In addition, the culture parameters are easily controlled. However, one of the disadvantages of SmF is that it uses expensive synthetic media (Lailaja and Chandrasekaran 2013; Niyonzima et al. 2013).

Table 5.1 Low-cost used substrates during industrial enzyme production by microorganisms

Substrate used	Enzymes produced	Microorganism	Reference
Chicken feathers	Keratinase	*Bacillus megaterium*	Saibabu et al. (2013)
Jamun leaves	Tannase	*Aspergillus* sp. GM4	Souza et al. (2015)
Molasses	Neutral invertase	*Aspergillus* sp	Lincoln and More (2017)
Orange peel moistened with molasses	Invertase	*Aspergillus sojae* JU12	Lincoln and More (2018)
Organic kitchen wastes	Amylase	*Chryseobacterium* sp. *Bacillus* sp.	Hasan et al. (2017)
Rice bran	Phytase	*Bacillus lehensis* MLB2	More et al. (2015)
Rice bran and wheat bran	Laccase	*Stereum ostrea*	Usha et al. (2014)
Sawdust	Cellulase	*Penicillium* sp.	Prasanna et al. (2016)
Sugarcane bagasse	Cellulase	*Aspergillus awamori*	Pachauri et al. (2018)
Sugarcane bagasse or straw, wheat bran, dry corn, sawdust	Cellulase	*Pseudomonas fluorescens*	Sethi et al. (2013)
Tea stalks	Tannase	*Aspergillus tubingensis* CICC 2651	Xiao et al. (2015)
Wheat bran	Keratinase	*Aspergillus niger/flavus*	Mini et al. (2012)
Wheat bran	Phytase	*Aspergillus niger*	Sandhya et al. (2015)
Wheat bran	Xylanase	*Sphingobacterium* sp. SaH-05	Ghasemi et al. (2014)

Solid-state fermentation is also used for production of some industrial enzymes (Lincoln and More 2018). It uses inexpensive substances like agro-industrial by-products and downstream process is not expensive. In addition, higher industrial enzyme yield is obtained in a brief period of time. However, the physicochemical parameters are not easily monitored and regulated (Wang et al. 2012). Some industrial enzymes were reported to be produced by both solid-state and submerged fermentation. For instance, invertase (β-D-fructofuranosidase) was produced by *Aspergillus sojae* JU12 using SSF and the fermentation was cost effective as orange peels moistened with molasses were used. The enzyme is very important in the production of alcoholic beverages (Lincoln and More 2018). Similarly, the abundant agricultural residue known as lignocellulosic biomass was used to produce cellulase in significant amount with *Aspergillus awamori* (Pachauri et al. 2018).

Batch and fed-batch fermentation, with constant and/or linear feeding, are also used for industrial enzyme production by microorganisms. Aishwarya et al. (2013) reported the production of a detergent protease by the bacterium *Alcaligenes* sp.

Table 5.2 Optimum conditions for industrial enzyme production by bacteria and fungi

Microorganism	Enzyme produced	pH	T (°C)	Agitation (rpm)	Inoculum size (%)	Incubation period	Good C source	Preferred N source	**Reference**
Bacterial species									
Acinetobacter calcoaceticus 1–7	Lipase	9	37	180	0.6	48 h	Starch, olive oil,	Corn flour bean flour	Wang et al. (2012)
Bacillus cereus	Protease	10	37	n s	2	48 h	Glucose and skim milk	Peptone and yeast extract	Beena et al. (2012)
Bacillus cereus	Pullulanase	6	50	200	ns	48 h	Soluble pullulan	Tryptone	Waleed et al. (2015)
Bacillus cereus GA6	Amylase	10	20	120	1	96 h	Glycerol	Ammonium acetate	Roohi et al. (2013)
Bacillus cereus FT9	L-glutaminase	7	37	150	ns	24 h	Malt extract	Peptone	Sinha and Nigam (2016)
Bacillus cereus LC13	L-Glutaminase	7	37	100	ns	24 h	Maltose	L-Glutamine	More et al. (2016)
Bacillus flexus MSBC 2	CGTase	10	37	ns	ns	120 h	Corn starch	Peptone with yeast extract	Shwetha et al. (2017)
Bacillus flexus XJU-1	Lipase	11	37	100	2	36 h	Cotton seed oil	Yeast extract	Niyonzima et al. (2013)
Bacillus halodurans	CGTase	10.5	37	ns	ns	48 h	Soluble starch	Yeast extract and peptone	More et al. (2012)
Bacillus halodurans	Pullulanase	10	37	ns	ns	72 h	Pullulan	Peptone	Asha et al. (2013)
Bacillus lehensis MLB2	Phytase	5.5	30	100	2	24 h	Rice bran	Potassium nitrate	More et al. (2015)
Bacillus licheniformis AS08E	Amylase	ns	45	200	ns	60 h	ns	Ns	Roy and Mukherjee (2013)

Bacillus licheniformis KBDL4	Protease	10	37	200	3	48 h	Casein	Yeast extract	Pathak and Deshmukh (2012)
Bacillus megaterium	Keratinase	10	37	ns	ns	72 h	Feather meal	Feather meal supplemented with yeast extract	Saibabu et al. (2013)
Bacillus megaterium BMS4	Cellulase	7	37	120	ns	24 h	Sucrose	LB medium BHM medium	Hussain et al. (2017)
Bacillus smithii BTMS 11	Lipase	8	28	170	3	24 h	Glucose, sesame oil	Soybean meal	Lailaja and Chandrasekaran (2013)
Bacillus sonorensis	Lipase	ns	40	150	1	48 h	Olive oil	Peptone and yeast extract	Nerurkar et al. (2013)
Bacillus sp.	Amylase	7	50	120	2	48 h	Soluble starch	Peptone	Hasan et al. (2017)
Bacillus sp. SMIA-2	Protease	8.5	50	150	n s	36 h	Starch	Whey protein and corn steep liquor	Rodrigues et al. (2013)
Bacillus sp., MSL2	Cellulase	6	50	200		48 h	Carboxymethylcellulose	Yeast extract	Sriariyanun et al. (2016)
Bacillus subtilis	Pectinase	8.5	37	ns	ns	24 h	Glucose	Yeast extract	Mathur et al. (2014)
Bacillus subtilis AS-SO1a	Amylase	6	35	150	ns	60 h	Starch	Beef extract	Roy et al. (2012)
Chryseobacterium sp.	Amylase	5	50	120	2	60 h	Glucose	Peptone	Hasan et al. (2017)

(continued)

Table 5.2 (continued)

Microorganism	Enzyme produced	pH	T (°C)	Agitation (rpm)	Inoculum size (%)	Incubation period	Good C source	Preferred N source	**Reference**
Pseudomonas fluorescens	Cellulase	10	40	ns	ns	24 h	Glucose	Ammonium sulfate	Sethi et al. (2013)
Pseudomonas stutzeri AS22	Amylase	8	30	200	ns	24 h	Potato starch	Yeast extract	Maalej et al. (2013)
Sphingobacterium sp. SaH-05	Xylanase	8	37	150	ns	24 h	Wheat bran	Ammonium sulfate	Ghasemi et al. (2014)
Staphylococcus arlettae JPBW-1	Lipase	8	35	100	10	48 h	Soybean oil	ns	Chauhan et al. (2013)
Streptomyces sp. A3	Amylase	9	45	200	2	168 h	Maltose	Peptone	Chakraborty et al. (2012)
Virgibacillus sp. EMB13	Protease	8	30	150	4	24 h	Mannitol	Peptone and yeast extract	Sinha and Khare (2012)
Fungal species									
Aspergillus niger	Pectinase	4.5	30	150	ns	7 days	Pectin	Peptone	Dhital et al. (2013)
Aspergillus niger	Phytase	5	30	200	ns	4 days	Glucose	Ammonium nitrate	Sandhya et al. (2015)
Aspergillus niger	Tannase	4	37	120	ns	4 days	Tannate	Yeast extract	Cavalcanti et al. (2017)
Aspergillus awamori	Cellulase	6	25	ns	ns	7 days	Carboxymethyl cellulose (CMC)	Peptone	Pachauri et al. (2018)
Aspergillus niger/flavus	Keratinase	8.5	37	120	ns	4 days	Keratin	Peptone	Mini et al. (2012)
Aspergillus sojae JU12	Invertase	8	37	ns	9	5 days	Orange peel moistened with molasses	Beef extract	Lincoln and More (2018)
Aspergillus sp. GM4	Tannase	4	30	ns	ns	4 days	Jamun leaves/tannate	Potassium nitrate	Souza et al. (2015)

Aspergillus terreus gr.	Protease	10	37	ns	2	5 days	Casein	Soybean meal	Niyonzima and More (2013a)
Aspergillus versicolor PF/F/107	Protease	9	35	150	n s	4 days	Wheat bran	Sodium nitrate	Choudhary (2012)
Fusarium sp. XPF-5	Xylanase	8	47	160	ns	4 days	Xylan	ns	Sharma et al. (2016)
Mucor circinelloides F6–3-12	Tannase	5.5	30	200	2	7 days	Green tea leaf powder	Sodium nitrate	El-Refai et al. (2017)
Mucor hiemalis	L-Asparaginase	7	30	ns	ns	4 days	Glucose	L-Asparagine	Thakur et al. (2014)
Penicillium nigricans	β-Amylase	5	60	ns	ns	ns	Starch	Sodium nitrite	Uday et al. (2013)
Penicillium sp.	Cellulase	5	30	ns	ns	7 days	Lactose	Yeast extract	Prasanna et al. (2016)
Saccharomyces cerevisiae	Invertase	7	30	ns	ns	2 days	Starch	Urea	Sivakumar et al. (2013)
Scopulariopsis sp.	Protease	9	37	n s	3	5 days	Casein	Tryptone	Niyonzima and More (2013b)
Scytalidium lignicola	Laccase	6	30	ns	ns	7 days	Sucrose	Sodium nitrate	Sidhu et al. (2017)
Stereum ostrea	Laccase	5.5	30	180	ns	12 days	Glucose	Peptone	Usha et al. (2014)
Trichoderma viride Pers: SF Grey	L-Asparaginase	6.5	37	ns	ns	3 days	Maltose	L-Asparagine and peptone	Lincoln et al. (2015)

ns not specified/not determined

(MTCC 9730) using fed-batch fermentation. Different statistical experimental designs are used to produce industrial enzymes by microorganisms in higher amount. For instance, a sequential statistical strategy, by Plackett-Burman design followed by steepest ascent method and response surface methodology, was used to maximally produce tannase from *Aspergillus* (Xiao et al. 2015). Plackett-Burman design was used in SSF to produce a tannase by *Aspergillus* sp. GM4 (Souza et al. 2015).

The production of some industrial enzymes is stimulated by various substances in the production medium. For instance, the production of the detergent lipase by *Bacillus flexus* XJU-1 was activated by Tween-80 and Triton X-100. Indeed, the present surfactants act by modifying plasma membrane, thereby activating the medium compound uptake, leading to the lipase release in a significant amount (Niyonzima et al. 2013; Niyonzima and More 2014a). Prasanna et al. (2016) reported the increase in extracellular cellulase secretion by *Penicillium* sp. when the surfactant Triton-X100 was supplemented to the fermentation medium. The production of laccase by a mushroom *Stereum ostrea* was stimulated by various inducers such as aromatic or phenolic compounds, copper, and surfactants. Indeed, the micronutrient copper activates laccase transcription and production, whereas surfactants favor enzyme production by stimulating spore growth and enhancing the availability of less soluble substrates for the microorganism (Usha et al. 2014).

The use of the inexpensive by-products or agricultural residues does not only make the fermentation and production cost effective, but also reduces the environmental pollution that may be caused by by-products or agricultural residue disposition. Indeed, if a cheap substrate is chosen carefully to grow the microorganism, one-third of process cost can be reduced. For instance, keratinase was produced in a significant amount by *Bacillus megaterium* when chicken feathers were used as both carbon and nitrogen sources (Saibabu et al. 2013). Similarly, Lincoln and More (2018) optimally produced invertase from *A. sojae* JU12 when orange peel moistened with molasses was used as the substrate. Likewise, wheat bran, a cheap agricultural substrate, was used by different *Aspergillus* species to produce phytase (Sandhya et al. 2015), keratinase (Mini et al. 2012), and laccase (Usha et al. 2014). The coconut cake was also reported to be the best substrate for cellulase secretion by *Pseudomonas fluorescens* (Sethi et al. 2013). Table 5.1 shows some by-products or agricultural residues which are used as substrates during industrial enzyme production by microorganisms.

To meet the present increased industrial enzyme demand, the concomitant production of some enzymes by microorganisms in a single cultivation medium with cheap substrates becomes necessary. Industrial enzymes have been concomitantly produced from microorganisms and used in industries. Indeed, if two, three, or more industrial enzymes are co-produced by bacteria or fungi under similar conditions, the whole process is not expensive. In addition, the stability among the simultaneously produced enzymes is assured. For example, protease, lipase, amylase, and cellulase are used in detergent industries to remove various stains; if they are produced by the same cultivation medium, the proteolysis of lipase, amylase, and cellulase by the protease is prevented. No amylase or lipase proteolysis was observed when protease, lipase, and amylase were co-produced together by *B. flexus*

XJU-1 in the same fermentation (Niyonzima and More 2014a, b, c, d). The amylases, cellulases, and lipases resistant to protease hydrolysis have an excellent applicability to the detergent formulation. The alkaline amylase of *B. megaterium* B69 was also not hydrolyzed by the alkaline protease when concomitantly produced together (Saxena and Singh 2014).

The agro-industrial byproducts were also used as inexpensive substrates for the simultaneous production of industrial enzymes in higher amounts. Shivakumar (2012) co-produced an important amount of amylase and protease by *Bacillus* sp. Y using cheap substrates under solid-state fermentation. Similarly, the agro-industrial waste mustard oilseed cake was the substrate of choice for amylase and protease coproduction by *B. megaterium* B69 (Saxena and Singh 2014). Likewise, the inexpensive substrate was employed for the concomitant production of lipase and amylase by *Bacillus subtilis* JPBW-9 (Anwar et al. 2011). The use of the agricultural residue or by-products as production substrates to produce industrial enzymes by microorganisms makes the fermentation process cost effective, and also avoids environmental pollution by using these wastes.

5.3 Optimization of Process Parameters

The industrial enzyme production by microorganisms is principally influenced by various factors such as incubation time, agitation/shaking, initial pH, inoculum concentration, incubation temperature, carbon source, metal ions, and nitrogen source (Table 5.2). The optimization of these factors has a significant role in enhancing the enzyme yield. The optimization of media components, cultural parameters, and fermentation conditions is therefore necessary to maximally produce the industrial enzymes in adequate amounts. The process factors are generally optimized one factor each time, holding all other factors unchanged, and the optimized condition/factor is considered in the subsequent experiments in sequential order. The advantage of the optimization of various nutritional parameters, physicochemical aspects, and fermentation factors is that it helps in designing a cost-effective fermentation process (Bora and Bora 2012; Pathak and Deshmukh 2012; Niyonzima and More 2013a, b; Roy and Mukherjee 2013).

The optimization of the process parameters by statistical methods is also used to increase the production of industrial enzymes in a shorter time. For instance, nutritional factors (C, N, and P sources) and physicochemical parameters (inoculum age, incubation time and temperature) were optimized by applying Plackett-Burman design and Box–Behnken design and the enhancement in tannase production by *Mucor circinelloides* F6–3-12 was observed (El-Refai et al. 2017). An adequate amount of lipase was produced by *Staphylococcus arlettae* JPBW-1 using response surface methodology (RSM) (Chauhan et al. 2013). Roy and Mukherjee (2013) used *B. subtilis* DM-03 and *Bacillus licheniformis* AS08E to maximally produce the amylases with statistical methods (Roy and Mukherjee 2013). The culture conditions for the production of protease by *Trichoderma estonicum* was optimized by a two-level factorial Plackett-Burman design followed by central composite design (Saravanakumar and Kathiresan 2012). A list of various microorganisms producing industrial enzymes and the optimum conditions for the industrial enzyme production are shown in Table 5.2.

5.3.1 Time Course Study

The incubation time plays an important role in the production of industrial enzymes by bacteria and fungi. The optimal time recorded for industrial enzyme production mainly ranges from 24 to 48 h for bacterial species (More et al. 2012; Pathak and Deshmukh 2012; Sinha and Khare 2012; Chauhan et al. 2013; Lailaja and Chandrasekaran 2013; Maalej et al. 2013; Nerurkar et al. 2013; Niyonzima et al. 2013; Rodrigues et al. 2013; Sethi et al. 2013; Ghasemi et al. 2014; Mathuret al. 2014; More et al. 2015; Waleed et al. 2015; Sinha and Nigam 2016; Sriariyanun et al. 2016; More et al. 2016; Hasan et al. 2017; Hussain et al. 2017). The shorter incubation time reported for most of the industrial enzyme production makes the fermentation process inexpensive. The incubation period of 60–96 h range was also reported for the production of industrial enzymes such as keratinase, protease, pullulanase, amylase, and lipase by bacteria (Asha et al. 2013; Roohi et al. 2013; Roy and Mukherjee 2013; Saibabu et al. 2013; Hasan et al. 2017). Higher incubation periods of 120 and 168 h were observed for cyclodextrin glycosyltransferase (or CGTase) production by *B. flexus* MSBC 2 (Shwetha et al. 2017) and amylase secretion by *Streptomyces* strain A3 (Chakraborty et al. 2012), respectively (Table 5.2). Therefore, the time period of industrial enzyme secretion by bacteria varies from one species to another. This may be ascribed to the genome difference.

The fungal industrial enzymes are secreted at optimal level at fourth or fifth day (Choudhary 2012; Mini et al. 2012; Niyonzima and More 2013a, b; Thakur et al. 2014; Sandhya et al. 2015; Souza et al. 2015; Sharma et al. 2016; Cavalcanti et al. 2017; Lincoln and More 2018) (Table 5.2). A low fermentation time of 2 and 3 days was observed for the production of the invertases from *Saccharomyces cerevisiae* (Sivakumar et al. 2013) and L-asparaginase by *Trichoderma viride* Pers: SF Grey (Lincoln et al. 2015), respectively. The higher incubation time of 7 days was also noted for the production of pectinases by *Aspergillus niger* (Dhital et al. 2013), cellulases by *Penicillium sp.* and *Aspergillus awamori* (Prasanna et al. 2016; Pachauri et al. 2018), tannase by *M. circinelloides* isolate F6–3-12 (El-Refai et al. 2017), and laccases by *Scytalidium lignicola* (Sidhu et al. 2017). Likewise, Usha et al. (2014) reported a higher incubation time of 12 days when laccase was secreted in an adequate amount by *Stereum ostrea*.

In general, as the incubation time increases, the enzyme secretion by microorganisms also increases. However, after optimum incubation period, a decline in enzyme production is observed. This decrease in enzyme production was attributed to the reduced availability of nutrients and the toxic metabolite secretion (Romero et al. 1998) or decomposition of enzyme by the protease (Anandan et al. 2007). For industrial lipase, the decrease was ascribed to the accumulation of fatty acids and glycerol resulted from lipolysis (Smith and Alford 1996). For most industrial enzymes like amylases, the decrease in production was due to the denaturation resulted from the enzyme produced and the medium component interaction (Niyonzima and More 2015d). The variation in industrial enzyme production by bacteria and fungi can be attributed to growth and metabolic activity variation (Bhavani et al. 2012; Niyonzima and More 2015b).

5.3.2 Influence of Initial pH of the Medium

The initial pH of the culture and fermentation medium is a major factor regulating industrial microbial enzyme secretion. It may influence the availability of nutrient substrates or the transport of various nutrient components across the bacterial or fungal membranes, which in turn stimulates the microbial growth and thus industrial enzyme production (Bhavani et al. 2012; Bora and Bora 2012; Niyonzima and More 2015b). The optimum initial pH range recorded for most of the industrial bacterial enzymes is 6–10 (Chakraborty et al. 2012; Pathak and Deshmukh 2012; Sinha and Khare 2012; Asha et al. 2013; Chauhan et al. 2013; Lailaja and Chandrasekaran 2013; Maalej et al. 2013; Nerurkar et al. 2013; Niyonzima et al. 2013; Roohi et al. 2013; Rodrigues et al. 2013; Roy and Mukherjee 2013; Saibabu et al. 2013; Sethi et al. 2013; Ghasemi et al. 2014; Mathur et al. 2014; Waleed et al. 2015; Sinha and Nigam 2016; Sriariyanun et al. 2016; More et al. 2016; Hussain et al. 2017; Shwetha et al. 2017). Low pH values of 5.0 and 5.5 were seen for amylase production by *Chryseobacterium sp.* (Hasan et al. 2017) and phytase by *Bacillus lehensis* MLB2 (More et al. 2015), respectively. A high pH of 10.5 was recorded for industrial CGTase production by *Bacillus halodurans* (More et al. 2012) (Table 5.2). The difference in genomes may also be the reason why the bacteria producing industrial enzymes have different initial pH requirements.

The optimum pH observed for fungal industrial enzyme production ranges from acidic to basic pH range, viz. pH 5 to 9 (Choudhary 2012; Mini et al. 2012; Niyonzima and More 2013b; Sivakumar et al. 2013; Thakur et al. 2014; Usha et al. 2014; Lincoln et al. 2015; Prasanna et al. 2016; Sharma et al. 2016; Cavalcanti et al. 2017; El-Refai et al. 2017; Sidhu et al. 2017; Lincoln and More 2018; Pachauri et al. 2018). Niyonzima and More (2013a) produced an alkaline protease active at pH 10.0 using *Aspergillus terreus* gr. Similarly, lower pH values of 4 and 4.5 were observed as optimum for the production of tannase, phytase, and pectinases by *Aspergillus* species (Dhital et al. 2013; Sandhya et al. 2015; Souza et al. 2015) (Table 5.2). The variation in industrial enzyme yields at different initial pH requirements may be due to the bacterial or fungal strain specificity. Any deviation from optimum initial pH resulted in low industrial enzyme secretion. This was attributed to the disruption of transport mechanisms through the bacterial or fungal membrane that prevents the industrial enzyme release (Padhiar et al. 2011).

5.3.3 Effect of Incubation Temperature

The incubation temperature is a vital environmental parameter for industrial enzyme secretion by microorganisms. Like initial pH, it may influence the growth of bacteria and fungi, and thus industrial enzyme production by these microbes. The optimum incubation temperature seen for bacterial industrial enzyme production is in 30–50 °C range. Temperature of 37 °C was noted as optimum in most cases (Beena et al. 2012; More et al. 2012, 2015; Pathak and Deshmukh 2012; Wang et al. 2012; Asha et al. 2013; Niyonzima et al. 2013; Saibabu et al. 2013; Ghasemi et al. 2014; Mathur et al. 2014; Waleed et al. 2015; Sinha and Nigam 2016; Hussain et al. 2017) (Table 5.2). Low optimum incubation temperatures of 20 and 28 °C were also observed for industrial amylase and lipase production, respectively, by *Bacillus* species (Lailaja and Chandrasekaran 2013; Roohi et al. 2013).

For fungal species, the optimum incubation temperature ranged from 25 to 47 °C (Choudhary 2012; Mini et al. 2012; Dhital et al. 2013; Niyonzima and More 2013a, b; Sivakumar et al. 2013; Thakur et al. 2014; Usha et al. 2014; Lincoln et al. 2015; Sandhya et al. 2015; Souza et al. 2015; Prasanna et al. 2016; Sharma et al. 2016; Cavalcanti et al. 2017; El-Refai et al. 2017; Sidhu et al. 2017; Lincoln and More 2018; Pachauri et al. 2018). However, 60 °C was the optimum fermentation temperature for β-amylase production by *Penicillium nigricans* (Uday et al. 2013). Similarly, Panosyan (2019) reported thermostable and active protease, amylase, and lipase from *Thermoactinomyces* isolated from hot springs. At elevated incubation temperature, the yield in industrial enzyme production is low due to the thermoliability of the industrial enzymes or the denaturing of industrial enzyme structure in the active site.

5.3.4 Effect of Inoculum Level

The concentration of inoculum is one of the key culture parameters for microbial growth and thus industrial enzyme production. Various inoculum concentrations ranging from 0.6 to 4% are optimum for bacterial industrial enzyme production by microorganisms (Beena et al. 2012; Chakraborty et al. 2012; Pathak and Deshmukh 2012; Sinha and Khare 2012; Wang et al. 2012; Lailaja and Chandrasekaran 2013; Nerurkar et al. 2013; Niyonzima et al. 2013; More et al. 2015; Hasan et al. 2017) (Table 5.2). However, Chauhan et al. (2013) reported a detergent lipase production by *S. arlettae* JPBW-1 when a higher inoculum size of 10% was used. Different inoculum levels were found to maximally produce the industrial enzymes by diverse fungi. For instance, the inoculum level of 2% was optimum for the tannase production by *M. circinelloides* isolate F6–3-12 (El-Refai et al. 2017) and protease production by *A. terreus* gr. (Niyonzima and More 2013a). Inoculum size of 3% was optimum for the secretion of a protease by *Scopulariopsis* sp. (Niyonzima and More 2013b). Lincoln and More (2018) obtained an industrial invertase from *A. sojae* JU12 using a higher inoculum level of 9%. The effect of inoculum on industrial enzyme secretion may thus depend on the type of microorganism, inoculum load, and bacterial or fungal size and type.

Generally, significant increase in industrial enzyme production by microorganisms correlated with an enhancement in inoculum concentration till optimum inoculum size is reached (Niyonzima and More 2013a, b) owing to rapid substrate degradation (Sarao et al. 2010). Indeed, the enzyme production by microorganisms is often high at lower inoculum levels; however, low enzyme yield is observed after increasing inoculum size (de Souza et al. 2001). Hesseltine et al. (1972) proposed that the decrease observed when an important inoculum level is used can be attributed to the faster bacterial or fungal growth and thus shortage of the nutrients. Likewise, Hasan et al. (2017) observed a low yield at higher inoculum level owing to the lack of total dissolved oxygen and nutrient supply to the microorganisms.

5.3.5 Effect of Carbon Source

The carbon sources serve as a primary energy source for bacterial and fungal growth and therefore industrial enzyme production. Carbon sources such as starch (Roy et al. 2012; Wang et al. 2012; Rodrigues et al. 2013; Hasan et al. 2017; Shwetha et al. 2017), glucose (Sethi et al. 2013; Mathur et al. 2014; Hasan et al. 2017), soluble pullulan (Asha et al. 2013; Waleed et al. 2015), malt extract (Sinha and Nigam 2016), maltose (Chakraborty et al. 2012; More et al. 2016), pullulan (Asha et al. 2013), sucrose (Hussain et al. 2017), and carboxymethylcellulose (Sriariyanun et al. 2016) are used for the production of bacterial enzymes. Soybean oil (Chauhan et al. 2013), olive oil (Nerurkar et al. 2013), and cotton seed oil (Niyonzima et al. 2013) are also used as carbon sources. In some cases, a mixture of carbon sources like corn flour and bean flour (Wang et al. 2012), glucose and skim milk (Beena et al. 2012), and glucose and sesame oil (Lailaja and Chandrasekaran 2013) are also used (Table 5.2).

For the production of fungal enzymes, starch (Sivakumar et al. 2013; Uday et al. 2013), glucose (Thakur et al. 2014; Usha et al. 2014; Sandhya et al. 2015), maltose (Lincoln et al. 2015), sucrose (Sidhu et al. 2017), lactose (Prasanna et al. 2016), carboxymethylcellulose (Pachauri et al. 2018), pectin (Dhital et al. 2013), and tannate (Cavalcanti et al. 2017) are employed as carbon sources. Various inexpensive substances such as wheat bran (Choudhary 2012), jamun leaves (Souza et al. 2015), green tea leaves (El-Refai et al. 2017), and orange peel moistened with molasses (Lincoln and More 2018) are preferred as carbon sources for industrial enzyme production by fungi. The industrial enzymes are generally produced by microorganisms with low carbon source concentration. This makes the production cost effective. In some cases, a carbon source repression is observed when it is used in a significant amount. When an important amount of carbon source is utilized, enzyme secretion decreases owing to limitation of oxygen transfer resulting in poor bacterial or fungal growth (Niyonzima et al. 2013).

5.3.6 Effect of Nitrogen Sources

The nitrogen sources serve as a secondary energy source for the microbial growth and thus for industrial enzyme production. They play an important role in most of the microorganisms to synthetize the cell-wall components, amino acids, peptides, proteins (including industrial enzymes), and nucleotides/nucleic acids (Gupta et al. 2002). For the production of bacterial enzymes, tryptone (Waleed et al. 2015), peptone (Chakraborty et al. 2012; Asha et al. 2013; Sinha and Nigam 2016; Hasan et al. 2017), yeast extract (Pathak and Deshmukh 2012; Maalej et al. 2013; Niyonzima et al. 2013; Mathur et al. 2014), and beef extract (Roy et al. 2012) are employed (Table 5.2). Sometimes, the optimum industrial enzyme production is observed when there is a combination of nitrogen sources, like peptone and yeast extract (More et al. 2012; Nerurkar et al. 2013; Rodrigues et al. 2013; Shwetha et al. 2017). The cheap organic nitrogen sources like corn flour and bean flour (Wang et al. 2012) and soybean meal (Lailaja and Chandrasekaran 2013) are also utilized to produce industrial enzymes by bacteria.

Like for bacteria, organic nitrogen sources such as peptone (Mini et al. 2012; Dhital et al. 2013; Usha et al. 2014; Pachauri et al. 2018), tryptone (Niyonzima and More 2013b), urea (Sivakumar et al. 2013), yeast extract (Prasanna et al. 2016; Cavalcanti et al. 2017), beef extract (Lincoln and More 2018), and L-asparagine (Thakur et al. 2014) are used in fungal enzyme production. In general, the industrial enzymes are generally produced at their optimum levels when organic nitrogen sources are incorporated in the production medium. The preference of organic nitrogen sources by industrial enzyme-producing microorganisms can be ascribed to the presence of some micro- and macronutrients, vitamins, amino acids and/or peptides, and growth factors present in them (Pathak and Deshmukh 2012; Sinha and Khare 2012; Niyonzima and More 2013b) (Table 5.2).

The feather meal serves as both carbon and nitrogen source for industrial keratinase production by *B. megaterium* (Saibabu et al. 2013). The nitrogen sources stimulate the industrial enzyme production up to a certain level beyond which metabolite repression is seen. Indeed, the complex organic nitrogen sources may show enzyme repression when employed in high amounts because they are rich in amino acids (such as glycine) and short peptides. A higher nitrogen source concentration is therefore, inhibitory to the enzyme secretion by bacteria or fungi.

Although the inorganic nitrogen sources are not generally found to increase the production of the industrial enzymes, there are exceptions as well. For instance, a significant industrial enzyme production was seen when ammonium acetate (Roohi et al. 2013), potassium nitrate (Niyonzima and More 2015a), and ammonium sulfate (Sethi et al. 2013; Ghasemi et al. 2014) were used for industrial production by bacteria. Similarly, ammonium nitrate (Sandhya et al. 2015), potassium nitrate (Souza et al. 2015), sodium nitrate (Choudhary 2012; El-Refai et al. 2017; Sidhu et al. 2017), and sodium nitrite (Uday et al. 2013) were inorganic nitrogen sources used to produce industrial enzymes by fungi.

5.3.7 Effect of Agitation

The industrial enzyme-producing microorganisms are usually grown under shaking conditions. The production of bacterial enzymes is often done with shaking in the range of 100 to 200 rev/min (Table 5.2). For instance, 100 rpm was optimum for the production of industrial enzymes by *Bacillus* species (Niyonzima et al. 2013; Niyonzima and More 2015c; More et al. 2016), while 200 rpm is also good for some bacterial species (Chakraborty et al. 2012; Pathak and Deshmukh 2012; Maalej et al. 2013; Roy and Mukherjee 2013; Waleed et al. 2015). Similarly, the production of industrial enzymes by fungi is done at 120–200 rev/min range. For example, 120 rpm (Mini et al. 2012; Cavalcanti et al. 2017), 150 rpm (Choudhary 2012; Dhital et al. 2013), 160 rpm (Sharma et al. 2016), 180 rpm (Usha et al. 2014), and 200 rpm (Sandhya et al. 2015; El-Refai et al. 2017) were found as optimal shaking conditions for different fungal species. The agitation of the culture flasks at a moderate rate allows a good availability of the nutrients to the microorganisms and a proper aeration, favoring the production of enzymes in an optimum amount.

5.4 Conclusion

In this chapter, an overview was given for the optimization of nutritional and physical parameters important for production of industrial enzymes. Since there is an increasing demand of industrial enzymes, the present information can help other researchers to optimize production of these metabolites cost effectively. Although various industrial enzymes have been produced under optimized conditions in the last decade, most of them are not marketed or reached optimal levels as yet. This was mainly attributed to the non-cost-effective production and lack of enzymes with desired properties. Hence, intensive research is required to obtain bacterial or fungal enzymes to meet the demand of the industries and market. Various enzyme based industries thus shouldwork together with researchers to strengthen this linkage.

References

Aishwarya M, Kumari S, Sivasubramanian A, Muthuraman MS (2013) Production, characterization and purification of alkaline protease from *Alcaligenes* sp. and its application in detergent industry. Asian J Pharm Clin Res 6:151–155

Anandan D, Marmer WN, Dudley RL (2007) Isolation, characterization and optimization of culture parameters for production of an alkaline protease isolated from *Aspergillus tamarii*. J Ind Microbiol Biotechnol 34:339–347

Anwar T, Mathur A, Mathur G, Chauhan RS (2011) Statistical optimization for coproduction and partial characterization of thermohalotolerant lipase and amylase from halotolerant *Bacillus subtilis* JPBW-9. World J Sci Technol 1:54–63

Asha R, Niyonzima FN, More SS (2013) Purification and properties of pullulanase from *Bacillus halodurans*. Int Res J Biol Sci 2:35–43

Beena AK, Geevarghese PI, Jayavardanan KK (2012) Detergent potential of a spoilage protease enzyme liberated by a psychrotrophic spore former isolated from sterilized skim milk. Am J Food Technol 7:89–95

Bhavani M, Chowdary GV, David M, Archana G (2012) Screening, isolation and biochemical characterization of novel lipase producing bacteria from soil samples. Int J Biol Eng 2:18–22

Bora L, Bora M (2012) Optimization of extracellular thermophilic highly alkaline lipase from thermophilic *Bacillus* sp. isolated from hot spring of Arunachal Pradesh, India. Braz J Microbiol 43:30–42

Cavalcanti RMF, Ornela PHO, Jorge JA, Guimarães LHS (2017) Screening, selection and optimization of the culture conditions for tannase production by endophytic fungi isolated from Caatinga. J Appl Biol Biotechnol 5:1–9

Chakraborty S, Raut G, Khopade A, Mahadik K, Kokare C (2012) Study on calcium ion independent α-amylase from haloalkaliphilic marine *Streptomyces* strain A3. Indian J Biotechnol 11:427–437

Chauhan M, Chauhan RS, Garlapati VK (2013) Evaluation of a new lipase from *Staphylococcus* sp. for detergent additive capability. Biomed Res Int. https://doi.org/10.1155/2013/374967

Choudhary V (2012) Production, isolation and characterization of alkaline protease from *Aspergillus versicolor* PF/F/107. J Acad Indus Res 1:272–277

de Souza DF, de Souza CGM, Peralta RM (2001) Effect of easily metabolizable sugars in the production of xylanase by *Aspergillus tamarii* in solid-state fermentation. Process Biochem 36:835–838

Dhital R, Panta OP, Karki TB (2013) Optimization of cultural conditions for the production of pectinase from selected fungal strain. J Food SciTechnol 8:65–70

El-Refai HA, Abdel-Naby MA, Hanan MMA, Amin MA, Salem HA (2017) Statistical optimization for tannase production by *Mucor circinelloides* isolate F6-3-12 under submerged and solid state fermentation. Curr Trends Biotechnol Pharm 11:167–180

Ghasemi Y, Sadeghi H, Gholami A, Mohkam M, Kargar M (2014) Isolation and identification of highly xylanase producing bacterium *Sphingobacterium* sp. SaH-05 from soil. Int J Sci Eng Res 5:205–209

Gupta R, Beg QK, Khan S, Chauhan B (2002) An overview on fermentation, downstream processing and properties of microbial alkaline proteases. Appl Microbiol Biotechnol 60:381–395

Hajji M, Rebai A, Gharsallah N, Nasri M (2008) Optimization of alkaline protease production by *Aspergillus clavatus* ES1 in *Mirabilis jalapa* tuber powder using statistical experimental design. Appl Microbiol Biotechnol 79:915–923

Hasan MM, Marzan LW, Hosna A, Hakim A, Azad AK (2017) Optimization of some fermentation conditions for the production of extracellular amylases by using *Chryseobacterium* and *Bacillus* isolates from organic kitchen wastes. J Genet Eng Biotechnol 15:59–68

Hesseltine CW (1972) Biotechnology report: Solid-state fermentations. Biotechnol Bioeng 14:517–532

Hussain AA, Abdel-Salam MS, Abo-Ghalia HH, Hegazy WK, Hafez SS (2017) Optimization and molecular identification of novel cellulose degrading bacteria isolated from Egyptian environment. J Gen Eng Biotechnol 15:77–85

Ire FS, Okolo NBN, Moneke AN, Odibo FJC (2011) Influence of cultivation conditions on the production of a protease from *Aspergillus carbonarius* using submerged fermentation. Afr J Food Sci 5:353–365

Kumar CG, Takagi H (1999) Microbial alkaline protease: from bioindustrial viewpoint. Biotechnol Adv 17:561–594

Lailaja VP, Chandrasekaran M (2013) Detergent compatible alkaline lipase produced by marine *Bacillus smithii* BTMS 11. World J Microbiol Biotechnol 29:1349–1360

Lincoln L, More SS (2017) Screening and enhanced production of neutral invertase from *Aspergillus* sp. by utilization of molasses—a by-product of sugarcane industry. Adv Biores 8:103–110

Lincoln L, More SS (2018) Purification and biochemical characterization of an extracellular β-d-fructofuranosidase from *Aspergillus* sp. 3 Biotech 8. https://doi.org/10.1007/s13205-018-1109-2

Lincoln L, Niyonzima FN, More SS (2015) Purification and properties of a fungal L-asparaginase from *Trichoderma viride* Pers: SF Grey. J Microbiol Biotechnol Food Sci 4:310–316

Maalej H, Hmidet N, Ghorbel-Bellaaj O, Nasri M (2013) Purification and biochemical characterization of a detergent stable α-amylase from *Pseudomonas stutzeri* AS22. Biotechnol Bioprocess Eng 18:878–887

Mathur A, Kushwaha A, Singh AK, Katiyar A (2014) Isolation, purification and characterization of alkaline pectinase from *Bacillus subtilis* isolated from soil. J Pharm Sci 5:1–6

Mini KD, Paul MK, Mathew J (2012) Screening of fungi isolated from poultry farm soil for keratinolytic activity. AdvApplSci Res 3:2073–2077

More SS, Niraja R, Evelyn C, Byadgi AM, Shwetha V, Mangaraj SD (2012) Isolation, purification and biochemical characterization of CGTase from *Bacillus halodurans*. Croat J Food Technol Biotechnol Nutr 7:26–33

More SS, Supreeth S, Anushree A, Arpitha C, Bhavya J, Niyonzima FN (2015) Purification and characterization of phytase from *Bacillus lehensis* MLB2. Biologia 70:294–304

More SS, Swamy R, Niveditha M, Navyashree M, Bhavya J, N. Niyonzima FN. (2016) Purification and characterization of anti-cancer l-glutaminase of *Bacillus cereus* strain LC13. Proc Nat Acad Sci doi. https://doi.org/10.1007/s40011-016-0808-x

Nelson DL, Cox MM (2008) Lehninger principles of biochemistry, 5th edn. W.H. Freeman and Company, New York

Nerurkar M, Manasi J, Sujata P, Ravindra A (2013) Application of lipase from marine bacteria *Bacillus sonorensis* as an additive in detergent formulation. J Surfact Deterg 16:435–443

Niyonzima FN, More SS (2013a) Screening and optimization of cultural parameters for an alkaline protease production by *Aspergillus terreus* gr. under submerged fermentation. Int J Pharm Bio Sci 4:1016–1028

Niyonzima FN, More SS (2013b) Optimization of fermentation culture conditions for alkaline protease production by *Scopulariopsis* sp. Appl Biol Res 15:66–69

Niyonzima FN, More SS (2014a) Concomitant production of detergent compatible enzymes by *Bacillus flexus* XJU-1. Braz J Microbiol 45:903–910

Niyonzima FN, More SS (2014b) Purification and properties of detergent compatible extracellular alkaline protease from *Scopulariopsis* sp. Prep Biochem Biotechnol 44:738–759

Niyonzima FN, More SS (2014c) Biochemical properties of the alkaline lipase of *Bacillus flexus* XJU-1 and its detergent compatibility. Biologia 69:1108–1117

Niyonzima FN, More SS (2014d) Detergent compatible bacterial amylases. Appl Biochem Biotechnol 174:1215–1232

Niyonzima FN, More SS (2015a) Purification and characterization of detergent compatible alkaline protease from *Aspergillus terreus* gr. 3 Biotech 5:61–70

Niyonzima FN, More SS (2015b) Detergent compatible proteases: Microbial production, properties and stain removal analysis. Prep Biochem Biotechnol 45:233–258

Niyonzima FN, More SS (2015c) Microbial detergent compatible lipases. JSIR 74:105–113

Niyonzima FN, More SS (2015d) Coproduction of detergent compatible bacterial enzymes and stain removal evaluation. J Basic Microbiol 55:1–10

Niyonzima FN, More SS, Muddapur U (2013) Optimization of fermentation culture conditions for alkaline lipase production by *Bacillus flexus* XJU-1. Curr Trends Biotechnol Pharm 7:793–803

Pachauri P, More S, Aranganathan V, Sullia SB, Deshmukh S (2018) Kinetic study and characterization of cellulase enzyme from isolated *Aspergillus niger* subsp. awamori for cellulosic biofuels. J Sci Ind Res 77:55–60

Padhiar J, Das A, Bhattacharya S (2011) Optimization of process parameters influencing the submerged fermentation of extracellular lipases from *Pseudomonas aeruginosa*, *Candida albicans* and *Aspergillus flavus*. Pak J Biol Sci 14:1011–1018

Panosyan H (2019) Thermoactinomycetes isolated from geothermal springs in Armenia capable of producing extracellular hydrolases. Env Sust. https://doi.org/10.1007/s42398-019-00066-0

Pathak AP, Deshmukh KB (2012) Alkaline protease production, extraction and characterization from alkaliphilic *Bacillus licheniformis* KBDL4: a Lonar soda lake isolate. Indian J Exp Biol 50:569–576

Prasanna HN, Ramanjaneyulu G, Reddy BR (2016) Optimization of cellulase production by *Penicillium* sp. 3 Biotech 6:162–172

Pundir RK, Rana S, Tyagi H (2012) Studies on compatibility of fungal alkaline protease with commercially available detergents. Int J Mod Biochem 1:41–56

Rao MB, Tanksale AM, Ghatge MS, Deshpande VV (1998) Molecular and biotechnological aspects of microbial protease. Microbiol Mol Bio Rev 62:597–635

Rodrigues PM, Andrade VVV, Martins MLL (2013) Stability and activity of the partially purified spray-dried protease from *Bacillus* sp SMIA-2 and its characterization as a laundry detergent additive. Int J Bioassays 2:562–567

Romero F, Garcia LA, Diaz M (1998) Protease production from whey at high concentration by *Serratia marcescens*. Resour Environ Biotechnol 2:93–115

Roohi R, Kuddus M, Saima S (2013) Cold-active detergent-stable extracellular α-amylase from *Bacillus cereus* GA6: Biochemical characteristics and its perspectives in laundry detergent formulation. J Biochem Technol 4:636–644

Roy JK, Mukherjee AK (2013) Applications of a high maltose forming, thermo-stable α-amylase from an extremely alkaliphilic *Bacillus licheniformis* strain AS08E in food and laundry detergent industries. Biochem Eng J 77:220–230

Roy JK, Rai SK, Mukherjee AK (2012) Characterization and application of a detergent-stable alkaline α-amylase from *Bacillus subtilis* strain AS-S01a. Int J Biol Macromol 50:219–229

Saibabu V, Niyonzima FN, More SS (2013) Isolation, partial purification and characterization of keratinase from *Bacillus megaterium*. Int Res J Biol Sci 2:13–20

Sandhya A, Sridevi A, Devi PS, Narasimha G (2015) Production and optimization of phytase by *Aspergillus niger*. Der Pharm Lett 7:148–153

Saracoglu ZN, Tanrisever D, Arikan B, Guvenmez HK (2013) Characterization of an oxidant and detergent stable alkaline protease produced from a novel isolate *Bacillus* sp. strain. J Appl Biol Sci 7:4–9

Sarao LK, Arora M, Sehgal V, Bhatia S (2010) Production of protease by submerged fermentation using *Rhizopus microsporus* Var Oligospous. Internet J Microb 9(1)

Saravanakumar K, Kathiresan K (2012) Statistical optimization of protease production by mangrove-derived *Trichoderma estonicum* and its potential on blood stain removal. Int J Biotechnol Mol Biol Res 3:15–21

Sarrouh B, Santos TM, Miyoshi A, Dias R, Azevedo V (2012) Up-to-date insight on industrial enzymes applications and global market. J Bioprocess Biotechniq S4:002. https://doi.org/10.4172/2155-9821.S4-002

Saxena R, Singh R (2014) Contemporaneous production of amylase and protease through CCD response surface methodology by newly isolated *Bacillus megaterium* strain B 69. Enzyme Res. https://doi.org/10.1155/2014/601046

Sethi S, Datta A, Gupta BL, Gupta S (2013) Optimization of cellulase production from bacteria isolated from soil. ISRN Biotechnol. https://doi.org/10.5402/2013/985685

Sharma P, Kaushik N, Sharma S, Kumar V (2016) Isolation, screening, characterization and optimization of xylanase production from thermostable alkaliphilic *Fusarium* sp. XPF5. J Biochem Technol 7:1089–1092

Shivakumar S (2012) Co-production of alkaline protease and amylase of *Bacillus* sp. Y in solid state cultivations. Res J Biotech 7:32–38

Shwetha V, More SS, Uday M (2017) Screening, optimization and production of a novel β- cyclodextrinase by *Bacillus flexus* MSBC 2. Curr Trends Biotechnol Pharm 11:136–143

Sidhu AK, Darade SB, Bhavsar PP, Gaikwad VB, Patil SN (2017) Isolation, screening and optimization for laccase production by *Scytalidium lignicola* pesante under submerged fermentation. Int J Curr Microbiol App L Sci 6:2477–2491

Sinha R, Khare SK (2012) Isolation and characterization of halophilic *Virgibacillus* sp. EMB13: Purification and characterization of its protease for detergent application. Indian J. Biotechnol 11:416–426

Sinha S, Nigam VK (2016) Production and characterization of L-glutaminase by *Bacillus* sp. Int J Pharm Sci Res 7:1620–1626

Sivakumar T, Ravikumar M, Prakash M, Shanmugaraju V (2013) Production of extracellular invertase from *Saccharomyces cerevisiae* strain isolated from grapes. Int J Curr Res Aca Rev 1:72–83

Smith JL, Alford JA (1996) Inhibition of microbial lipases by fatty acids. Appl Microbiol 14:699–705

Souza PNC, Maia NC, Guimarães LHS, de Resende MLV, Cardoso PG (2015) Optimization of culture conditions for tannase production by *Aspergillus* sp. gm4 in solid state fermentation. Acta Sci 37:23–30

Sriariyanun M, Tantayotai P, Yasurin P, Pornwongthong P, Cheenkachorn K (2016) Production, purification and characterization of an ionic liquid tolerant cellulase from *Bacillus* sp. isolated from rice paddy field soil. Electron J Biotechnol 19:23–28

Thakur M, Lincoln L, Niyonzima FN, More SS (2014) Isolation, purification and characterization of fungal extracellular L-asparaginase. J Biocatal Biotransf 2. https://doi.org/10.4172/2324-9099.1000108

Uday MM, Bennur RS, Veena SM, Niyonzima FN, More SS (2013) Isolation and characterization of β-amylase from *Penicillium nigricans*. BTAIJ 7:85–88

Usha KY, Praveen K, Reddy BR (2014) Enhanced production of ligninolytic enzymes by a mushroom *Stereum ostrea*. Biotechnol Res Int. https://doi.org/10.1155/2014/815495

Waleed M, Faiza AA, Nizar I (2015) Production optimization of pullulanase enzyme produced by *Bacillus cereus* isolated from Syrian sources. Int Food Res J 22:1824–1830

Wang H, Zhong S, Ma H, Zhang J, Qi W (2012) Screening and characterization of a novel alkaline lipase from *Acinetobacter calcoaceticus* 1–7 isolated from Bohai Bay in China for detergent formulation. Braz J Microbiol 43:148–156

Willey JM, Sherwood LM, Woolverton CJ (2008) Prescott Harley Klein's microbiology (7th Ed). McGraw Hill, Boston, USA

Xiao A, Huang Y, Ni H, Cai H, Yang Q (2015) Statistical optimization for tannase production by *Aspergillus tubingensis* in solid-state fermentation using tea stalks. Electron J Biotechnol 18:143–147

Commercial Applications of Microbial Enzymes

6

Adhithi Ramesh, P. Harani Devi, Sulagna Chattopadhyay, and M. Kavitha

Abstract

Enzymes are widely used nowadays in industrial and household catalysis. Enzyme-catalysed conversions occur in minutes or even in seconds. Further, they catalyse reactions which are difficult to perform by chemical methods, like the enantio- or regioselective hydrolysis or addition of chiral groups. In industries, enzymes are steadily replacing chemical reactions since they are greener and eco-friendly. Enzymes produce fewer by-products, consume less energy, reduce environmental pollution and add improved value to the products. Consequently, it is not surprising to notice the blooming global enzyme market. Major factors driving the market growth of enzymes are cost efficiency, stringent enforcement of environmental regulations in many countries and growing interest among end users. Microbes are an inexhaustible source of enzymes which have numerous advantages with regard to their use in industrial applications, as compared to conventional methods using chemicals. They are good catalysts, increase the rate of reactions and work optimally under given environmental conditions and scaling up of the production process is possible by genetic manipulation. Recombinant DNA technology and protein engineering open up the possibilities of obtaining novel products. Industries utilising microbial enzymes are food, pharmaceutical, detergent, leather, waste management and many others. This chapter discusses the commercial applications of various microbial enzymes and also highlights the sources and nature of industrial applications.

A. Ramesh · P. Harani Devi · S. Chattopadhyay · M. Kavitha (✉)
School of Biosciences and Technology, Vellore Institute of Technology, Vellore, Tamilnadu, India

N. K. Arora et al. (eds.), *Microbial Enzymes: Roles and Applications in Industries*, Microorganisms for Sustainability 11,
https://doi.org/10.1007/978-981-15-1710-5_6

6.1 Introduction

Enzymes are proteinaceous macromolecules (except ribozyme which is an RNase) produced by living organisms that are responsible for catalysing various biochemical reactions taking place inside the organism (Singh et al. 2016a, b, c, d). They are highly specific and accelerate the rate of reaction by lowering the activation energy of the reactants. They work well at atmospheric pressure and require milder conditions for catalytic action, making them highly advantageous for industries. Enzymes have been classified by the IUBMB (International Union of Biochemistry and Molecular Biology) into six major classes depending on their mode of action, oxidoreductases, transferases, hydrolases, lyases, isomerases and ligases (Singh et al. 2016a, b, c, d). Enzymes derived from microbial sources are currently the hub of all scientific research and applications in various fields, especially in the industrial sector. With the advent of the industrial revolution and technological bloom, processing and marketing of industrial products take much less time than was possible earlier using conventional methods; this is further leading to an increase in demand. This continuous cycle of supply and demand requires an efficient and continuous system to satisfy both ends, which are being realised through the use of microbial enzymes. Conventional methods using chemicals and other physical parameters are now being replaced by increased use of natural products or engineered ones from living organisms; this is owing to the drawbacks of the conventional chemical methods which are tackled by the advantages of the biological ones. Chemical methods have low catalytic efficiency and need harsh conditions for treatment (high temperatures, extreme pH, and high pressure) which are hazardous to the environment (Adrio and Demain 2014). Use of organic solvents leads to the production of wastes, further leading to pollution. These disadvantages and more are being countered by enzymes, especially microbe-derived biocatalysts.

Industrial enzymes are widely accepted in food and beverage applications, owing to their functional properties. They play an important role in determining the desired attributes (in the products) such as taste, texture, appearance, and flavour and are also used in the production of biofuels. The extensive requirement of carbohydrases in food processing, brewing, baking and biofuel manufacturing has been one of the key drivers of the industrial enzyme market. Proteases accounted for the largest share in the industrial enzyme market due to their wide range of applications in the food and beverage, detergent and biofuel industries. The production of industrial enzymes is a billion-dollar industry in the global market, where sales in the USA reached $5.1 billion in 2009 (Sarrouh et al. 2012). According to the report "Industrial Enzymes Market," the global industrial enzyme market is projected to reach US $6.30 billion by 2022 in terms of value, at a CAGR of 5.8% from 2017.

Microbes have been utilised by humans since antiquity (Singh et al. 2016a, b, c, d). They are ubiquitous, are easily available and have a fast growth rate, hence ideal for industrial use. They also ensure a regular and abundant supply of the product of interest (Mienda et al. 2014), have the ability to be commercialised by ease of scaling up of production and are non-toxic and non-pathogenic. Enzymes derived from microbes are highly desirable for industrial processes, as they have a broad spectrum of chemical and physical conditions for optimal activity, higher and superior performance and ability to scale up production by gene manipulation (Singh et al. 2016a, b, c, d). Additionally, microbial enzymes have longer shelf life enabling long-term storage without significant loss of activity, and ease of downstream processing due to the fact that many enzymes are extracellular in nature (Porta et al. 2010). In this chapter, we discuss the commercial applications of major microbial enzymes like amylases, proteases, lipases and cellulases with a brief analysis on the industrial applications of other enzymes. And we also explore various microbial sources of enzymes and the nature of industrial applications.

6.2 Amylases

Amylases are a group of extracellular enzymes that hydrolyse the α-1,4- and α-1,6-glycosidic bonds present in starch and glycogen to give diverse products as dextrins and progressively smaller polymer composed of glucose units (Fig. 6.1). The first enzyme to be produced industrially was an amylase from a fungal source in 1894, which was used as a pharmaceutical aid for digestive disorders (Mojsov 2012). Based on the action specificity towards the α-glucan chains, amylases are classified into the following: (1) Endoamylases, which cleave α-1,4-glycosidic linkage between adjoining glucose units in the product chain retaining the anomeric carbon configuration in the product. α-Amylases belong to this class; they cleave α-1,4 bonds and result in α-anomeric products (Rana et al. 2013). α-Amylases play a central role in the hydrolysis of starch, both in nature and in the starch industry. They can hydrolyse the α-1,4 linkages but cannot hydrolyse the α-1,6 linkages. (2) Exoamylases, which act at the non-reducing ends of polysaccharides and produce low-molecular-weight products. They cleave α-1,4 or α-1,6 bonds of the external glucose residues resulting in α- or β-anomeric products (β- and γ-amylases) (Sivaramakrishnan et al. 2006). β-Amylases hydrolyse the α-1,4 linkage next to the non-reducing end of α-glucan, which successively yields maltose in a β-configuration. γ-Amylases, also called as glucoamylase or amyloglucosidase, catalyse the hydrolysis of successive α-1,4 linkages in the

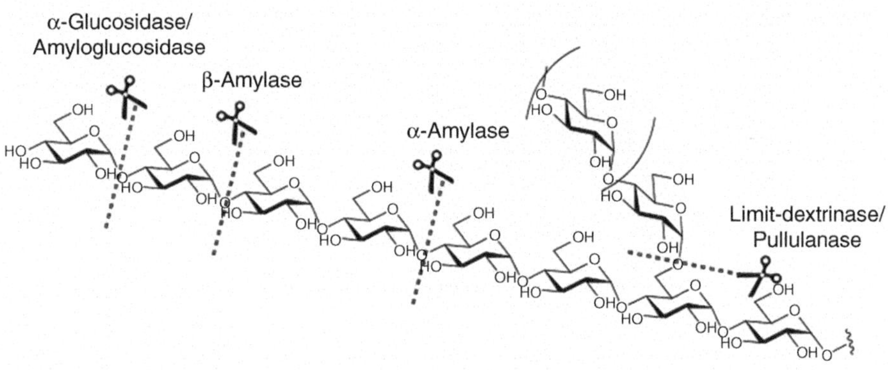

Fig. 6.1 Mode of action of amylases

non-reducing ends of glucans, thereby producing β-D-glucose as the hydrolysis product. It also hydrolyses the α-1,6 linkage, although at a rate lower than that of the α-1,4 linkage hydrolysis (Taniguchi and Honnada 2009).

The amylase has a three-dimensional structure capable of binding to substrate and by the action of highly specific catalytic groups promotes the breakage of the glycoside links (Fig. 6.2). Its 3D structure is composed of three domains called A, B and C. Domain A has eight identical pairs of β-sheet and α-helix. Domain B is composed of three β-sheets and is inserted between the domains A and C. Domain C has eight antiparallel β-sheets and an independent domain with unknown function. The catalytic site of the enzyme is located at the catalytic cleft formed between the carboxyl domains A and B (de Souza and Magalhaes 2010). This cleft has a size that can accommodate just seven glucose units of α-glucans. About 20 amino acid residues are aligned on the surface of this catalytic cleft. Two catalytic amino acid residues, Asp and Glu in the fourth and fifth β-sheets, respectively, are located near the third glycosidic linkage of the substrate (Taniguchi and Honnada 2009).

Amylases are present in plant, animal and microbial cells. Microorganisms are commonly used in the large-scale production of amylases. Commercially exploited amylases are obtained from both bacteria and fungi. *Bacillus licheniformis*, *Bacillus stearothermophilus*, *Bacillus amyloliquefaciens*, *Bacillus subtilis*, *Bacillus polymyxa*, *Bacillus megaterium*, *Chromobacter* sp., *Haloarcula hispanica*, *Halomonas meridiana*, *Rhodothermus marinus*, *Corynebacterium gigantea*, *Geobacillus thermoleovorans*, *Lactobacillus fermentum*, *Lactobacillus manihotivorans* and *Pseudomonas stutzeri* (de Souza and Magalhaes 2010; Gopinath et al. 2017) are the common bacterial sources of amylases. The common fungal sources are *Aspergillus oryzae*, *Aspergillus niger*, *Aspergillus awamori*, *Aspergillus kawachii*, *Aspergillus flavus*, *Penicillium brunneum*, *Penicillium expansum*, *Penicillium chrysogenum*, *Penicillium roqueforti*, *Penicillium janthinellum*, *Penicillium camemberti*, *Penicillium olsonii*, *Streptomyces rimosus*, *Thermomyces lanuginosus*, *Cryptococcus flavus* and *Mucor* sp. (Gopinath et al. 2017).

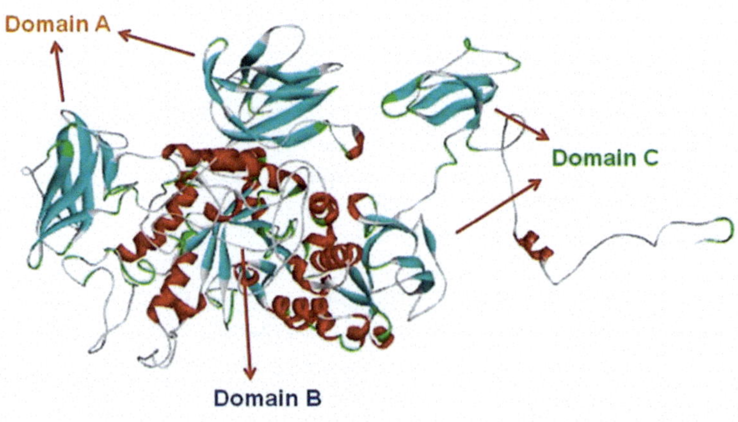

Fig. 6.2 Structure of amylase with three distinct domains A, B and C

6.2.1 Commercial Applications of Amylases

6.2.1.1 Detergent Industry

Earlier detergents used to contain surfactants which were active at high temperatures and considerable amounts of energy are used to heat laundry water, particularly in cold countries. Apart from that when they were released into the environment, they caused toxic effects in the ecosystem. Moreover, at present consumers are preferring to use cold water and mild conditions for washing their clothes, which requires the detergent to work in those conditions. Enzymes can be used as alternatives in this regard. They have the capability to degrade the stains at low washing temperatures and hence are used as supplements in detergents (Jegannathan and Nielson 2012). Amylases are also used in the formulation of detergents, and 90% of all liquid detergents contain these enzymes to remove tough stains (Gopinath et al. 2017). α-Amylase is used in detergents and in automatic dishwashing, to digest the starch-containing food particles such as potatoes, gravies, custard and chocolate, etc. (Mojsov 2012). At lower temperatures removal of starch from porcelain becomes more problematic because starch can attract other dirt particles and hence detergents with amylases which optimally work at lower temperatures and alkaline pH can help to overcome this problem (Rana et al. 2013). Under the action of amylases starch is degraded into smaller food particles or into water-soluble oligosaccharides or dextrins by catalysing the hydrolysis of glycosidic linkages in stains and eliminating the starchy glue that combines with other stains and dirt. The hydrolysing activity of these enzymes also restricts binding of swollen starch to other stains and dirt (Singh et al. 2016a, b, c, d).

The presence of Ca^{2+} up to a certain level is required for the good activity of the amylases becuase these ions act as stabilisers and protect against denaturation. Optimal activity lies in the range of 40–60 °C and pH of 5–8. The oxidative stability of amylases is one of the most important criteria for their use in detergents where the washing environment is very oxidising (Gopinath et al. 2017). Washing environment plays a vital role in the action of amylases. At low temperatures of washing, the effect of amylases is very high. Amylases are being used for the improvement of laundry bleach composition and bleaching without the colour. Enzyme addition stabilises the bleach agent and preserves the effectiveness of the bleach in laundry detergent bar composition (Saini et al. 2017).

6.2.1.2 Food Industry

Dextrin with a DP 3-25 is commercially produced and used as a food ingredient in various applications, such as to improve the viscosity or to be a filler or an ingredient of food. To produce glucose or oligosaccharides, raw starch is gelatinised to create a paste to which amylases are added to hydrolyse starch into low-molecular-weight dextrins (Taniguchi and Honnada 2009). Amylases from the bacterial sources are added to the starch slurry, and the mixture is heated quickly around 100 °C for a suitable period to obtain dextrins. In bread baking, amylases are added to dough that results in enhancing of the rate of fermentation, in the reduction of viscosity and in the degradation of the starch in the flour into smaller dextrins, which are utilised by yeast during fermentation (Rana et al. 2013). Increase in fermentable sugars eventually causes increase in loan volume. Another added role of amylases is to produce surplus amounts of maltose in the dough, some of which lasts after baking and prevents baked bread from becoming stale by interfering with the starch molecules on the shelf (Taniguchi and Honnada 2009). It also improves the taste, crust colour and toasting qualities of the bread (de Souza and Magalhaes 2010).

In beer and juice clarification, amylases are used together with pectinases (de Souza and Magalhaes 2010). The most widespread application of amylases in the starch industry is the hydrolysis in the starch liquefaction process which converts starch into fructose and glucose syrups. Cocoa slurries are treated with amylases to produce a product which does not tend to layer in storage, eliminates appreciable stiffening or setback and gives rise to a product with an improved flavour and solubility in milk, which is known as chocolate syrup, in which chocolate starch is dextrinised and syrup does not become thick. The stabilised syrups which are cocoa flavoured are added at room temperature to conventional non-acid confection mixes for use in the production of quiescently frozen chocolate-flavoured confections (Saini et al. 2017). Jellies made from apple, quince and crab are hazy in appearance because of the high starch content. Treating the jelly with amylase for 1 h at 80–95 °F and filtering produce a clear paste suitable for making a sparkling jelly.

Maltose is often known as a sweetener to improve the taste and quality of the food. Part of maltose is even used in the medical field. For the production of maltose, liquefied starch is treated with amylases to yield maltose (Taniguchi and Honnada 2009). High-fructose-containing syrup is prepared by enzyme isomerisation. Amylases are used extensively in the preparation of dried baby foods and cereal products. The cereal to be treated is heated to a very high temperature and fungal amylases are added to digest the starch. Milk contains no starch and low levels of oligosaccharides and contains unusual monosaccharides with glycosidic linkages. Therefore, it seems that amylases are used in the hydrolysis of oligosaccharides in milk. Amylases also aid in the production of cakes (Mojsov 2012). Amylases are also used in the pretreatment of animal feed to improve the digestibility of the fibre (Saini et al. 2017).

6.2.1.3 Paper and Pulp Industry

Amylases are used in this industry for the modification of starches for coated paper that is for the production of low-viscosity, high-molecular-weight starch. The coating treatment improves the quality of the finished product (smoothness) and enhances stiffness, elasticity and writing quality of the paper (Mojsov 2012; de Souza and Magalhaes 2010). The viscosity of the natural starch is very high for the sizing of paper, which is altered by partially degrading the polymer with amylases in a batch or continuous process. Starch is added to the paper in the size press and paper picks up the starch by passing through the rollers which are used for transferring of the slurry. The temperature lies around 45–60 °C. A constant viscosity of the starch is needed for good results at this stage. The mill also has the flexibility of varying the viscosity of the starch for various paper grades (Gupta et al. 2003). The size enhances the strength in the paper (Rana et al. 2013). Amylases are used in this industry to protect it from mechanical strain during processing (de Souza and Magalhaes 2010). Cold-active amylases are used mostly as they reduce the viscosity of the paper.

Paste of starch when used as a mounting adhesive and modified with additives such as protein glue or alum, frequently, causes damage to the paper as a result of its embrittlement. In such cases starch-digesting enzymes such as α-amylases, in immersion or as gel poultice, are applied to facilitate its removal (Saini et al. 2017). Amylases are often used for deinking depending on the type of paper and ink, and drainage improvement (Singh et al. 2016a, b, c, d). Treatment of multi-print furnishes with cellulases and amylases at pH 7–7.5 improved the brightness of the pulp. The ink particles released on treating with amylases appeared to be more hydrophobic than ink particles released on treating with cellulases. Using more specific amylases during the procedure of mill trail, the brightness of the paper was significantly improved. The ash content was also greatly reduced after flotation and washing, resulting in a change of the final pulp characteristics (Bajpai 2010).

Introduction of amylases during the process of deinking also increased the roughness of the paper. The fibre surface was attacked by amylase for the release of particles of ink from the surface due to starch solubilisation. Mostly α-amylases catalyse the hydrolysis of internal α-1,4-glycosidic links in starch in a random manner. The atomic force micrograph of fibres treated with amylases confirmed the roughness of the paper with more irregularities (Dutt et al. 2012). Amylases along with mixtures of cellulases and xylanase are used for degrading the starch layer on the paper surface. The toner particles adhering to the surface of the paper were released by enzymatic treatment and subjected to subsequent separation. This concoction is often used for hydrolysing vegetable oil-based ink binders (Dutt et al. 2012).

6.2.1.4 Textile Industry

Amylases are used in the textile industry for desizing process. Starch, which is a sizing agent, is applied to yarn before fabric production to ensure a fast and secure weaving process because without starch cloth tends to break the threads making up wraps. In this industry, starch, which is cheap and available in almost all regions in the world and can be removed quite easily, is applied for textile wrapping which gives strength to textile at weaving (Pandi et al. 2016). These enzymes are mainly preferred because of the activity under high temperature and pH, thereby increasing the use under the harsh conditions of textile wet processing. After weaving the starch is removed usually by utilising amylase. This process is called as desizing. Amylases are employed to cleave starch particles randomly into water-soluble components and can be removed by washing (Sundarram and Murthy 2014). The cloth is first saturated with enzyme solution in a desizing bath at approximately 50 °C and held until all the size has been solubilised. The enzymatic desizing of cotton is being done for many decades. Bio-desizing is mostly appreciated because of its high efficiency and specific action. Not only cotton but also ubiquitous jeans are desized after mashing (Saini et al. 2017). Amylases bring about complete removal of the size without any harmful effects on the fabric besides eco-friendly behaviour. The amylases do a selective work, remove particularly the size and do not attack the fibres (Mojsov 2012). Amylases from *Bacillus* sp. have been employed in this industry since long (Rana et al. 2013).

6.2.1.5 Leather Industry

Amylases are primarily used in the extraction of collagen from animal skin. α-Amylases are used to open up the skins in leather processing. Liming and reliming achieve the objective of splitting up of the fibre bundles and enabling them for tanning agents, dyes, fat liquors and other materials to diffuse into the matrix. So, fibre-opening enzymes especially α-amylases are used as a pivotal tool in replacing the hazardous chemicals involved in this process (Pandi et al. 2016; Aravindhan et al. 2017). Dehairing is another stage where amylases are used. Use of amylases along with proteases increases the efficiency of the procedure and also increases the outputs of the leather (Choudhary et al. 2004; Kyaw et al. 2010).

6.2.1.6 Biofuel

Amylases are one of the most used groups of enzymes for the production of biofuels. Enzymes from bacterial and fungal sources are being used for decades at commercial level, using both conventional (dry grinding and wet milling) and non-conventional (starch hydrolysis or granular starch hydrolysis) processes. For ethanol production, the commonly used substrate is starch due to its availability all around the world. This procedure involves two main steps in order to obtain fermentable sugars. The bioconversion of starch into ethanol involves liquefaction and saccharification, where starch is converted into sugar using amylases which is followed by fermentation, where the conversion of sugar takes place. The conversion of sugar to ethanol is done by specific yeast, like *Saccharomyces cerevisiae* (Mojsov 2012).

The industrial procedure involves cooking of starch granules at high temperatures in order to solubilise the starch molecules, followed by adding of starch-degrading enzymes such as amylases. The high-temperature cooking contributes to the consumption of energy of the fermentation process, thus reducing the total energy in the industrial plant. Thus, a yeast strain, usually *S. cerevisiae*, which is genetically engineered and is capable of expressing a hydrolysing enzyme usually α-amylase, would greatly reduce the processing cost of the production of bioethanol. This whole procedure is referred to as the cold starch hydrolysis process. Genetically engineered barley α-amylase expressed in *S. cerevisiae* is used in the conversion of starch to bioethanol which lowers the cost of production at a large scale (Taniguchi and Honnada 2009). When it comes to the fungal sources, *A. oryzae* is mostly used with rice flake's waste as a substrate.

6.2.1.7 Petroleum Industry

The main function of amylases in petroleum refining is to remove the filter cakes that are formed on the petroleum wells. Starch is a major component of the filter cake and hence its removal is an important step concerning the production and injection in wells. Amylases are employed in recent days because these enzymes are capable of hydrolysing the starch molecules into dextrins or smaller oligosaccharides. This results in a rapid decrease in the viscosity of the gelatinised starch. Thus, these enzymes have a great potential in filter cake removal applications. The problem is the interference in enzymatic activity due to the operating conditions of the petroleum industry like high temperatures, high-salt condition and high pressures. Genetic modifications were done to produce heat-stable enzymes from *Bacillus* sp. which could work at increased concentrations of salt and pressure (Kyaw et al. 2010).

6.2.1.8 Other Applications

The use of amylases is increasing day by day in all the industries. It is being used in the pharmaceutical industry as a digestive aid. Pure amylases are being required in clinical sectors. Amylases are being used in the cosmetic industry for haircare (shampoos, oils and styling) and toiletries (bathing soaps and liquids) and oral care. Amylases are generally used to treat starch-processing waste water. This waste is mainly produced from the food processing industries. In those areas before releasing the effluent, the water is treated with amylases to remove the starch. This kind of treatment produces very useful products like microbial biomass protein and also purifies the effluent (Mojsov 2012).

To conclude, pollution-free processes are gaining importance all over the world. Enzymes are being employed in all the industries to make it eco-friendly. The use of α-amylase has been prevalent in industries for many decades. Though there are three classes of amylases, α-amylase is used more extensively in industries. They can be obtained from plants, animals and microbes. However, enzymes from fungal and bacterial sources are gaining importance because the conditions required for the fermentation are very conducive and the production rate is very high. Other reasons for this are that the production and downstream processing cost is low and the yield rate is very high. Production is generally carried out using submerged fermentation, but solid-state fermentation is gaining importance and will have a very promising future especially in applying agro-industrial residues as substrate. In recent days, it is capturing importance and researchers are trying new experiments to increase their use in the industrial sector. The properties of amylases such as stability, pH profile and Ca^{2+} ion independency encourage their increasing use in industries. Amylases are mainly used in industries for the preparation of fermented foods. Apart from the food industry their applications are spreading widely into other industries such as paper and pulp, textile and leather. It is used in biofuel production with starchy waste as raw material. The commercial applications of amylases are summarised in Table 6.1.

6.3 Proteases

Proteases, term used interchangeably with proteinases or peptidases, are a large class of enzymes catalysing the hydrolysis of peptide bonds linking two adjacent amino acid residues in a protein. They might be specific to a single protein (angiotensin-converting enzyme) or non-specific, acting against a wide spectrum of proteinaceous compounds (proteinase K) (Lopez-Otín and Bond 2008). Specificity subsites, each of which attaches to a side chain of a single amino acid residue of the substrate, flank one or both sides of the catalytic site of proteases. In serine proteases,

Table 6.1 Industrial applications of amylases

Industry	Applications
Food industry	Production of dextrins, high-fructose corn syrup and maltose which can be used as an artificial sweetener Hydrolysis of starch causing turbidity due to insolubility in fruit juices Production of bread with sufficient loaf volume and softness Production of candy with desired softness Hydrolysis of oligosaccharides in milk For the digestion of few fibres and increase in nutritional value of the feed For the manufacture of chocolate syrup by using cocoa slurries For the manufacture baby foods especially cereal-based foods
Detergent industry	Additive to remove starch-based dirt Increase oxidative stability upon the use of these enzymes which make the washing environment very oxidising
Pharmaceutical industry	Digestive aid
Cosmetic industry	In shampoos, oils and oil care
Paper industry	Reduction of viscosity of starch for appropriate coating of paper, deinking
Textile industry	Wrap sizing of textile fibres, used in removal of sizing agent from woven fabric
Petroleum industry	Remove the filter cakes on petroleum wells
Fuel industry	Ethanol from starchy waste materials
Leather industry	Splitting up of the fibre bundles and enabling it for tanning agents
Medicine field	Diagnosis of peptic ulcers and other diseases
Clinical chemistry	Detection of higher order oligosaccharides
Wastewater treatment	Treatment of starchy water released from food industries

the catalytic triad Asp-Ser-His carries out catalysis (Salleh et al. 2006). Proteases are broadly classified into endopeptidases and exopeptidases based on the cleavage site, where the former cleaves proteins at sites away from the either terminal to produce mono-, di-, tri- or polypeptides, whereas exopeptidases cleave proteins from the N- or the C-terminal. Proteolytic enzymes are also categorised into six types based on the type of amino acid residue present at their active site; they are serine, cysteine, threonine, aspartic, glutamic and metalloproteases (Mótyán et al. 2013) (Fig. 6.3). A seventh protease was added in 2011, asparagine peptide lyase, where cleavage occurs not through hydrolysis but via amidine lyases (Rawlings et al. 2011).

Proteolytic enzymes are obtained from plant, animal and microbial sources. However, for most industrial applications microbial sources are widely used as they prove to be more advantageous. This is due to their wide range of physical and chemical characteristics, large-scale production and lower cost of production and possibility of genetic manipulation. Moreover, plants and animals as sources are influenced by factors such as availability of agricultural land, certain climatic conditions and livestock for slaughter, which are difficult to control (Sawant and Nagendran 2014). Acidic proteases are more commonly found in fungi like

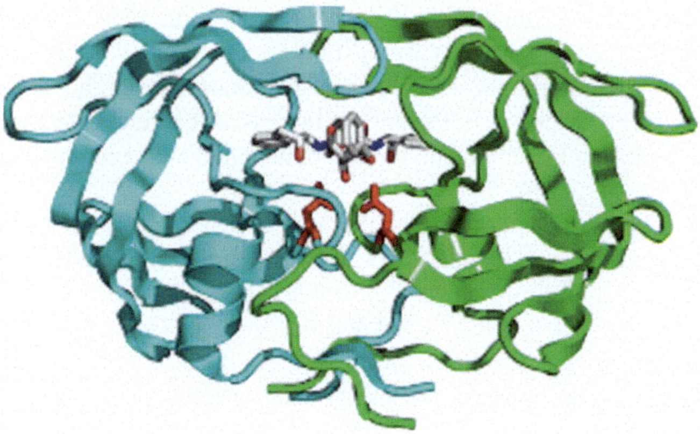

Fig. 6.3 Structure of protease

Mucor miehei, *Mucor pusillus* and *Mucor hiemalis* for rennet-like proteases and *Aspergillus* sp. and *Rhizopus* sp. for pepsin-like proteases. Neutral proteases are present in both bacteria and fungi. *Clostridium histolyticum*, *Streptococcus* sp., *B. subtilis*, *Bacillus cereus*, *B. megaterium* and *B. stearothermophilus* are some of the bacterial sources, whereas *A. oryzae*, *Aspergillus sojae* and *Penicillium* sp. are the fungal sources (Sumantha et al. 2006). Alkaline proteases are well studied due to their immense use in the detergent industries and are obtained from fungal sources like *Aspergillus candidus*, *Aspergillus fumigatus*, *A. flavus* and *Cephalosporium* sp. and bacterial sources like *Bacillus alcalophilus, Bacillus proteolyticus, B. subtilis*, *Pseudomonas* sp. and *Streptomyces* sp. (Singh et al. 2016a, b, c, d).

6.3.1 Commercial Applications of Proteases

6.3.1.1 Detergent Industry

Proteases are used for the breakdown and removal of organic stains (human sweat, egg, blood, etc.) and in dishwasher detergents for removing protein-containing food particles (Singh et al. 2016a, b, c, d). Out of the total protease sales worldwide, 89% is contributed by detergent enzymes with *Bacillus* sp. contributing a major share producing subtilisins and alkaline proteases (Jisha et al. 2013). Enzymes stable to work in alkaline conditions and high temperatures are utilised because of the harsh nature of detergent formulations. Hence, alkalophilic *Bacillus* sp. producing alkaline proteases is the most used in the detergent industry. In fact, *Bacillus* strains are the sole source of all proteases which are currently in the market, although fungal alkaline proteases have also proven advantageous due to simplified downstream processing (Salleh et al. 2006). The first microbial protease to be added in detergents was subtilisin Carlsberg from *B. licheniformis*. From then on, enzymes in detergent formulations have been increasingly preferred. This is because they are environmentally more compatible than other non-biological components. For instance, using lower temperatures for washing process and avoiding phosphates as a detergent component reduce environmental contamination. It was shown that when proteases were added to detergents, it increased the detergent's overall effectiveness, also enabling lower washing temperatures and environment-friendly components (Salleh et al. 2006).

Ideally, for maximum effectiveness, minimal use of enzymes is desired (around 0.4–0.8%). Additionally, extended shelf life and compatibility with other components in the detergent mixture are also desirable. These qualities were found in proteases acquired from *Vibrio fluvialis* and *Nocardiopsis* sp. Two commercial mixtures Savinase T and Esperase, produced from alkalophilic *Bacillus* sp., had their pI at 11.1 enabling them to be stable at higher pH ranges (Salleh et al. 2006). Continuous modification and development of the detergent formulation are necessary to maintain effectiveness and efficiency. This could be seen when the flourishing usage of enzymes in detergents developed a hitch in the 1970s, due to some workers in detergent factories developing allergies against certain enzyme preparations. The detergent preparation was then modified to make it dust free by encapsulating the protease in an inner core (containing inorganic salts and sugars as preservatives) and coating this core with inert waxy materials which could disperse during the washing step. This prevented dust formation and enabled the protection of the enzyme from degradation by other components used in the detergent preparation (Kumar et al. 2008).

6.3.1.2 Leather Industry

Microbial proteases find uses in the leather industry in processes like soaking, dehairing and tanning. Earlier, hazardous chemicals were used in these processes which gave rise to major environmental concerns regarding effluent disposal and pollution. For instance, chemicals used for dehairing, although a more cheap and fast method, produced increased amounts of hydrogen sulphide constituting a health hazard for the workers, while also being environmentally unfriendly. Hence a shift to the use of enzymes was highly desired (Salleh et al. 2006). Proteases are mainly used for removal of non-fibrillar proteins (albumins, globulins) from the skin and hides of animals and for the selective hydrolysis of the non-collagenous constituents of the same. Alkaline proteases possess elastolytic and keratinolytic properties, enabling their use in dehairing as they react with hair follicle proteins allowing easy hair removal; the alkaline environment produced as a result swells up the hair roots, speeding up the dehairing process, thus proving to be a cheap alternative (Mojsov 2012). Another advantage is the reduction of soaking time, since alkaline proteases soak up water at an increased rate. Waste disposal can also be reduced by using a mixture of proteases with lime and sodium chloride during dehairing and dewooling (Gopinath et al. 2017).

Major protease that has shown maximum usage in this industry is alkaline protease extracted from alkalophilic *Bacillus* sp. Improved results have been obtained upon its usage: for instance, destruction of unwanted leather pigment, increase in leather surface and removal of fine hair, so that the bating time can be highly reduced or will not be required at all. Physiological conditions used for the same were temperature range of 20–30 °C, pH of 12–12.5 and production time of 18–24 h. Additionally, sources like *A. flavus*, *Streptomyces* sp., *B. amyloliquefaciens* and *B. subtilis* are also used for alkaline proteases used in leather tanning. Aquaderm, NUE and Pyrase are three commercial proteases manufactured by Novo Nordisk for applications in soaking, dehairing and bating, respectively (Salleh et al. 2006).

6.3.1.3 Dairy Industry

Cheese manufacturing uses microbial proteases extensively, where specific peptide bonds are hydrolysed to yield casein and macro peptides, the building blocks of cheese (Gopinath et al. 2017). Casein is present in conjunction with calcium phosphate in the form of casein micelles with casein on the surface (Ismail and Nielsen 2010). The enzyme rennet (chymosin) hydrolyses this casein and induces aggregation of casein micelles which helps in coagulation of milk that is the first stage in the cheese production process. Some microbial sources which produce rennet-like proteases are *A. oryzae*, *Rhizomucor pusillus*, *Rhizopus miehei*, *Endothia parasitica* and *Irpex lactis* (Afroz et al. 2015). Casein micelles can be further degraded by other enzymes. For example, plasmin (when converted to its active form from its inactive plasminogen form) degrades casein. This plasmin-induced proteolysis can prove to be either beneficial or detrimental depending on the extent of hydrolysis and the type of product on which it occurred. For instance, proteolysis can induce the formation of the required texture during the ripening process and also produce flavour in cheesemaking, and on the other hand it causes undesirable gelation in processed milk (pasteurised and ultra-high temperature). This proteolysis is caused by both native enzymes and enzymes produced by psychrotrophic microbes during cold storage of milk (refrigeration). Microbes like *Pseudomonas fluorescens* and *B. polymyxa* have been shown to produce heat-stable metalloproteases which can destabilise casein micelles by hydrolysing casein, leading to reduced quality of milk and cheese if stored at low temperatures for long time periods (Ismail and Nielsen 2010).

Lactic acid bacteria are also widely used for the production of curd as a starter culture. These bacteria need a high nutritional diet for optimal yield, so as to meet the requirements for the same, they contain a complex system of proteases and peptidases to supply the essential amino acids while also allowing rapid growth in proteinaceous food like milk (Mathias et al. 2017). Proteases from lactic acid bacteria that belong to the genera *Lactobacillus*, *Streptococcus* and *Lactococcus* also aid in degradation of major allergens in milk, like the whey proteins α-lactalbumin and β-lactoglobulin (Atanasova et al. 2014).

6.3.1.4 Food Industry

Proteases from microbial sources find numerous applications in the food and feed industries. Their protein-degrading ability results in imparting advantageous properties like increasing its nutritional value, solubility and digestibility, while also decreasing factors causing adverse side effects like presence of allergenic compounds. Modification of some of the properties of food such as coagulation, foaming and gel formation can also be done using microbial proteases. Bakery and beverage industries find the most extensive use of proteases, as well as other fields (Singh et al. 2016a, b, c, d).

Bakery items such as bread and cookies get many of their desirable properties such as their characteristic flavour and texture from the action of proteases. They additionally ensure dough homogeneity, maintain gluten strength in breads and reduce dough consistency (Singh et al. 2016a, b, c, d). Fungal acid proteases from *A. oryzae* are used to soften the bread, making it easier to pull and knead by modifying its gluten content which is made to undergo partial hydrolysis; the quality of bread and dough rheology also improves as the gluten network is affected, resulting in greater mobility, extensibility and crispiness (Salleh et al. 2006; Miguel et al. 2013). Effect of proteases on gluten is also exploited in making pastries, biscuits and cookies, where gluten elasticity, as well as shrinkage of dough or paste after sheeting or moulding, is reduced. Hydrolysis of gluten proteins improves the spread ratio of the cookies (Miguel et al. 2013).

Due to their gluten-modifying action, addition of proteases in the baking process also helps reduce gluten-related disorders, which is the intolerance or sensitivity to gluten-containing food products (Heredia-Sandoval et al. 2016). Such disorders produce several gastrointestinal symptoms like diarrhoea, constipation, vomiting and bloating. Administration of proteases helps to reduce immunogenicity due to the gluten content while maintaining its baking properties. Certain selected *Lactobacillus* sp. in conjunction with fungal and/or malt proteases were recently shown to reduce the residual gluten sequences when fermentation time was increased. Gluten contains 10–15% proline residues which are targeted by these microbial enzymes; they act by cleaving the peptide bonds next to these proline residues. The resulting small peptides have lower immunogenicity. Fungal proteases from *A. niger* have shown to effectively degrade immunogenic gluten in vitro while also being resistant to stomach acids.

In breweries, a zinc metallo-endoprotease from *B. amyloliquefaciens*, marketed as Neutrase, has been used for extracting proteins from barley and malt, improving yeast growth and obtaining appropriate concentrations of nitrogenous nutrients (Sumantha et al. 2006). Lactic acid produced by some lactic acid bacteria, mainly from the genus *Lactobacillus* like *L. delbrueckii* sp. *delbrueckii*, *L. delbrueckii* sp. *lactis* and *L. fermentum*, induces natural acidification of malt and helps in adjusting the pH of the mash (Mathias et al. 2017). Addition of bio-acidified malt with base malt is also in keeping with the Beer Purity Laws in some countries. Microbial protease also are useful in reducing the waste from breweries; these are of four types: the brewer's spent grain (from grain processing), hot trub (coagulated protein during boiling of wort), residual brewer's yeast (from the starter culture) and diatomaceous earth (resulting from the clarification stage of beer production).

L-Glutaminase is used as a flavour enhancer, producing the taste of umami especially in oriental-style dishes. Proteases are also used in the tenderisation of meat by cleavage of collagen fibres and also in meat recovery. The production of soy sauce and other soy products also use microbial proteases (Singh et al. 2016a, b, c, d). For instance, Kojizyme, a complex of endopeptidases and exopeptidases from *A. oryzae*, is used in soy sauce fermentation. The degradation of the turbid complex contributed by fruit juice proteins and alcohol-based liquor is done by aspergillopepsin I, an acid protease from *Aspergillus saitoi*; it is additionally used to produce seasoning powders from proteinaceous food materials (Sumantha et al. 2006). The hydrolysing property of proteases is exploited in processes such as hydrolysis of soy protein, casein and whey protein, gelatin and fish protein. Ultra-sonication in conjunction with the addition of proteases has been found to be useful in the extraction of rice starch.

6.3.1.5 Pharmaceutical Industry

Proteases and regulation of proteolysis of numerous molecules in the body contribute in maintaining normal physiological conditions and functioning. For instance, the concentration and production of growth factors, cytokines, cellular receptors and chemokines are regulated by both activation and inactivation of certain proteases (Craik et al. 2011). Improper regulation of the same may lead to adverse effects in the body manifesting as diseases like cancer, various inflammatory disorders and others. Proteases additionally provide numerous advantages for their use in therapeutics. Their specificity and high catalytic activity translate to lower and less frequency of doses, higher efficacy and lower costs (Li et al. 2013). Higher specificity means lower number of substrates, which can be redesigned to further narrow down the range, thus eliminating the possibility of unwanted activation of non-specific molecules. Hence, proteases present profound potential in the field of therapeutics for the treatment of such diseases. Generally therapeutic proteases available in the market nowadays are serine proteases, with the exception of the zinc protease and bacterial botulinum toxins (Craik et al. 2011).

Primary application of proteases in medicine lies in the degradation of blood clots. Plasminogen activators (PA) convert inactive plasminogen to active plasmin, which then degrades the blood clots. This occurs by lysis of fibrin, which constitutes the fibrin meshwork making up the clot. Such clots may produce blockage in the vascular system and other such vessels leading to complications like constricted or reduced blood flow and inflammation. Many proteases target fibrin which helps in dissolving such harmful clots. Serrapeptase (or serratiopeptidase) obtained from *Serratia marcescens* and *Serratia* sp. E 15 commercially is one such enzyme which digests blood clots, arterial plaques and other such dead protein debris (Chanalia et al. 2011). It acts as an anti-inflammatory agent by helping to drain the fluid from an inflamed area and inhibiting the release of pain-inducing amines. Hence it targets many inflammatory disorders like rheumatoid arthritis and osteoarthritis. Serrapeptase also breaks down mucus deposits in the respiratory tract associated with inflammation and swelling in the area, therefore also becoming a treatment option for diseases like sinusitis and bronchitis.

In addition to microbes being used to produce the actual therapeutic enzyme, they have also been used as an expression system to produce an increased amount or a modified form of a therapeutic protease. Gram-negative bacteria like *Escherichia coli* and Gram-positive bacteria like *Clostridium botulinum* are being currently used for the industrial production of therapeutic enzymes (Craik et al. 2011). For instance, a type of PA, tissue-type plasminogen activator (t-PA), was engineered to bind specifically to fibrin molecules in the meshwork of blood clot and degrade it. The first t-PA molecule, marketed as Alteplase, was approved by the FDA for treatment of acute myocardial infarction, stroke and catheter clearing. Variants of t-PA, reteplase and tenecteplase, were produced using *E. coli* as the host which resulted in the molecule lacking glycosylation. Such engineered proteases showed increased half-life in plasma as well as reduced time and expense of administration. Reteplase has been proved to be effective in clinical trials and tenecteplase has been approved by the FDA.

L-Asparaginase from *E. coli* and *Erwinia chrysanthemi* and L-glutaminase from various bacteria, yeast and fungi have been obtained for use as an anticancer agent (Chanalia et al. 2011). Tumour cells are unable to synthesise L-asparagine and depend on extracellular asparagine for their growth and proliferation; the use of L-asparaginase cuts off their main supply of extracellular asparagine, inhibiting their growth, by hydrolysing extracellular asparagine to ammonia and aspartic acid. Moreover, L-asparaginase also works selectively on malignant cells as it is able to differentiate the metabolic activities of tumour cells and normal cells. L-Glutaminase, on the other hand, attacks cancer growth by degrading glutamine; this is because glutamine is needed for the metabolism of energy and also that of purines, pyrimidines and other proteins. *Pseudomonas* sp. has been used for the production of recombinant glutaminase which has been patented.

Type A botulinum toxin is widely used for cosmetic and medical uses (Craik et al. 2011). It is a neurotoxin, inhibiting the release of acetylcholine at the presynaptic cholinergic nerve terminals of the PNS and at the ganglionic nerve terminals of the ANS, thus disrupting neurotransmission. Type A has the longest duration of action. The toxin A Botox was approved by the FDA in 1989 for the treatment of blepharospasm associated with dystonia and also strabismus. The FDA later also approved type B for the treatment of cervical dystonia.

Some bacteria have the ability to produce factors which are inhibitory to the growth of yet another species of bacteria which cause disease. These factors can be extracted from these microbial sources and purified to make a therapeutic agent against those diseases. For instance, *Staphylococcus simulans* produces a protease, lysostaphin, which has shown inhibitory activity to the growth of *Staphylococcus aureus* as well as other *Staphylococcus* sp. such as *S. carnosus*, *S. epidermidis* and *S. saprophyticus* among others (Chanalia et al. 2011). *S. aureus* causes various diseases, the likes of which are septicaemia, endocarditis, abscesses (localised as well as systematic) and septic emboli. It also has the ability to develop resistance against antibiotic agents like methicillin, hence becoming methicillin-resistant *S. aureus* (MRSA), to which people working in hospitals are frequently exposed. These bacteria also produce toxins and form layers of biofilms on the surfaces of plastics and glass. Lysostaphin has shown effectiveness against MRSA (in synergism with other antimicrobial agents and membrane-active agents) and causes destruction of bacterial biofilms by disruption of the extracellular matrix. It is a glycylglycine endopeptidase, cleaving the bond between adjacent glycine residues found especially in the peptidoglycan membrane of staphylococcal species. Hence it has a specific activity against staphylococcal species, resulting in perforation of the cell wall and increasing the roughness of the cell surface, ultimately leading to cell death.

Streptokinase produced by *Streptococcus* sp. is a plasminogen activator and can lyse human blood clots (Chanalia et al. 2011). It is currently being used in a therapeutic formulation for coronary thrombosis and myocardial infarction. It is also used in combination with streptodornase, a DNase used for cleaning wounds containing necrotic tissue, purulent exudates and blood clots, and is sold as Varidase (streptokinase-streptodornase). Clostridial collagenases are also used in wound debridement. Recombinant lysostaphin showed therapeutic activity against aortic valve endocarditis. *C. histolyticum* produces two monomeric collagenases which are purified and marketed as Xiaflex, used against Dupuytren's contracture with a palpable cord. Proteases can also be used as a digestive aid for lytic enzyme deficiency syndromes. Proteases from *A. oryzae, B. polymyxa* and *Beauveria bassiana* are used in digestive disorders and as antitumour agents. A serine protease, Nattokinase, derived from *B. subtilis*, is used for treating cardiovascular disease by reducing blood clotting factors and lipids associated with it (Mane and Tale 2015).

6.3.1.6 Textile Industry

The textile industry makes use of proteases from fungal and bacterial sources like *A. niger* and *B. subtilis* for the removal of wool fibre scales and for the degumming of silk (Singh et al. 2016a, b, c, d). Wool processing uses microbial proteases in the processes of scouring and bleaching (Ammayappan 2013). Scouring removes natural impurities like wax, fat and sweat adhered to the material. Wax and grease are hydrophobic impurities attached to the epicuticle surface membranes which makes the surface of raw wool hydrophobic (Araujo et al. 2008). Conventionally, harsh chemicals like sodium carbonate and potassium permanganate in a hot detergent solution were used for scouring, leading to use of increased chemical and heat energy as well as consumption of a large amount of water (Araujo et al. 2008). Alkaline proteases provide a better alternative for the same by decreasing the amount of water uptake. It also resulted in the improvement of dye penetration and softness.

Wool develops natural yellow coloration (called canary coloration) upon light exposure, alkali or microbial degradation. To prevent this, bleaching is done by hydrogen peroxide. It has been shown that overall whiteness improves upon the addition of proteases, which causes partial removal of cuticle cells during peroxide bleaching. Novolin L, a protease derived from a genetically modified *Bacillus* sp., when applied on wool, improved shrink resistance of the clothes manufactured from it. Serine proteases of subtilisin type, when applied in the pretreatment of wool fibres, have shown to improve anti-shrinkage properties, remove impurities and increase subsequent dyeing affinity (Araujo et al. 2008). Degumming of silk is done to remove sericin, a serine-rich gum gluing fibroin filament in silk (Freddi et al. 2003). Removal of sericin gives silk its characteristic soft and smooth texture, shiny appearance and elegance. Genetically modified *B. subtilis* producing an oxidative-stable endopeptidase, genetically modified *Bacillus lentus* producing an alkaline protease, and *A. saitoi* producing a pepsin were shown to have appropriate silk degumming properties.

6.3.1.7 Waste Treatment and Reduction

Many industrial wastes comprise proteins as their component(s), which lend the waste its toxic properties. These proteinaceous components are degraded by proteases, many of which are of microbial origin. In wastewater treatment plants using activated sludge process, sludge flocs contain 70–80% of organic matter (apart from the biomass made of aggregated microbes) produced when cells are lysed to release substrates and nutrients, forming an autochthon substrate for microbial metabolism (lysis-cryptic growth) (Guo and Xu 2011). This microbial metabolism and some of the carbon contents, being released as a product of respiration, lower the overall biomass production and subsequently the amount of sludge produced. Microbial proteases in conjugation with other hydrolases (amylase, lipase, etc.) have been found to increase the lysis efficiency of activated sludge, which reduces investment and operational costs and also helps to optimise sewage treatment systems. It was found that major components of sludge are carbohydrates and proteins, hence the increase in lysis efficiency. Bacterial species and filamentous fungi have been used for this purpose, as well as *Raoultella* sp. and *Pandoraea* sp. belonging to the Acinetobacter family.

Proteases have been found to work better in combination with other enzymes rather than in a solo effort. For instance, increased efficiency of sludge lysis has been reported when administered with lipase and endoglucanase and also with amylase and cellulase. Another study showed 80% reduction of solids in sludge when pronase E, a protease from *Streptomyces griseus*, was added with cellulose as a mixture, but only 46% reduction when just protease was added (Roman et al. 2006). In the same study, COD reduction was 93% when enzyme mixture was added and 59% in control. Proteases have been used for the treatment of waste from other industries and households also. Poultry and livestock industries produce tonnes of chicken feather and animal wastes such as hides, hairs and hoofs, respectively, worldwide annually (Verma et al. 2016). Alkaline protease from *B. subtilis* has been used for keratin degradation in waste feathers from poultry slaughterhouses (Gupta et al. 2002) with the processed feathers forming a nutrient source for the food and feed industry. Hydrolysis by keratinases can also help convert poultry waste into biogas (Verma et al. 2016). Keratinase from *B. licheniformis*, marketed as Verazyme, is the first commercial keratinase and it guarantees value addition of chick feathers as well as converts feather protein into biodegradable plastics.

6.3.1.8 Cosmetics

The most well-known usage of microbes in cosmetics is Botox therapy. It is done by the application of botulinum toxin produced not only by *C. botulinum* but also by other *Clostridium* species like *C. baratii* and *C. butyricum* (Walker and Dayan 2014). The said compound is a neurotoxin which, as mentioned earlier, is responsible for the disruption of the release of acetylcholine, a presynaptic neurotransmitter. Botox toxin is a protease which cleaves one or more SNARE (soluble *N*-ethyl-maleimide-sensitive factor attachment protein receptor) proteins on the presynaptic vesicle so that the neurotransmitter is not able to fuse with the membrane for its synaptic release. Botox A was approved by FDA for the treatment of facial rhytides of the upper face. It is also an optiontional treatment for dynamic lines, namely glabellar lines, horizontal forehead wrinkles and crow's feet. Keratinases have been shown to work as a component of a skin cosmetic composition with the presence of amino acids (Cho et al. 2007). Novolin T obtained from a genetically modified *Bacillus* sp. possesses a high keratinase activity and also acts as a skin moisturiser with whitening effects.

6.3.1.9 Silver Recovery

The preparation of photographic and X-ray films uses silver halides along with gelatin. There is about 1.5–2% (w/w) silver present in these films; hence waste from photographic and X-ray films produces more silver than that from other sources (Shankar et al. 2010). Silver is an invaluable metal having many applications, viz. catalysis of chemical reactions, construction of mirrors, use in microbicides and disinfectants (in the form of silver nitrate) and supplementation by antibiotics (Parpalliwar et al. 2015). Conventional methods for silver recovery from photographic and X-ray films suffered from many disadvantages. The most primitive method is to burn the films directly (Al-Abdalall and Al-Khaldi 2016); this causes environmental pollution, generates foul odour and produces a polyester layer which makes the silver extraction process difficult. Hence, microbial proteases offer many advantages for the same. The gelatin layers on the X-ray film are hydrolysed by these enzymes exposing the silver for recovery and also the polyester base which can be recycled; the latter cannot be done by conventional methods. Moreover, enzymes are more specific in their action and avoid damage to the polyester membrane. The duration of action is also less, with the gelatin layer being removed in only a few minutes. Mostly bacterial proteases are used for silver recovery from X-ray/photographic films and only one fungal protease has been reported (Shankar et al. 2010). Proteases from *B. subtilis*, *Conidiobolus coronatus* and *Streptomyces avermectinus* have been used for obtaining silver (Al-Abdalall and Al-Khaldi 2016).

To summarise, proteases derived from microbial sources are exploited in a variety of industrial applications. Major industries to employ such enzymes are the food and pharmaceutical industries, where these enzymes enhance the flavour and overall quality of food products, and they are also used for the treatment of numerous diseases. Alkaline proteases are more favoured among the above industries than others. In addition to meeting the increasing demands of the current market, these enzymes are also promoting a green environment by reducing the use of harsh chemicals needing even harsher physiological conditions. Novel proteases are also being produced and utilised thanks to recombinant technology and protein engineering, enabling the effectiveness and optimal action of these enzymes. Proteases have also been reported to act better in concert with other enzymes than as a solo effort, as is evident in their use in waste treatment. However, not all combinations produce a positive result, raising concerns about compatibility with other components in a mixture. Also, the stability of novel proteases needs to be tested before applying in a large scale, which can be done by conducting pilot-scale studies. Nevertheless, microbial proteases represent a huge potential of application in many industries at present and in future with the advancement of technologies. Table 6.2 represents the applications of microbial proteases in various industries.

Table 6.2 Industrial applications of proteases

Industry	Applications
Detergent industry	Removal of proteinaceous stains
Leather industry	Dehairing, destruction of unwanted leather pigment, increase in leather surface
	Leather tanning
Food industry	
Dairy industry	Casein micelle aggregation, coagulation of milk Curdling of milk, degradation of major allergens in milk
Bakery	Reduce gluten content, softening of bread, improved dough rheology and quality, greater mobility, extensibility and crispiness
Brewery	Extract proteins from barley and malt, improve yeast growth and obtain appropriate concentrations of nitrogenous nutrients
Others	Flavour enhancement Soy sauce fermentation Degradation of the turbid complex in fruit juice proteins and alcohol-based liquor
Pharmaceuticals	Lyse human blood clots; treatment for coronary thrombosis, myocardial infarction Inhibiting tumour growth by hydrolysing extracellular asparagine
Textile industry	Removal of wool fibre scales, degumming of silk Improved shrink resistance of wool
Waste treatment	Reduction of sludge content Reduction of solids, reduction of COD Degradation of keratin in chick feathers, conversion to biogas and biodegradable plastics
Silver recovery	Hydrolyse gelatin layer in X-ray/photographic film, recycle polyester base

6.4 Lipases

Lipases are one of the most adaptable and versatile biocatalysts and the third most important group of enzymes (Kavitha 2016). These are a subclass of esterases and belong to the family of hydrolases that act on the carboxylic ester bonds. Lipases are lipolytic enzymes, which catalyse a wide range of reactions such as the hydrolysis of triacylglycerol to form free fatty acids and glycerol by a process called lipolysis, interesterification, esterification, acidolysis, alcoholysis and aminolysis (Aravindan et al. 2007). A unique characteristic of lipase is its ability to hydrolyse heterogeneous substrates and synthesise fatty acid esters in an aqueous and non-aqueous medium. Lipases also possess interfacial activation; that is, it is activated by the presence of an interface or there is a sharp increase in the activity of lipases when the substrates start to form an emulsion.

Though there are different types of lipases, their basic three-dimensional backbone remains the same (Fig. 6.4). They are characterised by α/β-hydrolase folding, which is a specific sequence of α-helix and β-sheets. Also, most of them contain a helical segment called as the lid which covers the active-site residues in a closed conformation. In case of true lipases the active site of the α/β-hydrolase folding has three catalytic residues, nucleophilic residues (serine, cysteine and aspartate), a catalytic acid residue (aspartate or glutamate) and a histidine residue (Andualema and Gessesse 2012). As a result, lipases are also known as serine hydrolases.

Lipolytic enzymes can be classified into eight different families based on the difference in the amino acid sequences within or around their active site (Ramnath et al. 2017). Family I comprises true lipases which should possess the property of interfacial activation and contain the lid structure. This family is further divided into six subfamilies and predominantly catalyses hydrolysis of the long acyl chains that are its substrates. The members of this family II have a modified GDS(L) motif that contains the active-site serine residue. Family III lipases contain a conserved GXSXG motif, with a canonical α/β-fold and a catalytic triad. Family IV lipases are similar to mammalian hormone-sensitive lipase (HSL), with a motif structure of His-Gly-Gly [HGG]. Family V lipases have a conserved motif structure of His-Gly-Gly-Gly [HGGG] upstream to the pentapeptide GDSAG motif sequence. Members of family VI, VII and VIII predominantly contain esterases.

Lipases are found abundantly in nature and are produced by microorganisms, plants and animals, but for most of the commercial applications, they are obtained from microbial sources. The lipase-producing microorganisms are found in habitats like industrial wastes, soil polluted with oilseeds and/or oil, vegetable-oil processing factories and dairy plants (Kiran et al. 2016). The common bacteria used in lipase production are *Bacillus pumilus*, *Bacillus coagulans*, *B. alcalophilus*, *B. subtilis*, *Bacillus psychrosaccharolyticus*, *B. stearothermophilus*, *Chromobacterium viscosum*, *Acinetobacter* sp., *Aeromonas hydrophila*, *Burkholderia cepacia*, *Burkholderia multivorans*, *Colwellia psychrerythraea*, *Desulfotalea psychrophile*, *Janibacter* sp., *Micrococcus roseus*, *Microbacterium phyllosphaerae*, *Corynebacterium paurametabolum*, *Microbacterium luteolum*, *Moritella* sp., *Pelagibacterium halotolerans*, *Photobacterium* sp., *Pseudoalteromonas* sp., *Pseudomonas* sp., *Psychrobacter* sp.,

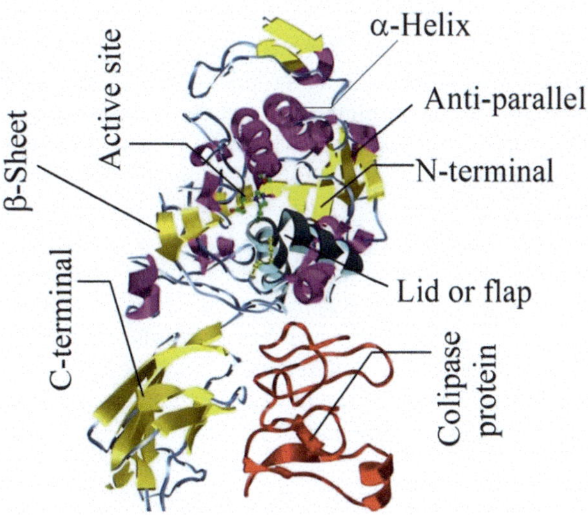

Fig. 6.4 Structure of lipase

Staphylococcus caseolyticus, *Staphylococcus xylosus*, *S. marcescens* and *Streptococcus lactis* (Shelatkar and Padalia 2016). Fungal species that produce lipases include *Rhizopus* sp., *Aspergillus* sp., *Penicillium* sp., *Mucor* sp., *Geotrichum* sp., *Rhizomucor* sp., *Colletotrichum gloeosporioides*, *S. griseus*, *C. antarctica*, *Candida albicans*, *Humicola lanuginosa*, *Rhizopus chinensis*, *Rhizopus homothallicus*, *Geotrichum candidum*, *Rhizopus oryzae*, *Candida utilis*, *Candida rugosa*, *Candida cylindracea*, *Fusarium solani*, *Rhizopus arrhizus* and *Trichosporon laibachii* (Kavitha 2016; Kiran et al. 2016; Shelatkar and Padalia 2016; Ugo et al. 2017). Though for most of the industrial applications bacterial or fungal sources are used, yeasts are also a potent source of lipase production. They include *Yarrowia lipolytica*, *Aureobasidium pullulans*, *S. cerevisiae*, *Williopsis californica* and *Rhodotorula mucilaginosa* (Kiran et al. 2016).

6.4.1 Commercial Applications of Lipases

6.4.1.1 Detergent Industry

The detergent industry is one of the commercially most important fields of application for microbial lipases, constituting about 32% of total lipase sales. Any detergent lipase should fulfil the following criteria: (1) should have a low substrate specificity, (2) should be stable under washing condition and (3) should exhibit resistance from the oxidising and chelating agents used, in the form of active oxygen bleach and builders (Salleh et al. 2006). Due to their ability to attack lipids, lipases are able to remove and decompose fatty stains and residues like butter, oil and sauces. This helps in increasing the efficiency of stain cleaning as well as fabric care. Use of enzymes in detergents also helps to improve its detergency, prevent scaling and reduce the time period, agitation and temperature that are needed for washing, therefore extending the lifespan of fabrics. Using lipases aids in reducing the quantity of surfactants used (Ugo et al. 2017).

The first commercial lipase was identified and isolated by Novo Nordisk, from *Humicola* strain. Later, several strategies were applied for increasing the enzyme efficiency, which lead to the production of the first commercial recombinant lipase, called as Lipolase which contained the lipase sequence of the fungus *T. lanuginosus*, expressed in *A. oryzae*. This paved a way for further exploration in the field of using lipases as a component in detergents. Alkaline lipases produced by *Bacillus* sp. B207, *Pseudomonas paucimobilis* and *Streptomyces* sp. can be used as additives during detergent formulation. Lipase from *Bacillus* sp. and *Pseudomonas* sp. shows an excellent stability at normal washing conditions (Salihu and Alam 2012). Bacterial lipase from *Staphylococcus arlettae* JPBW-1, found in rock salt mine, showed a good stability towards surfactants and oxidising agents, removing around 62% of olive oil from cotton fabrics. Alkaline lipase from *B. cepacia* RGP-10 exhibited a better stability towards commercial detergents and oxidising agents, as compared to Lipolase (Prakasan et al. 2016).

Nowadays, researchers are focusing on the use of cold-active lipases (CLPs) in detergents due to their ability to act under low-water and cold conditions, which results in decreased energy consumption as well as lesser wear and tear of cloth fibres. CLPs obtained from *Pseudoalteromonas* sp. NJ 70, *M. phyllosphaerae*, *Bacillus sphaericus* and *Pichia lynferdii* were shown to be effective for the same purpose (Kavitha 2016). Immobilisation of lipases onto various surfaces or carriers eases the oil removal from fabrics, alters the wettability of fabric surfaces, enhances stability to denaturation and prevents heat deactivation. Lipase obtained from *Pseudomonas putida* ATCC 53552 was adsorbed on fabric, forming a fabric-lipase complex, for the removal of oil stain (Hasan et al. 2006). Using lipases along with other oxidoreductases allows further decrease in the quantity of surfactants used, thereby making them more eco-friendly (Ugo et al. 2017).

6.4.1.2 Food Industry

Chocolates, toffees and other bakery goods contain cocoa butter, which is needed for giving them crystallising and melting characteristics. But cocoa butter is highly expensive. As an alternative, lipases from *M. miehei* and *C. antarctica* can be used to produce cocoa butter equivalences from palm olein and distillate from palm oil refinery (Hasan et al. 2006).

Lipases are mainly used to enhance the flavour content of bakery products due to their ability to liberate short-chain fatty acids by esterification, prolong the shelf life and render an improved texture and softness to the bakery goods. For example, artificially expressed lipases in *A. oryzae* and recombinant yeast with *Geotrichum* LIP 2 gene were found to reduce the initial firmness, increase specific volume of breads and give a more uniform crumb structure. Increased butter flavour can be produced through the lipolysis of butterfat. Lipase M 300 LF can be used to increase the volume of bread and bakery items. Lipases from *A. niger*, *R. oryzae* and *C. cylindracea* are used in the making of bakery products (Hasan et al. 2006). Phospholipases can substitute or supplement traditional emulsifiers as they degrade wheat lipids to produce emulsifying lipids in situ. It can be applied for egg yolk treatment to hydrolyse egg lecithin and iso-lecithin, to produce lyso-phospholipids. This helps to improve the emulsion and heat stability and can be useful in custard, mayonnaise, baby food, dressing and dough preparations (Aravindan et al. 2007; Singh et al. 2016a, b, c, d). Lipases like DSM Maxapal A2 can be cloned and expressed in *A. niger* for the same purpose.

In the fat and oil industry, lipases are mainly used for two main purposes, degumming of vegetable oils in refineries and production of oils and fats of a higher commercial as well as nutritional value from low-value substrates. Altering the location of fatty acid chains or replacing one or more fatty acids in a glyceride by esterification or interesterification reactions can result in modification of lipid properties. This may help to convert relatively inexpensive and less desirable fat to a higher value fat. For instance, lipolysis of salmon oil resulted in a 2.5-fold increase in its omega-3 polyunsaturated fatty acids (PUFA). Lipases from *C. rugosa*, *C. cylindracea*, *Mucor javanicus* and *A. niger* are used for the enzymatic hydrolysis of sardine oil to increase the content of omega-3 PUFAs by 10–35%

(Prakasan et al. 2016). Phospholipase can also be used for degumming purposes, which involves the removal of phospholipids. These are hydrolysed to a more water-soluble form, to facilitate washing out (Aravindan et al. 2007). Lipases can also be used to recycle and reuse the waste; for example, phospholipases can be used to recover additional oil from oil seeds (Hasan et al. 2006).

In dairies, lipases are mainly used to hydrolyse milk fats and render the characteristic flavour to cheese. These act as flavouring agents by production of short-chain fatty acids (C_4 and C_6) or medium- or large-chain fatty acids (C_{12} and C_{14}). The former gives a sharp and tangy mouthfeel and is preferred for cheese production, while the latter renders a soapy taste. The free fatty acids produced also take part in simple chemical reactions, which leads to production of flavour ingredients like acetoacetate, beta-keto acids, methyl ketones, flavour esters and lactones. Lipases produced by microorganisms like *Rhizomucor miehei*, *A. niger* and *A. oryzae* are used in cheesemaking (Hasan et al. 2006). For instance, extracellular lipase produced by *Cryptococcus flavescens* 39-A releases short-chain fatty acids from milk fat to produce a desirable taste during the mozzarella cheesemaking process. The lipolytic activity of *P. roqueforti* gives a characteristic flavour and appearance (i.e. a blue-green veined) to blue cheese. Lipases help to enhance the flavours of natural milk fats to produce volatile flavouring compounds (Prakasan et al. 2016). These are also involved in faster cheese preparation and production of customised milk products. Human milk fat substitute (HMFS) production also uses lipases and can be obtained by sn-1,3 lipase-catalysed acidolysis of tripalmitin, palm stearin, butterfat, palm oil or lard with free fatty acids. Nowadays, enzyme-modified cheese is being produced by incubating the cheese in the presence of enzymes at elevated temperature. This gives rise to concentrated flavour by lipase catalysis and can be used in other products, such as dips, sauces, soups and snacks. Interesterification and hydrogenation are used to prepare glyceride products for the production of butter and margarine (Hasan et al. 2006).

In the meat and fish processing industries, lipases can be used to develop flavour as well as for biolipolysis, i.e. removal of fats during meat and fish processing. Biolipolysis results in production of leaner and fat-free meat and fish (Hasan et al. 2006; Aravindan et al. 2007). Lipase produced by *R. miehei* is used in tea industries to promote enzymatic degradation of lipid membrane for enhanced characteristic flavour in black tea (Hasan et al. 2006). Lipases can act on wheat flour to produce variations in quality, and therefore, can be used to produce noodles and pasta, giving them an even and intense colour with reduction in stickiness when cooked (Prakasan et al. 2016). Immobilised lipases from *C. antarctica* and *Lactobacillus reuteri* are used to produce nutraceuticals. Regioselective hydrolysis of octaacetyl sucrose by lipases is used for the production of sucralose, an artificial sweetener (Aravindan et al. 2007). Since many of the micro- and macronutrients in food are heat sensitive in nature, the reactions that take place during food processing are preferred to be at low temperatures. Cold-active lipase from *P. fluorescens* P38, at low temperatures, synthesises a flavouring compound, butyl caprylate in *n*-heptane. Esterification of functionalised phenols to form antioxidants, to be used in sunflower oils, can be carried out with lipases from *C. antarctica* (CALB), *C. cylindracea* AY30, *Hansinuela lanuginosa*, *Pseudomonas* sp. and *G. candidum* (Kavitha 2016).

6.4.1.3 Medicine and Pharmaceutical Industry

The ability of lipases to hydrolyse its substrates in an enantioselective manner is exploited in pharmaceutical industries. By changing the production conditions of drugs, like temperature and pH, the enantiomeric value and stereo-preference of lipases can be altered, which makes them useful in the manufacturing of single isomers of chiral drugs by trans-esterification or direct esterification in non-aqueous medium and resolving of racemic mixtures of drugs (Momsia and Momsia 2013). *A. niger* lipase can be used to resolve ibuprofen and ketoprofen. Lipases obtained from *C. rugosa*, *C. antarctica* and *R. miehei* are efficient in resolving the enantiomers of ibuprofen, naprofen and suprofen (Ugo et al. 2017). Pure form of (s)-enantiomer of ibuprofen was synthesised using lipase as a biocatalyst (Aravindan et al. 2007). Lipase from *Pseudomonas* sp. AK kinetically resolved the chiral silane reagents for the synthesis of epothilone A, a potent antitumour agent (Prakasan et al. 2016). Lipases can also be used to hydrolyse polyester alcohols, which can be used as optically active intermediates in the synthesis of drugs. When 3-phenylglycidic acid ester is asymmetrically hydrolysed by *S. marcescens* lipase, an intermediate in the synthesis of diltiazem hydrochloride (a calcium antagonist used as a coronary vasodilator) is formed (Kiran et al. 2016). Lipase from *C. antarctica* or *Pseudomonas* sp. acts on stereospecific N-acylamines, leading to the formation of optically active amines that act as intermediates in the various pharmaceutical preparations (Kavitha 2016).

Lipases have also been found to be useful in the regioselective modifications of polyfunctional organic compounds and have been used to regioselectively modify castanospermine (drug used to treat AIDS) (Hasan et al. 2006). Some cold-active lipases such as CAL-B can be used to manufacture and segregate a number of nitrogenated compounds for the synthesis of pharmaceuticals (Kavitha 2016). Currently, lipases are used for kinetic resolution processes that are efficient for the preparation of optically active intermediates for synthesising of drugs such as nikkomycin-B and NSAIDs like ibuprofen, suprofen, ketoprox and naproxen, and lamivudine, an antiviral agent. Also, these can be used for the enantiospecific synthesis of a wide range of antibiotics, vitamins, alkaloids, antitumour, anti-arteriosclerotic and antiallergic agents (Aravindan et al. 2007).

Lipases obtained from *Bacillus* sp., *Candida lipolytica* and *A. oryzae* can be used to synthesise digestive aids, and therefore, can be used to treat gastrointestinal disturbances, dyspepsia and other gastrointestinal disorders. It can also be used in the treatment of malignant tumours as lipases are activators of tumour necrosis factor (Momsia and Momsia 2013; Pogaku et al. 2017). Lipase from *C. rugosa* and immobilised on a nylon scaffold produced lovastatin, a drug used to decrease the level of cholesterol in serum, by regioselective acylation of a diol lactone precursor with 2-methylbutyric acid, in organic solvents (Prakasan et al. 2016). Polyunsaturated fatty acids (PUFAs) obtained by using microbial lipases and immobilised lipases can be used as nutraceuticals due to their metabolic benefits (Kiran et al. 2016).

Lipases have also been used as a component of topical anti-obese creams and oral administration (Hasan et al. 2006).

Lipases can be used as diagnostic tools for the detection of diseases. Serum triglyceride levels can be determined by hydrolysis using lipases, where the enzyme hydrolyses triglycerides to generate glycerol, following which enzyme-linked colorimetric reactions can be carried out to quantitate the same. The level of blood serum lipases can be used as a means of diagnosis of conditions like acute pancreatitis or pancreatic injury (Hasan et al. 2006). Immobilisation of lipases onto pH/oxygen electrodes in combination with glucose oxidase can help to determine the triglyceride and blood cholesterol levels (Aravindan et al. 2007).

6.4.1.4 Cosmetics

Lipases play an important role in the production of aroma compounds and show surfactant activity. These properties can be exploited to produce cosmetics and perfumeries. In organic solvents, lipases catalyse esterification, thus forming esters which can be used to produce personal care products, such as skin creams, suntan creams and bath oils. These products have shown to be of a higher quality as compared to those formed by acid catalysis. Examples include the use of isopropyl myristate, isopropyl palmitate and 2-ethylhexyl palmitate to produce personal care products, by employing *Rhizomucor miehei* lipase as a biocatalyst. Lipases are also used as components in hair waving preparation. Lipases can also be used as additive in shampoo formulations to enhance the removal of excess oil from hair and reduce dandruff, thereby improving the overall hair quality (Salleh et al. 2006; Ugo et al. 2017).

Lipase from *C. antarctica* catalyses the trans-esterification between retinol and L-methyl lactate for synthesis of retinyl L-lactate and between ascorbic acid and L-methyl lactate to produce ascorbyl L-lactate. Both retinyl and ascorbic esters are of a great use in cosmetic production. These also mediate trans-esterification between olive oil and ascorbic acid, to synthesise ascorbyl oleate that can be used as antioxidant (Prakasan et al. 2016). Lipases are also used in synthesising natural dyes. For instance, lipases catalyse the release of indoxyl from isatan B, which can be combined with isatan C to produce indigo dye (Prakasan et al. 2016). Other commonly used microbes for lipases in the cosmetic industry are *A. oryzae*. Lipase produced by *C. cylindracea* is used in the commercial production of soap (Pogaku et al. 2017).

6.4.1.5 Textile Industry

The use of lipase in the textile industry serves two main purposes, desizing of cotton fabrics and denim, where lubricants from fabrics are removed, to provide a higher absorbency and a better levelness while dyeing and providing of a good finish to denim by reducing the frequency of streaks and cracks. In most of the cases, lipases and α-amylase are used in conjugation for desizing (Kiran et al. 2016). Lipases isolated from *C. antarctica* are commonly used in the textile industry (Pogaku et al. 2017). In textiles, polyester plays a key role in imparting fabrics high strength, soft hand, machine wash, stretch, strain abrasion and wrinkle resistance. Polyesterases can also be employed to promote polyester fabric to take up chemical compounds, such as cationics, fabric finishing compositions, dyes, antimicrobial compounds or deodorants. Lipases produced by *Pseudomonas* sp. can be used to degrade polymers of aliphatic polyethylene. An example of a commercial lipase used to improve the fibre texture without losing strength is JP 5344897 A (Hasan et al. 2006).

6.4.1.6 Leather Industry

Leather industry is one of the major contributors of environmental pollutants due to the chemicals they use and contributes to 80–90% of the total pollution. The process of degreasing or removal of fats is an essential part of manufacture as insufficient removal of natural fat prevents the chemicals from penetrating into the leather leading to a poor quality, in terms of internal softness, appearance (a stained appearance is seen due to chrome soap formation) and a bad odour (Kavitha 2016). Conventional methods of degreasing involve the application of organic solvents and surfactants which gives rise to environmental pollutants like volatile organic compound (VOC) emissions. Use of lipase is a more environment-friendly method for removing fats and grease from skins and hides. Lipase also acts in a more efficient manner by giving a uniform colour and a cleaner appearance, improving the hydrophobicity of leather, giving it a more waterproof character and a low-fogging stock (Hasan et al. 2006).

Since fat is present in sebaceous glands, hair follicles, and between collagen fibres and connective tissue fibres, to increase the process efficiency, alkaline stable proteases are employed. Proteolysis helps to emulsify natural fat by hydrolysing the membranes surrounding the fat cell and making the fat accessible for lipases to act upon. This increases the efficiency of degreasing (Ugo et al. 2017). Lipases stable at alkaline as well as acidic conditions can be used in skin and hide degreasing. For instance, acid lipase isolated from *Rhizopus nodosus* and a commercial degreaser were used for this process (Kavitha 2016). For degreasing of suede clothing leathers from sheepskins *R. nodosus* lipase can be used (Hasan et al. 2006). *B. subtilis* lipase was used in the degreasing process through an 8–12 h of incubation of leather and enzyme. This was also found to maintain the natural skin colour. Examples of commercially available lipases for use in leather industries include NovoLime, (a protease/lipase blend) and NovoCor AD (an acid lipase) (Prakasan et al. 2016).

6.4.1.7 Paper and Pulp Industry

Pulp and paper industries mainly use lipases for the process of depitching, a process by which the hydrophobic components of wood from the pulp are removed during papermaking. These hydrophobic components are called as pitch or resin stickies and include triglycerides and waxes. The presence of pitch may lead to holes and spots in the final paper product and also reduce the production rates. Lipases hydrolyse these hydrophobic components into glycerol and free fatty acid and enhance the pitch control. Since these products are water soluble, they can be easily washed away from the machines. Lipases also help to increase the rate of pulping, whiteness intensity of paper and equipment life, and reduce composite cost and chemical usage, thereby decreasing the pollution level (Ugo et al. 2017). Lipases obtained from *C. antarctica* are commonly used for depitching (Pogaku et al. 2017).

Immobilised lipase, isolated from *T. lanuginosus* on a resin coated with chitosan along with pectinase, resulted in a reduction of pitch deposits by 74%. Fungal lipase isolated from *C. rugosa* is used by Nippon Paper Industries, to hydrolyse wood triglycerides up to 90%. Alkaline lipases are used to efficiently remove pitch from recycled fibre pulping waste water (Prakasan et al. 2016). Lipases are also used for deinking wastepaper. Lipase from *Pseudomonas* sp. KWI-56 deink composition for ethylene oxide–propylene oxide adduct stearate resulted in reduction of residual ink spots on paper and improved its whiteness (Hasan et al. 2006).

6.4.1.8 Fine Chemical Synthesis

The ability of lipases to catalyse reactions such as alcoholysis, acidolysis, hydrolysis and glycerolysis is exploited in the oleochemical industry. When these reactions are carried out by conventional methods, they are energy intensive and require high temperature and pressure. However, use of lipases reduces the energy consumption. Currently, immobilised lipases are used to initiate these reactions using mixed substrates. This ensures a high productivity (Salihu and Alam 2012). In the polymer industry, lipases carry out polycondensation and ring-opening polymerisation of lactones and carbonates. CAL-B was used in the catalysis of a ring-opening polymerisation of epsilon-caprolactone in cellulose fibre surface (Pogaku et al. 2017).

Some cold-active lipases such as CAL-B can also be used to synthesise optically active alcohols. Lipase obtained from *P. fluorescens* P38 was used to synthesise butyl caprylate in the presence of *n*-heptane at low temperatures. Esterification of docosahexaenoic acid to form ethyl docosahexaenoate was done in an organic solvent-free system using CAL-B. Asymmetric synthesis of amino acids/amino esters was carried out using CAL-A that shows chemoselectivity for amine groups (Kavitha 2016). Another novel application of lipase is its use as a biocatalyst in the production of useful biodegradable compounds. For instance, lipases can be used to catalyse esterification of butanol and oleic acid to produce 1-butyl oleate, which is used in biodiesel to decrease its viscosity (Momsia and Momsia 2013). Using lipases, glucoside esters with surfactant properties are synthesised (Salleh et al. 2006). With the help of lipases, pesticides, insecticides, fungicides, herbicides and/or their precursors can also be produced. Lipases produced by *Achromobacter* sp. can be used to hydrolyse acetic acid ester of a racemic mixture of a-cyano-3-phenoxybenzyl alcohol (CPBA), in an enantioselective manner to give (S)-CPBA, which is an active stereoisomer of insecticide (Kuhad et al. 2011).

6.4.1.9 Environmental Applications

Biodiesel is produced by trans-esterification of triglycerides with methanol/ethanol. This serves as a renewable, biodegradable and nontoxic source of fuel. Lipases can convert oil, in the presence of alcohols, to short-chain alcohol esters by a single trans-esterification reaction (Salihu and Alam 2012). Therefore, lipases, immobilised within a biomass support particle, can be used as biocatalysts (Salleh et al. 2006). CAL-B can be immobilised and used for Jatropha biodiesel production, while lipase obtained from *Microbacterium* sp. can be immobilised to produce biodiesel from algal oil (Ugo et al. 2017). Immobilised *Pseudomonas cepacia* lipase carries out trans-esterification of soybean oil with methanol and ethanol. Using Novozym 435 trans-esterification of crude soybean oils can be carried out in a solvent-free medium, for the biodiesel production. Fatty acid esters can be produced from palm kernel oil and coconut oil, using PS30 lipase, by trans-esterification with different alcohols. In this, palm kernel oil conversion with ethanol resulted in highest conversion and was found to be 72% (Hasan et al. 2006). Methanolysis was observed in residual oil obtained from soybean, rapeseed and palm oil refining waste when it had been subjected to hexane by *R. oryzae* lipase. The use of palm oil resulted in highest conversion to methyl esters, with a yield of 55% (Salihu and Alam 2012). Biodiesel was synthesised from degummed soybean oil, using CAL-B (Kavitha 2016). In contrast to conventional methods that use acid or alkali catalyst for the production, lipase-catalysed esterification as well as trans-esterification of triacylglycerols yields cleaner and more environment-friendly by-products (Hasan et al. 2006).

Industrial effluents and domestic waste are high in chemical commodities such as phenols, aromatic amines and nitriles, which are toxic to the environment. Microbial enzyme alone or in combinations carries out the bioconversion of these toxic compounds to non-toxic or less toxic products. For instance, *A. oryzae* and *Candida tropicalis* lipases may be used to degrade crude oil hydrocarbons. Commercial mixtures containing lipase, cellulase, protease, amylase, inorganic nutrients, wheat bran, etc. are used to degrade organic debris present in sewage, holding and septic tanks, grease traps, etc. (Pogaku et al. 2017). Apart from chemicals, there may be presence of high amount of fats, mainly as triglycerides in waste water. The employment of bacterial as well as immobilised lipases for the breakdown of fats present in domestic sewage and wastewater treatment has become an alternative method to the conventional technique. For instance, during anaerobic digestion, lipases can be used to break down lipids into simpler low-molecular-weight components. During the activated sludge process, lipases are used for the removal of thin layers of fats that deposit on the surface of aerated tanks, which is required for efficient transport of oxygen in order to maintain the biomass (Salleh et al. 2006). Lipases isolated from organisms such as *C. rugosa* can be used to effectively break down fats and prevent fat blockage. Effluent and lipid-rich wastewater treatment in various industries can be carried out by employing lipases such as those obtained from *Pseudomonas aeruginosa* LP602 cells. Lipases such as LG-1000 can be used to hydrolyse and reduce the size of fat particles in slaughterhouse waste water (Hasan et al. 2006).

Bioremediation in oil-contaminated areas, biodegradation of petroleum and other hydrocarbons can also be carried out using lipases. For instance, the ability of *P. putida, Acinetobacter* sp. and *Rhodococcus* sp. to degrade *n*-alkanes, *P. putida* xylE to degrade aromatic hydrocarbons and *P. putida* ndoB and *Mycobacterium* sp. strain PYR-1 nidA to degrade polycyclic aromatic hydrocarbons was determined in 12 areas contaminated with petroleum hydrocarbons (Hasan et al. 2006). In cold conditions, these processes can be carried out using cold-adapted microorganisms that produce cold-active lipases. Lipase, isolated from *Acinetobacter* sp., can be used in bioremediation of oil-contaminated soil due to its ability to hydrolyse triglycerides at 4 °C (Kavitha 2016).

6.4.1.10 Biosensors

Immobilised lipases can be used as biosensors to accurately and efficiently quantitate the levels of triacylglycerol, which can serve to be of a great use in the food and pharmaceutical industries and for clinical diagnosis. This exploits the ability of lipase to generate glycerol from the triglycerides present in the analytical sample. The amount of glycerol released can be quantified by either chemical or enzymatic means. For instance, the amount of organophosphorous pesticides and dichlorvous residues in the root, stem and blades of Chinese cabbage can be quantified by lipase hydrolysis, using a surface acoustic wave impedance sensor (Aravindan et al. 2007).

To conclude, the demand for lipases and other enzymes has increased in the twenty-first century due to an enormous demand for green products. Microbes serve as a factory for production of enzymes in order to fulfil these demands. Lipases are currently being used in a wide range of industries, such as food processing industries like the dairy and beverage, pharmaceuticals, detergent making, textile, and cosmetics. The most recent synthetic routes used in the production of a wide range of compounds have been optimised with lipases in order to yield a better quality of chemicals, drugs and other products. Yet there is a need for developing a more cost-effective and efficient process for screening, isolation, purification and/or immobilisation of lipases, so as to fulfil the demands of various industries. Application of novel biotechnological process like genetic engineering, protein engineering, mutagenesis and other rDNA technologies can modify certain features of lipases such as the enzyme's adapting properties or quantitative improvements can be made so as to enhance and elevate the production and efficiency of lipases.

Currently, the use of lipases is encircled by limitations such as inadequate enantioselectivity and limiting activity of lipases, challenges in recycling of the enzyme and inherent limitations of the kinetic resolution due to a maximum possible conversion limit to only 50%, during practical applications. The studies of production of microbial lipases and the role of inducers in lipase production are scanty. Also, many genes of the enzymes that may be associated with some unique features remain unexplored. Understanding of the structure-functional relationship (of lipases) will enable researchers to tailor novel lipases. This opens up a scope for further research and makes this field fascinating for future. This may pave new ways to solve biotechnological bottlenecks that industries are currently facing and also may serve useful in solving the environmental problems. Table 6.3 presents the industrial applications of lipases.

6.5 Cellulases

Cellulases are the fourth most commercially important enzymes. They are hydrolases which hydrolyse the β-1,4-glycosidic linkages present in cellulose, a linear polysaccharide containing monomers of the glucose, into fermentable sugars and hence are also known as glycosyl hydroxylases. Cellulose in most cases is not present in pure form but is associated with hemicelluloses (20–35% plant dry weight) and lignin (5–30% of plant dry weight) (Zhang and Zhang 2013; Behera et al. 2017). Cellulases are inducible in nature, as the enzyme is synthesised when the microbes grow on cellulosic materials and form independently folding units that are structurally and functionally discrete, and are called as cellulase domains or modules (Behera et al. 2017). Based on the amino acid sequences and crystal structure of these domains or modules, cellulases are classified into various families (Golan 2011).

Cellulases produced by microorganisms can exist as either free enzymes, as in case of aerobic bacteria, or complexed with polysaccharide forming enzyme-substrate complex, called as cellulosome, which is located on the cell surface, as seen in anaerobic bacteria (Ducros et al. 1995). Cellulase is made of two domains, namely a non-catalytic cellulose-binding module that is located at the N- or C-terminus of a catalytic module and a catalytic domain, joined together by a short and flexible poly-linker (rich in serine and threonine residues) at the N-terminus (Behera et al. 2017) (Fig. 6.5). The crystal structure of cellulase catalytic domain has a resolution of approximately 2.4 Å. The catalytic centre located at the β-strands of C-terminal end,

Table 6.3 Industrial applications of lipases

Industry	Applications
Detergent industry	Increases the efficiency of stain cleaning and fabric care
Food industry	
Dairy industry	Hydrolyse milk fats, render the characteristic flavour and texture to cheese
Bakery and confectionary	Production of chocolates, toffees and some bakery goods Enhance the flavour content of bakery, renders improved texture and softness
Fat and oil industry	Degumming of vegetable oils in refineries and production of oils and fats of a higher commercial and nutritional value
Medicine and pharmaceutical industry	Manufacture of single isomers of chiral drugs and preparation of optically active intermediates for synthesising of drugs Synthesis of digestive aids
Textile industry	Desizing of cotton fabrics and denim, imparting high strength, soft hand, machine washability, stretch, strain abrasion and wrinkle resistance to fabrics
Paper and pulp industry	Depitching and deinking of paper
Fine chemical synthesis	Polycondensation and ring-opening polymerisation of lactones and carbonates Synthesis of optically active alcohols, used as a biocatalyst in the production of useful biodegradable compounds, precursor for the production of pesticides, insecticides, fungicides and herbicides
Cosmetic industry	Production of aroma compounds from esters which are used to produce personal care products Additive in shampoo formulations to enhance the removal of excess oil from hair and reduce dandruff Synthesis of natural dyes
Leather industry	Removal of fats and grease from skins and hides
Environmental applications	Trans-esterification to produce biodiesel Wastewater treatment

with Glu170 as the proton donor and Glu307 as the nucleophile and the aromatic residues, forms the substrate-binding site (Singh et al. 2016a, b, c, d).

There are a wide variety of cellulase-degrading enzymes that are found in nature but generally the complete degradation of cellulose to β-glucose occurs as a result of complex and synergistic interaction between cellulolytic enzymes. Initially, there is a synergistic action of endoglucanases and exoglucanases, which gives cellobiose or cellodextrin residues which are further hydrolysed to β-glucose by β-glucosidases. There are two major types of cellulases, endoglucanases and exoglucanases, including cellobiohydrolases and β-glucosidase (Ducros et al. 1995; Golan 2011). Endoglucanases cleave the β-1,4-glycosidic bonds of celluloses randomly and generate cellodextrins or cellobiose units. The structure of their active sites is cleft/grove shaped. Exoglucanases act on reducing or non-reducing ends of the celluloses in a progressive manner to yield cellobiose or glucose. A topological feature of their catalytic module is the presence of a tunnel structure (formed due to two surface loop structures), which covers either the entire or a part of the active site of exoglucanases and enables it to hydrolyse cellulose in a processive manner. Examples include *Clostridium jusui, Clostridium cellulovorans* and *Clostridium thermocellum*. β-Glucosidases do not act on insoluble cellulose but hydrolyse soluble cellodextrins and cellobiose residues to produce glucose. β-Glucosidases are characterised by a pocket-shaped active site. This enables them to bind to the non-reducing ends of the cellodextrin or cellobiose.

Cellulases are produced naturally by a wide range of microorganisms such as fungi, bacteria and actinomycetes, in the presence of cellulosic materials, like wood. They can be aerobic or anaerobic, and mesophilic or thermophilic in nature. Out of all the microorganisms, fungi account for around 80% of cellulases and are the major producers. Cellulases produced by aerobic fungi are preferred over others as they produce cellulases extracellularly (Behera et al. 2017; Sukumaran et al. 2005). Among fungi employed in the production of cellulases, *Actinomucor* sp., *Trichoderma reesei, Trichoderma longibrachiatum, Trichoderma harzianum, Trichoderma atroviride, Penicillium brasilianum, Penicillium occitanis, Penicillium decumbans, Penicillium funiculosum, P. janthinellum, Humicola insolens, Humicola grisea, A. niger, Aspergillus nidulans, A. oryzae, Aspergillus terreus, F. solani* and *Fusarium oxysporum* are the chief producers. Other cellulase-producing fungi include *Candida thermophilum, Melanocarpus albomyces, Neurospora crassa, Paecilomyces inflatus, Paecilomyces echinulatum, Coniophora puteana, Thermoascus aurantiacus, Mucor circinelloides, Lenzites trabea, Sporotrichum thermophile, Trametes versicolor* and *Agaricus arvensis* (Shah 2013; Behera et al. 2017).

Though fungi are the major producers of cellulases, some bacteria like *Clostridium acetobutylicum, C. cellulovorans, C. jusui, Clostridium stercorarium, C. thermocellum, Acinetobacter junii, Acinetobacter amitratus, Pseudomonas cellulose, Acidothermus cellulolyticus, B. subtilis, B. pumilus, B. amyloliquefaciens, B. licheniformis, Bacillus circulans, Bacillus flexus, Cytophaga* sp., *Bacteroides* sp., *Acetivibrio cellulolyticus, Butyrivibrio fibrisolvens* and *Cellvibrio gilvus* are

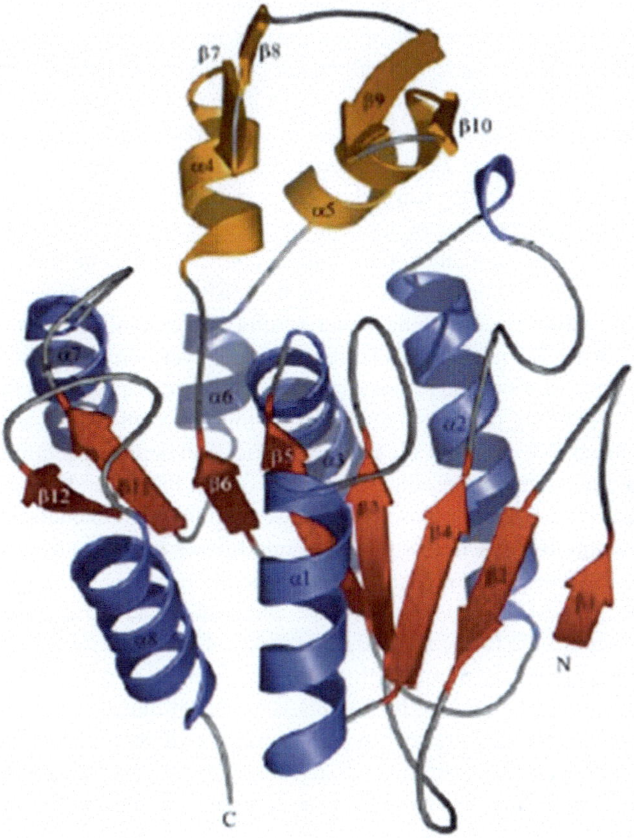

Fig. 6.5 Structure of cellulase

involved in the production of the cellulases (Shah 2013; Behera et al. 2017). Actinomycetes including *Streptomyces drozdowiczii*, *Streptomyces lividans*, *Cellulomonas fimi*, *Cellulomonas uda*, *Cellulomonas bioazotea*, *Thermomonospora fusca* and *Thermomonospora curvata* are actively involved in the production of cellulases (Shah 2013).

6.5.1 Commercial Applications of Cellulases

6.5.1.1 Textile Industry

Application of enzyme in textile processing industries began around 1850s and was mainly used to split starch during desizing of fabric. Nowadays, novel enzymes are being developed for the same purpose (Sajith et al. 2016). One of the most successful enzymes currently being used in the textile industry are cellulases. These are mainly being used for the purpose of biopolishing, carbonisation, stone wash, bio-finishing of cellulose-based fabrics and bio-stoning of jeans in the textile processing sectors. The overall effect of cellulase is an improved fabric appearance. Use of a cellulase-based approach will be advantageous as it is eco-friendly, less work intensive, saves energy, minimises water usage and lowers the fibre damage, resulting in an increased productivity. Thermostable cellulases are mainly obtained from *A. niger, H. insolens, Trichoderma viride, T. harzianum* and *Trichoderma reseei* (Sukumaran et al. 2005; Karmakar and Ray 2011).

Depending upon the purpose of application and the types of fabric, the type of cellulases, pH, temperature and other conditions can be varied. For instance, pure lyocell garments can be treated with acid cellulases, mixed-lyocell garments require the use of a neutral cellulases for obtaining better results while alkaline cellulases such as those obtained from *H. insolens* can be used to selectively break down the cellulose present in the fibre interiors and remove soil and dirt particles present in the inter-fibril spaces. Also, depending on the stage of processing, the content of various types of cellulases is varied. For example, cleaning of fabrics and depilling require the use of a mixture of cellulases whereas production of aged or soft fabrics and biopolishing need the use of either pure endoglucanase or cellulase mixtures rich in endoglucanase. The use of cellulases in the presence of proteases and lipases results in a better finish and appearance of the fabrics (Singh et al. 2016a, b, c, d).

During the process of bio-scouring (the removal of added as well as natural impurities present in the fabric surface), a mixture of cellulase and pectinase can be used, where the pectinaceous connections between cuticle structure and the body of the cotton fabric can be degraded by pectinase while cellulases penetrate the cuticle and degrade the cellulose present in the primary cell-wall structures that exist beneath the cotton cuticle (Singh et al. 2016a, b, c, d). Use of cellulases gives a lower degree of reduction in brightness as compared to other chemical methods (Sajith et al. 2016). Biopolishing is a stage in the wet processing of fabrics and includes desizing, scouring, bleaching, dyeing and finishing. Cellulase mixtures with a high endoglucanase content are the best suited for this stage. Here, cellulases act on the microfibrils that protrude from the surface of the fabric and aid in the enhancement of softness, hydrophilicity and moisture absorbance of fabrics, thereby giving a cleaner surface, enhanced look, feel, colour, brightness and a glossy appearance of fabrics and the reducing surface fuzziness and pill formation during processing (Sajith et al. 2016; Behera et al. 2017). For instance, the use of cellobiohydrolase I, obtained from *T. reesei*, can be used to reduce the crystallinity, thereby producing a more amorphous material having a higher water affinity and a reduced energy demand by up to 40% (Mojsov 2011).

At the stage of washing, cellulases assist in the removal of soil and other dirt particles from in between the microfibrils, while during the pretreatment of bast fibres, cellulases degrade the surface cellulosic materials, thereby facilitating other enzymes to act upon the other components present in the inner layers of the fibre (Sajith et al. 2016). In the process of bio-stoning or bio-stonewashing, cellulases such as those obtained from *H. insolens* and *Trichoderma* sp. can be employed as a substituent to pumice stone and can be used to remove the small fibrous ends or the dirt particles trapped within microfibrils present on the fabric surface, thereby giving a soft texture. Application of cellulases for denim finish minimises the utilisation of stone and the damage of denim. They help to loosen and remove the excess dye and facilitate the removal of mechanical abrasion during washing (Karmakar and Ray 2011; Behera et al. 2017).

6.5.1.2 Paper and Pulp Industry

The use of cellulases in the paper processing industry has considerably increased in the last decade. These are mainly employed for biomechanical pulping processes, including refining and grinding the woody raw materials, recycling the printed papers and deinking of different paper waste types, and as a co-additive in bleaching of pulp and/or drainage improvement (Singh et al. 2016a, b, c, d). Use of cellulase is an eco-friendly and energy-saving process, resulting in the production of paper pulps having a good bulk content and stiffness with improved hand-sheet strength, tensile strength and fibre brightness. During the deinking and toner removal processes, ink/toner is released from the paper surface by partially hydrolysing the carbohydrate residues present, which results in the breakdown of the fibrils and bundles, thereby lowering the affinity for the ink particles and also preventing the alkaline yellowing of the product. Use of cellulases leads to the removal of excess fibrils and other dissolved and colloidal matters on the fibre surface, thereby helping in the reduction of drainage. For instance, when cellulase isolated from *Trichoderma* sp. was used, a decrease in the turbidity of pulping filtrates was observed (Behera et al. 2017).

Out of all the cellulases, endoglucanases play a major role in the paper processing industries due to their ability to lower down the viscosity of pulp produced, at reduced levels of hydrolysis, as well as improve the bleachability of pulp. In certain cases, a mixture of endoglucanases and hemicellulases can also be used, which can further enhance the paper sheet density, drainage, runnability and beatability of pulp. For instance, the marginal modification of coarse mechanical pulp can be brought about by using cellulase and hemicellulase obtained from *Trichoderma* sp. Pergalase-A40, a commercial preparation of cellulase-hemicellulase mixture, can be used for the production of printing papers (Behera et al. 2017). Application of cellulases decreases defibrillation and reduces fibre coarseness on various fractions of douglas fir kraft pulp. Cellulases can also be used for the modification of cellulose fibres of kraft or sulphate pulp, which results in improved physical properties. They are being employed during the production of biodegradable cardboards, paper towels and other soft paper types and also during recycling of magazines, books and newspaper (Karmakar and Ray 2011; Zhang and Zhang 2013).

6.5.1.3 Food Industry

Nowadays, there is an increased demand for organic foods amongst consumers, mainly due to the ill effects of chemical additives on human health. The use of enzymes can help in the production of chemical-free food products that are much safer. Cellulases find a wide range of applications in food sectors as well (Ducros et al. 1995). During the production of fruit juices, its quality depends on the efficiency extraction, clarification and stabilisation methods used. Cellulosic enzymes make up an essential part of macerating enzyme complex, along with xylanases and pectinases. Maceration using enzymes enhances the efficiency of the above-mentioned processes, thereby improving the quality as well as the nutritive value of juices. Cellulases can be used to increase the yield and cloud stability, reduce the processing time and improve the texture of fruit and vegetable juices during the extraction and clarification process. They are used to decrease the viscosity of purees obtained from fruits including apricot, mango, papaya, peach, pear and plum.

Cellulases, mainly β-glucosidases, may aid to reduce the excessive bitterness present in citrus fruits, thereby improving its texture, flavour and aroma (volatile) characteristics. For instance, vacuum infusion of β-glucosidases into fruits serves as a potential method for the alteration of sensory properties such as aroma, flavour and texture. Cellulases are also used to extract phenolic compounds from grape pulp. In addition, cellulase may also be used to release antioxidants from fruit and vegetable pulp, to improve the yields of starch and proteinaceous substances during extraction, to enhance colour extraction of fruits and vegetables and to clarify fruit juices, in combination with pectinases, for the enhancement of nutritive value of fermented foods, homogeneous absorption of water by dried vegetables and cereals and low-calorie oligosaccharide production (Ducros et al. 1995). Cellulases isolated from *Aspergillus* sp., *Trichoderma* sp., *Bacillus* sp. and *Paenibacillus* sp. are primarily used to clarify juices (Sukumaran et al. 2005). Cellulases can also be used during the making of soy sauce, miso and other fermented soybean food products as they aid in the removal of external soybean coat, isolation of soybean and coconut proteins, and enhancement of the efficiency of starch isolation from potato and sweet potato, by enhancing the soaking efficacy and promoting the homogeneous water absorption (Ducros et al. 1995).

Cellulases are mainly used in bakery to render a good texture and quality in items such as bread. Carboxymethyl cellulases are mostly used for bread making due to its ability to aid in the hydrolysis of pentosans. Though pentoses are a minor part of wheat flour, because of its high water-holding capacity, they are a major determinant of dough rheology and quality of bread. The more the soluble pentosans are, the greater is the dough elasticity. Carboxymethyl cellulase has also shown significant farinographic and mixographic effects (Ducros et al. 1995). Cellulases promote the release of simple sugars, which makes them suitable to be used during the production of alcoholic beverages such as wine and beer. During wine production, cellulases are used to hydrolyse the polysaccharides present in the plant cell wall, which in turn enhance the colour extraction of grapes, skin maceration, quality, stability, clarification and aroma of wine. Macerating enzyme complexes containing cellulases also improve the pressability and juice yields of grapes during fermentation of wine (Zhang and Zhang 2013). Therefore, in the wine industry and breweries, they are added during the fermentation process in order to enhance the yield and quality of the end products. Endoglucanases can be included during either mashing or primary fermentation in order to promote the hydrolysis of glucan, which results in a decrease in the viscosity of wort, thereby enhancing its filterability. For instance, a decrease in the degree of polymerisation and wort viscosity due to the addition of endoglucanase and exoglucanase isolated from *Trichoderma* sp. was observed. They hydrolyse β-1,3- and β-1,4-glucans in low-grade barley that can be carried out using cellulases. β-Glucosidases also modify the glycosylated precursors and therefore, aid in improving the sensory properties of wine, such as aroma. Cellulosic enzymes during malting aid in improving aroma, yield and stability of liquor. Cellulases, along with hemicellulases and pectinases, are added during clarification and filtration of wine and aid to stabilise and improve its colour during extraction. This aids to improve the overall efficiency of the production process (Sukumaran et al. 2005; Karmakar and Ray 2011; Zhang and Zhang 2013). The performance of Cytolase 219, an enzyme mixture containing cellulase, pectinase and xylanase, during wine production was evaluated on three varieties of white grapes and reported an increase by 10–35% and 70–80%, in the first wine must and must filtration rate, respectively. It was observed that there is a 30–70% reduction in must viscosity, 20–40% energy saving during fermenter cooling and an enhancement in wine stability. Also, the pressing time was reduced by 50–120 min during wine production (Golan 2011).

The dietary fibres in animal feed are rich in polysaccharides such as cellulose, β-glucan and other oligosaccharides, which serve as antinutritional components for many of the animals. Cellulases, when added, eliminate these antinutritional components by degradation of certain cereal constituents, thereby improving the digestibility and nutritive value of monogastric animal feeds. For instance, β-glucanases, when added to animal feed, aid in the hydrolysis of β-glucan and decrease its intestinal viscosity, which in turn helps to release the nutrients and improve its digestibility and absorptivity. Cellulases obtained from *B. subtilis* are used to degrade the soya bean hull, thereby enhancing its nutritional value. Also, in the large intestine of animals, cellulases ferment the undigested cereal components and produce propionic acid, which due to its bacteriostatic effect prevents the colonisation by pathogenic bacteria. Addition of cellulases has also proven to improve the milk yield and body weight gain by ruminants (Sukumaran et al. 2005; Zhang and Zhang 2013).

6.5.1.4 Agriculture

Enzyme mixtures containing cellulases, hemicellulases and pectinases seek applications in the agriculture industry and can be employed to improve the growth of various crop varieties and soil quality. Soil supplemented with exogenous cellulase accelerated the decomposition of cellulose present in soil and straw decomposition, in turn increasing the soil fertility. As the result of this, the dependence on mineral fertilisers can be reduced. Some fungal cellulases and related cellulolytic enzymes, such as β-glucanases, control plant diseases, by disintegrating the cell wall of pathogens that enter the crops. Hence, they can be used as biocontrol agents. β-1,3-Glucanase isolated from *T. harzianum* CECT 2413, inhibits the growth of phytopathogenic *Rhizopus solani* and *Fusarium* sp. and hence is involved in controlling plant diseases. *Pythium* sp. is a plant pathogen that affects cucumber seedlings. Use of hypercellulolytic mutant isolated from *T. longibrachiatum* reduces the occurrence of the disease. The enzymes β-1,3-glucanase and *N*-acetylglucosaminidase isolated from *T. harzianum* P1 prevent the spore germination and elongation of germ tube of *Bacillus cinerea*, in a synergistic manner. Cellulases obtained from *Trichoderma* sp., *Gliocladium* sp., *Chaetomium* sp. and *Penicillium* sp. are useful in this industry due to their potential to improve seed germination, plant growth, flowering ability, root system and crop yields. Other cellulolytic enzymes used in the agriculture industry are mainly obtained from genera *Aspergillus*, *Chaetomium*, *Trichoderma* and actinomycetes (Zhang and Zhang 2013).

6.5.1.5 Carotenoid and Flavonoid Extraction

Carotenoids are organic pigments that include any class of pigments, such as carotene, that render colour to plant parts such as ripened tomatoes. Their desirable properties such as natural origin, negligible toxicity, ability to produce a range of colours, high versatility in terms of solubility and biological roles such as provitamin A activity, anticancer properties and role in lipid oxidation demand a higher rate of carotenoid production. Use of conventional methods such as solvent extraction can cause the dissociation of pigments from the proteins and render them water insoluble and cause their oxidation. In contrast, the enzyme-mediated extraction retains the natural and protein-attached status of these pigments, thus enhancing its stability and solubility. A mixture of cellulolytic enzymes and pectinases speeds up the hydrolysis rates and results in complete liquefaction. They disintegrate the cell-wall fruits and vegetables, such as orange peel, sweet potato and carrot, resulting in the release of carotenoids (Behera et al. 2017). Cellulases also assist in the extraction of the flavonoids from seeds and flowers. In contrast to conventional methods of extraction, such as acid or alkali treatment and/or solvent extraction, use of enzymes improves the yield and lowers the processing time and heat damage of the final product (Sukumaran et al. 2005).

6.5.1.6 Olive Oil Extraction

The current international market has an increasing demand for olive oil due to its high health benefits. The production of high-grade olive oil depends upon the quality of the substrates and fruits and the physical conditions provided during the extraction process. For instance, under cold-pressing conditions, a clean, freshly picked and slightly immature fruit produces a high quality of olive oil while at temperatures higher than ambient, a fully ripened fruit produces oil having high acidity and degree of rancidity and poor aroma. Cellulase and hemicellulase ensure an intense and quick cell-wall and membrane disintegration in olive fruits and favour the entry of substances such as polyphenols and aromatic precursors into the final products obtained, in turn strengthening the polyphenol extraction from olive fruits and causing a reduction of viscosity of olive. They also help to optimise the extraction process to a greater extent, generate a better centrifugal fractionation, decrease wastewater oil content and increase the overall efficiency of the process. For instance, Olivex, a commercial preparation containing pectinase, cellulase and hemicellulase, obtained from *Aspergillus aculeatus*, can be employed to enhance this process. Use of macerating enzymes also promotes the production of antioxidants and vitamin E in olive oil, which lowers down the degree of rancidity (Behera et al. 2017).

6.5.1.7 Detergent Industry

A recent addition to the application of cellulases is in the laundry and detergent industries. Cellulases are mostly used in conjugation with lipases and proteases in detergents. Their ability to remove small and fuzzy fibrils penetrating from the surface of fabric is exploited in this area of application, leading to an improved performance of detergents; modification of the cellulosic fibrils, thereby improving the colour, brightness and texture; enhancement of the quality of fabric; restoration of fabric softness; and removal of the dirt present in cotton garments. Alkaline cellulases obtained from *H. insolens* or *H. grisea* are used as additives in detergents to remove soil particles present in interfibrillar spaces due to their ability to specifically and selectively interact with cellulose in the fibre interiors, in the presence of other conventional ingredients. In order to enhance the stability of cellulases in liquid detergents, anionic or non-ionic surfactants, citric acid or water-soluble salt, proteolytic enzymes and mixtures of propanediol and boric acid or its derivatives can be used (Zhang and Zhang 2013; Behera et al. 2017). Currently, detergents use endoglucanases and its variants, particularly those isolated from *T. reesei*, *T. viride*, *T. harzianum* and *A. niger* as additives. The detergent industry needs alternative cellulase preparations as cellulases currently used are considered expensive, though they constitute only around 0.4% of the total detergent cost (Shah 2013).

6.5.1.8 Medicine and Pharmaceutical Industry

In the field of medicine, fungal cellulases can be used as a cure for phytobezoars and trapped masses of indigestible plant components in the gastrointestinal tract and aid in the hydrolysis of cellulose, hemicellulose and β-glucan polymers present in food. For instance, digestin, a digestive enzyme product, has cellulase and helps in improving digestion. Bacterial cellulases are potential in curing blinding keratitis and granulomatous amoebic encephalitis due to their cell wall-degrading action against pathogens, like *Acanthamoeba* sp. They also serve as antibiofilm agents due to their ability to degrade the cellulosic constituents of biofilm, which in turn aids in restricting pathogen's distribution within the body and increases the accessibility of drugs to them (Sukumaran et al. 2005; Zhang and Zhang 2013).

6.5.1.9 Biofuel

In recent years, one of the major environmental concerns has been global warming due to an increased level of greenhouse gases. These are aggravated in response to continuous and increased use of fossil fuel. Continuous use will also lead to the depletion of fossil fuel. Hence, there is a need for an alternative fuel source and biofuels, especially bioethanol from renewable resources, serve as a potential candidate for the same. Currently, enzymatic saccharification for the bioconversion of lignocellulose wastes such as agricultural residues like sugarcane bagasse, rice straw, wheat straw, corncob and *Prosopis juliflora* and forestry residues for the production of higher value products such as biofuel is being investigated.

In general, degradation of lignocellulose biomass is a costly process and includes three steps, physiochemical pretreatment, enzymatic hydrolysis and fermentation.

During delignification, the lignocellulose is degraded and cellulose and hemicellulose are released. These renewable cellulosic materials are hydrolysed by cellulases, following which these can be fermented to produce simple and free sugars, like glucose, which serve as substrates for bioethanol production. For instance, cellulases obtained from *C. thermocellum* hydrolyse cellulose and aid in the fermentation of sugar to yield ethanol. Cellulases isolated from a variety of filamentous fungi such as *Aspergillus* sp., *Trichoderma* sp. and *Penicillium* sp. can be exploited for the production of biofuel. As compared to chemical processes, such as the use of alkali or acids, use of enzymatic methods for the same serves to be a specific, low-energy-intensive, natural and an eco-friendly process and hence is preferred. Use of cellulases also reduces the cost by 40% (Sukumaran et al. 2005; Karmakar and Ray 2011).

6.5.1.10 Environmental Applications

As cellulases are one of the major components in plant biomass, wastes, such as those generated from agricultural fields, agro-industries and forestry, are rich in unutilised cellulose and are one of the causes of environmental pollution. Cellulases can be used to produce valuable products such as biofuels, alcohols, enzymes,

Table 6.4 Commercial applications of cellulases

Industry	Applications
Textile industry	Bio-souring of fabrics Bio-stoning of jeans
Paper and pulp industry	Biomechanical pulping Deinking and recycling of printed papers
Food industry	
Food processing	Enhance colour extraction of fruits and vegetablesClarify fruit juices, in combination with pectinases Enhance the viscosity of fruit purees Release antioxidants from fruit and vegetable pulp
Bakery	To render a good texture and quality in bakery items such as bread
Wine and beverage industry	Enhancing the process of malting, mashing, pressing and colour extraction of grapes Enhance the primary fermentation, beer quality, viscosity stability and filterability of wort
Animal feed	Improve the digestibility
Agriculture	Improve soil quality and decrease dependence on mineral fertilisers Inhibit the growth of plant pathogens
Detergent industry	Remove the small fibres that extend out and improve the colour, brightness and texture of fabrics Enhance the fabric quality and restore of fabric softness Remove dirt Enhance performance of detergents
Biofuel	Convert renewable cellulosic materials into glucose and fermentable sugars, which in turn serve as substrates for biofuel production
Environmental applications	Produce value-added products from agricultural and forestry wastes Degrade celluloses present in waste water

Table 6.5 Commercial applications of other enzymes

S. no.	Enzyme name	Microbial sources	Industry	Application
1.	Pectinase	*Aspergillus niger* *Penicillium frequentans* *Sclerotium rolfsii* *Rhizoctonia solani* *Mucor pusillus* *Bacillus polymyxa* *Bacillus pumilus* *Bacillus stearothermophilus* *Bacillus subtilis* *Xanthomonas campestris* *Penicillium italicum* *Aspergillus carbonarius* *Streptomyces lydicus* *Aspergillus giganteus* *Aspergillus kawachii* *Aspergillus niger* *Fusarium moniliforme* *Mucor flavus* *Penicillium frequentans* *Rhizopus oryzae* *Thermoascus aurantiacus*	Textile industry	1. Bio-scouring and bio-polishing of fabrics in conjunction with other enzymes such as amylases, lipases, cellulases and hemicellulases 2. Removal of sizing agents from cotton fabrics and cotton softening
			Food industry	To increase the efficiency of stain removal and fabric care 1. Removal of mucilage coat from coffee beans; accelerate the fermentation and improve the foaming ability of tea 2. Depectinisation and clarification of juices
			Animal feed	1. A component of the enzyme cocktail; aids in decreasing its viscosity and improves the nutrient absorption
			Waste water treatment	1. Pretreatment of wastewater high in pectin
			Paper and pulp industry	1. Depolymerises pectin and helps in bleaching
2.	Phytase	*Aspergillus niger* *E. coli* *Bacillus* sp. *Xanthomonas oryzae*	Animal feed	1. Helps in the utilisation of natural phosphorous-bound phytic acid in cereals
			Food industry	1. Degrades phytate present in food and increases the absorption of dietary minerals 2. In bread making, reduces phytate level and enhances the softness and crumb texture
3.	Papain	Expressed as recombinant protein in *E. coli* *Saccharomyces cerevisiae* *Pichia pastoris*	Medicine	1. Analgesic and anti-inflammatory against acute allergic sinusitis 2. Wound healing 3. Additive in toothpaste and mouthwash to enhance the whitening and removal of plaque

(continued)

Table 6.5 (continued)

S. no.	Enzyme name	Microbial sources	Industry	Application
4.	Bromelain	Expressed as recombinant protein in *E. coli*	Food industry In bakery	1. In bakeries to enhance dough relaxation and even raising of dough 2. Meat tenderisation 3. Inhibits the phenol oxidation and prevents browning of fruits
			Textile industry	1. Removes scale and impurities from wool and silk fibres; enhances dyeing
			Cosmetics	1. Tooth whitening and removal of stains, plaque and debris 2. Used in treatment of acne, wrinkles, dry skin and post-injection bruising and swelling
5.	Mannanase	*Acinetobacter* sp. *Bacillus amyloliquefaciens* *Bacillus circulans* *Bacillus subtilis* *Cellulosimicrobium* sp. *Chryseobacterium indologenes* *Klebsiella oxytoca* *Paenibacillus* sp. *Aspergillus niger* *Aspergillus flavus* *Aspergillus oryzae* *Trichoderma reesi* *Streptomyces* sp. *Penicillium occitanis* *Scopulariopsis candida*	Paper and pulp industry	1. Degrades glucomannan to improve paper brightness 2. Used in bio-bleaching
			Detergent industry	1. Aids in hydrolysis of mannan and boosts stain removal
			Food industry	1. Hydrolyses the mannan present in coffee extract and reduces its viscosity 2. Prevents proteolysis during fish processing
			Animal feed	1. Hydrolyses the mannan and release of encapsulated nutrients 2. Enhances the adsorption and decreases the digestion viscosity
6.	Xylanase	*Aspergillus* sp. *Bacillus* sp. *Bacillus licheniformis* *Aspergillus niger*	Animal feed Paper and pulp	1. Enhances digestibility of starch and depolarises xylans for better digestion 2. Enhances bleaching and augments the value of pulp
			Fuel	1. Production of bioethanol from lignocellulose

(continued)

Table 6.5 (continued)

S. no.	Enzyme name	Microbial sources	Industry	Application
7.	Laccase	*Bacillus subtilis* *Trametes hirsute* *Trametes versicolor* *Pseudomonas aeruginosa*	Paper and pulp	1. Used in bleaching and delignification.
			Cosmetics	1. Used in the preparation of hair dye and pigments
			Textiles	1. Production of textile dyes
			Food	1. Flavour enhancer
			Waste management	1. Degrades waste containing olefin unit, polyurethane and phenolic compounds
			Therapeutics	1. Detoxification
8	Catalase	*Streptomyces* sp. *Corynebacterium glutamicum*	Dairy	1. Cheese processing
			Therapeutics	1. As antioxidants
9.	Gluco-oxidase	*Aspergillus niger* *Penicillium chrysogenum*	Bakeries	1. Dough strengthening
			Beverages	1. Oxygen removal from beer
10.	Isomerase	*Corynebacterium* sp.	Food industry	1. Production of high-fructose corn syrup
11.	Superoxide dismutase	*Corynebacterium* sp. *Lactobacillus* sp.	Cosmetics	1. Free radical scavenging and prevents skin damage
			Therapeutics	1. Used in sunscreen lotions 2. Anti-inflammatory and antioxidant.
12.	Peroxidase	*Corynebacterium glutamicum*	Therapeutics	1. Antioxidant
			Cosmetics	1. Free radical scavenging to reduce erythema; used in sunscreen lotions 2. Hair dyeing
			Textile	1. Bio-bleaching
	Manganese peroxidase and lignin peroxidase	*Phanerochaete chrysosporium* *Coprinus cinereus*	Waste management	1. Degradation of phenolic compounds
	Lacto peroxidase	*Lactobacillus* sp.	Dairy	1. Enhances shelf life of dairy products
	Glutathione peroxidase	*Corynebacterium glutamicum*	Therapeutics	1. Antioxidant

sugars, enhanced animal feeds and human nutrients and can be used as cheap energy sources for fermentation, by supplementation of cellulase and other enzyme mixtures, thereby reducing the environmental load (Mojsov 2011). The waste water generated from industries such as the paper and pulp industry can be treated with cellulases in order to break down the celluloses present (Ducros et al. 1995).

To conclude, cellulases are currently being used in a variety of industrial contexts, such as in textile industry, paper and pulp industry, food and feed industry, pharmaceutical industry, detergent and laundry industry, research and development, waste management and biofuel generation. Cellulosic enzymes in combination with hemicellulases and pectinases serve as a potential tool for research in the areas of plant biology. Cellulases along with other suitable enzyme mixtures can be used for the purpose of controlling plant disease or to generate protoplasts by degrading the cell wall (Mojsov 2011). Cellulases can be exploited for the development of analytical and purification processes. Cellulose-binding domains can be used as an affinity tag for protein purification. Cellobiohydrolase I gene has a strong promoter which can be used to trigger gene expression at high levels. Using cellobiohydrolase I promoter, heterologous proteins, enzymes and antibodies can be expressed. For example, chymosin, glucoamylase, phytase, acid phosphatase, lignin peroxidase, laccase, endochitinase, antibody Fab fragment, single-chain antibodies, IL-6 and human lysozyme can be expressed in *T. reesei* (Zhang and Zhang 2013). Though production of cellulases occurs at an enormous rate, more cost-effective and efficient process for screening, isolation, purification and/or immobilisation is needed and optimisation of the same is also required, so as to fulfil the demands of various industries. Currently, one of the major drawbacks of cellulase production is the low yield and high production cost. Using biotechnological tools, improvements can be made in the efficiency of production, at a competitive cost. Table 6.4 displays various commercial applications of cellulases.

6.6 Other Enzymes

In addition to the above-discussed major enzymes, several other enzymes are also utilised in industrial applications, which are summarised in Table 6.5 (Lei et al. 2007; Pedrolli et al. 2009; Amri and Mamboya 2012; Chauhan et al. 2012; Jegannathan and Nielson 2012; Sharma et al. 2012; Arshad et al. 2014; Tapre and Jain 2014; Dahiya 2016; Kumar et al. 2017). Pectinases are used widely in food and textile processing. Xylanases are used in the paper industry and in fuel production. Laccases are used in the paper industry and as therapeutics. Catalase and peroxidases are used as antioxidants.

6.7 Conclusion

Microbes are an inexhaustible source of enzymes which have numerous advantages with respect to their use in industrial applications, as compared to conventional methods (using chemicals). They are good catalysts, increase the rate of reactions and work optimally under given environmental conditions and scaling up of the production process is possible by genetic manipulation. The global market for the production of microbial enzymes is a billion-dollar industry and is ever-increasing. Recombinant DNA technology and protein engineering open up the possibilities of obtaining novel products. Industries utilising microbial enzymes are food, pharmaceuticals, detergents, leather, waste management and many others. Some applications require the use of only a single type of enzyme preparation, whereas other applications require a mixture of enzymes for optimal efficiency. Care should be taken that all components of the enzyme mixture are compatible with each other and there is no formation of undesired products. Microbial enzymes have the potential to meet the demands of the ever-increasing population as well as reduce factors contributing to environmental hazards, thus making it a green method for building a sustainable future.

References

Adrio JL, Demain AL (2014) Microbial enzymes: tools for biotechnological processes. Biomol Ther 4:117–139. https://doi.org/10.3390/biom4010117

Afroz QM, Khan KA, Ahmed P, Uprit S (2015) Enzymes used in dairy industries. Int J Appl Res 1(10):523–527

Al-Abdalall AH, Al-Khaldi EM (2016) Recovery of silver from used X-ray film using alkaline protease from Bacillus subtilis sub sp. subtilis. Afr J Biotechnol 15(26):1413–1416

Ammayappan L (2013) Application of enzyme on woolen products for its value addition: a review. J Text Apparel Technol Manage 8(3):1–12

Amri E, Mamboya F (2012) Papain, a plant enzyme of biological importance: a review. Am J Biochem Biotechnol 8(2):99–104. https://doi.org/10.3844/ajbbsp.2012.99.104

Andualema B, Gessesse A (2012) Microbial lipases and their industrial applications. Biotechnology 11(3):100–118

Araujo R, Casal M, Cavaco-Paulo A (2008) Application of enzymes for textile fibres processing. Biocatal Biotransformation 26(5):332–349

Aravindan R, Anbumathi P, Viruthagiri T (2007) Lipase applications in food industry. Indian J Biotechnol 6:141–158

Aravindhan R, Monika V, Balamurugan K, Subramanian V, Rao JR, Thanikaivelan P (2017) Highly clean and efficient enzymatic dehairing in green solvents. J Clean Prod 140:1578–1586

Arshad ZIM, Amid A, Yusof F, Jaswir I, Ahmad K, Loke SP (2014) Bromelain: an overview of industrial application and purification strategies. Appl Microbiol Biotechnol 98:7283–7297. https://doi.org/10.1007/s00253-014-5889-y

Atanasova J, Moncheva P, Ivanova I (2014) Proteolytic and antimicrobial activity of lactic acid bacteria grown in goat milk. Biotechnol Biotechnol Equip 28(6):1073–1078

Bajpai PK (2010) Solving the problems of recycled fiber processing with enzymes. Bioresources 5(2):1311–1325

Behera BC, Sethi BK, Mishra RR, Dutta SK, Thatoi HN (2017) Microbial cellulases—diversity & biotechnology with reference to mangrove environment: a review. J Genet Eng Biotechnol 15:197–210

Chanalia P, Gandhi D, Singh JD (2011) Applications of microbial proteases in pharmaceutical industry: an over view. Rev Med Microbial 22(4):96–101

Chauhan PS, Puri N, Sharma P, Gupta N (2012) Mannanases: microbial sources, production, properties and potential biotechnological applications. Appl Microbiol Biotechnol 93:1817–1830. https://doi.org/10.1007/s00253-012-3887-5

Cho SA, Cho JC, Han SH (2007) Cosmetic composition containing enzyme and amino acid. Amorepacific Corporation (11/990):431
Choudhary RB, Jana AK, Jha MK (2004) Enzyme technology applications in leather processing. Indian J Chem Technol 11:659–671
Craik CS, Page MJ, Madison EL (2011) Proteases as therapeutics. Biochem J 435(1):1–16. https://doi.org/10.1042/BJ20100965
Dahiya S (2016) Industrial applications of phytases. Int J Appl Res 2(2):95–98
de Souza PM, Magalhaes PO (2010) Application of microbial alpha-amylases in industry: a review. Braz J Microbiol 41:850. https://doi.org/10.1590/S1517-83822010000400004
Ducros V, Czizek V, Belaich A, Gaudin C, Fierobe HP, Belaich JP, Davies GJ, Haser R (1995) Crystal structure of the catalytic domain of cellulose. Structure 3(9):939–949
Dutt D, Tyagi CH, Singh RP, Kumar A (2012) Effect of enzyme concoctions on fiber surface roughness and deinking efficiency of sorted office paper. Cellul Chem Technol 46:611–623
Freddi G, Mossotti R, Innocenti R (2003) Degumming of silk fabrics with several proteases. J Biotechnol 106:101–112
Golan AE (2011) Cellulase: types and action, mechanism and uses; biotechnology in agriculture, industry and medicine. Nova Science, New York, p c2011
Gopinath SC, Anbu P, Arshad MK, Lakshmipriya T, Voon CH, Hashim U, Chinni SV (2017) Biotechnological processes in microbial amylase production. Biomed Res Int 1272193:9. https://doi.org/10.1155/2017/1272193
Guo JS, Xu YF (2011) Review of enzymatic sludge hydrolysis. J Bioremed Biodegr 2:130. https://doi.org/10.4172/2155-6199.1000130
Gupta R, Beg QK, Lorenz P (2002) Bacterial alkaline proteases: molecular approaches and industrial applications. Appl Microbiol Biotechnol 59:15–32. https://doi.org/10.1007/s00253-002-0975-y
Gupta R, Gigas P, Mohapatra H, Goswani VK, Chauhan B (2003) Microbial alpha-amylases: a biotechnological perspective. Process Biochem 38:1599–1616
Hasan F, Sha AA, Hameed A (2006) Industrial applications of microbial lipases. Enzyme Microb Technol 39:235–251
Heredia-Sandoval NG, Valencia-Tapia MY, de la Barca AMC, Islas-Rubio AR (2016) Microbial proteases in baked goods: modification of gluten and effects on immunogenicity and product quality. Foods 5:59. https://doi.org/10.3390/foods5030059
Ismail B, Nielsen SS (2010) Invited review: plasmin protease in milk: current knowledge and relevance to dairy industry. Dairy Sci 93:4999–5009. https://doi.org/10.3168/jds.2010-3122
Jegannathan KR, Nielson PH (2012) Environmental assessment of enzyme use in industrial production—a literature review. J Clean Prod 42:228–240
Jisha VN, Smitha RB, Pradeep S, Sreedevi S, Unni KN, Sajith S, Priji P, Josh MS, Benjamin S (2013) Versatility of microbial proteases. Adv Enzyme Res 1:39–51. https://doi.org/10.4236/aer.2013.13005
Karmakar M, Ray RR (2011) Current trends in research and application of microbial Cellulases. Res J Microbiol 6(1):41–53
Kavitha M (2016) Cold active lipases—an update. Front Life Sci 9:226. https://doi.org/10.1080/21553769.2016.1209134
Kiran S, Arshad Z, Nosheen S, Kamal S, Gulzar T, Majeed MS, Jannat M, Rafique MA (2016) Microbial lipases: production and applications: a review. J Biochem Biotechnol Biomater 1(2):7–20
Kuhad RC, Gupta R, Singh A (2011) Microbial Cellulases and their industrial applications. Enzyme Res 2011:280696. https://doi.org/10.4061/2011/280696
Kumar D, Savitri, Thakur N, Verma R, Bhalla TC (2008) Microbial proteases and applications as laundry detergent additive. Res J Microbiol 3(12):661–672
Kumar D, Kumar SS, Kumar J, Kumar O, Mishra SV, Kumar R, Malyan SK (2017) Xylanases and their industrial applications: a review. Biochem Cell Arch 17(1):353–360

Kyaw N, de Mesquita RF, Kameda E, Neto JC, Langone MA (2010) Characterization of commercial amylases for the removal of filter cake on petroleum wells. Appl Biochem Biotechnol 161:171–180. https://doi.org/10.1007/s12010-009-8773-y

Lei XG, Porres JM, Mullaney EJ, Brinch-Pedersen H (2007) Phytase: source, structure and application. In: Polaina J, MacCabe AP (eds) Industrial enzymes. Springer, Dordrecht, pp 505–529. https://doi.org/10.1007/1-4020-5377-0_29

Li Q, Yi L, Marek P, Iverson BL (2013) Commercial proteases: present and future. FEBS Lett 587:1155–1163

Lopez-Otín C, Bond JS (2008) Proteases: multifunctional enzymes in life and disease. J Biol Chem 283:30433–30437

Mane P, Tale V (2015) Overview of microbial therapeutic enzymes. Int J Curr Microbiol App Sci 4(4):17–26

Mathias TRS, Paula A, de Almeida e Silva JB, de Mello PPM, Eliana S (2017) Brewery waste reuse for protease production by lactic acid fermentation. Food Technol Biotechnol 55:218–224. https://doi.org/10.17113/tb.55.02.17.4378

Mienda BS, Yahya A, Galadima IA, Shamsir MS (2014) An overview of microbial proteases for industrial applications. Res J Pharm Biol Chem Sci 5(1):388–396

Miguel ASM, Martins-Meyer TS, da Costa Figueiredo EV, Lobo BWP, Dellamora-Ortizet GM (2013) Enzymes in bakery: current and future trends. Food Industry Innocenzo Muzzalupo, Intech Open. https://doi.org/10.5772/53168

Mojsov K (2011) Application of enzymes in the textile industry: a review. In: II International congress “Engineering, ecology and materials in the processing industry”, Bosnia and Herzegovina

Mojsov K (2012) Microbial alpha-amylases and their industrial applications: a review. Int J Manage IT Eng 2(9):272–295

Momsia T, Momsia P (2013) A review on microbial lipase-versatile tool for industrial applications. Int J Life Sci Biotechnol Pharma Res 2(4):1–16

Mótyán JA, Tóth MF, Tőzsér J (2013) Research applications of proteolytic enzymes in molecular biology. Biomol Ther 3:923–942. https://doi.org/10.3390/biom3040923

Pandi A, Ramalingam S, Rao JR, Kamini NR, Gowthaman MK (2016) Inexpensive alpha amylase production and application for fiber splitting in leather processing. Royal Soc Chem 6:33170. https://doi.org/10.1039/c5ra26373e.

Parpalliwar JP, Patil PS, Patil ID, Deshannavar UB (2015) Extraction of silver from waste X-ray films using protease enzyme. Int J Adv Biotechnol Res 6(2):220–226

Pedrolli DB, Monteiro AC, Gomes E, Carmona EC (2009) Pectin and pectinases: production, characterization and industrial application of microbial pectinolytic enzymes. Open Biotechnol J 3:9–18

Pogaku P, Bhavani P, Manjusha T, Saritha K (2017) Lipases in environmental management. Int J Curr Res Aca Rev 5(4):106–109. https://doi.org/10.20546/ijcrar.2017.504.016

Porta R, Pandey A, Rosell CM (2010) Enzymes as additives or processing aids in food biotechnology. Enzyme Res 2010:436859. https://doi.org/10.4061/2010/436859

Prakasan P, Sreedharan S, Faisal PA, Benjamin S (2016) Microbial lipases-properties and applications. J Microbiol Biotechnol Food Sci 6(2):799–807. https://doi.org/10.15414/jmbfs.2016.6.2.799-807

Ramnath L, Sithole B, Govinden R (2017) Classification of lipolytic enzymes and their biotechnological applications in the pulping industry. Can J Microbiol 63:179–192. https://doi.org/10.1139/cjm-2016-0447

Rana N, Walia A, Gaur A (2013) Alpha amylases from microbial sources and its potential applications in various industries. Natl Acad Sci Lett 36:9–17. https://doi.org/10.1007/s40009-012-0104-0

Rawlings ND, Barrett AJ, Bateman A (2011) Asparagine peptide lyases, a seventh catalytic type of proteolytic enzymes. J Biol Chem 286:38321–38328. https://doi.org/10.1074/jbc.M111.260026

Roman HJ, Burgess JE, Pletschk BI (2006) Enzyme treatment to decrease solids and improve digestion of primary sewage sludge. Afr J Biotechnol 5(10):963–967

Saini R, Saini HS, Dahiya A (2017) Amylases: characteristics and industrial applications. J Pharmacogn Phytochem 6(4):1865–1871

Sajith S, Priji P, Sreedevi S, Benjamin S (2016) An overview on fungal cellulases with an industrial perspective. J Nutr Food Sci 6:461. https://doi.org/10.4172/2155-9600.1000461

Salihu A, Alam MZ (2012) Production and applications of microbial lipases: a review. Sci Res Essays 7(30):2667–2677. https://doi.org/10.5897/SRE11.2023.

Salleh AB, Rahman RNZRA, Basri M (2006) New lipases and proteases. Nova Science Publishers, Inc., Hauppauge, NY. ISBN: 978-1-60876-518-8

Sarrouh B, Santos TM, Miyoshi A, Dias R, Azevedo V (2012) Up-to-date insight on industrial enzymes applications and global market. J Bioprocess Biotechniq 10:4–002 https://doi.org/10.4172/2155-9821.S4-002

Sawant R, Nagendran S (2014) Protease: an enzyme with multiple industrial applications. World J Pharm Pharm Sci 3:568–579

Shah SR (2013) Chemistry and applications of cellulase in textile wet processing. Res J Eng Sci 2(7):1–5

Shankar S, More SV, Laxman RS (2010) Recovery of silver from waste X-ray film by alkaline protease from Conidiobolus coronatus. Kathmandu Univ J Sci Eng Technol 6(1):60–69

Sharma N, Rathore M, Sharma M (2012) Microbial pectinase: sources, characterisation and applications. Rev Environ Sci Biotechnol 12(1):45–60. https://doi.org/10.1007/s11157-012-9276-9

Shelatkar T, Padalia U (2016) Lipase: an overview and its industrial applications; international journal of engineering science and. Computing 6(10):2629–2631

Singh R, Kumar M, Mittal A, Mehta PK (2016a) Microbial cellulases in industrial applications. Ann Appl Biosci 3(4):2016

Singh R, Kumar M, Mittal A, Mehta PK (2016b) Microbial enzymes: industrial progress in 21st century. 3 Biotech 6:174. https://doi.org/10.1007/s13205-016-0485-8

Singh R, Mittal A, Kumar M, Mehta PK (2016c) Amylases: a note on current applications. Int Res J 5(11):27–32

Singh R, Mittal A, Kumar M, Mehta PK (2016d) Microbial proteases in commercial applications. J Pharm Chem Biol Sci 4(3):365–374

Sivaramakrishnan S, Gangadharan D, Nampoothiri KM, Soccol CR, Pandey A (2006) Alpha amylases from microbial sources- an overview on recent developments. Food Technol Biotechnol 44(2):173–184

Sukumaran RK, Singhania RR, Pandey A (2005) Microbial cellulase- production, applications and challenges. J Sci Ind Res 64:832–844

Sumantha A, Larroche C, Pandey A (2006) Microbiology and industrial biotechnology of food-grade proteases: a perspective. Food Technol Biotechnol 44(2):211–220

Sundarram A, Murthy TPK (2014) Alpha amylase production and applications: a review. J Appl Environ Microbiol 2:166–175

Taniguchi H, Honnada Y (2009) Amylases. Elsevier Inc.

Tapre AR, Jain RK (2014) Pectinases: enzymes for fruit processing industry. Int Food Res J 21(2):447–453

Ugo AK, Amara AV, Igwe CN, Kenechuwku U (2017) Microbial lipases: a prospect for biotechnological industrial catalysis for green products: a review. Ferment Technol 6(2):144. https://doi.org/10.4172/2167-7972.1000144.

Verma A, Singh H, Anwar S, Chattopadhyay A, Tiwari KK, Kaur S, Dhilon GS (2016) Microbial keratinases: industrial enzymes with waste management potential. Crit Rev Biotechnol 37(4):476–491. https://doi.org/10.1080/07388551.2016.1185388

Walker TJ, Dayan SH (2014) Comparison and overview of currently available neurotoxins. J Clin Aesthet Dermatol 7(2):31–39

Zhang XZ, Zhang YHP (2013) Cellulases: characteristics, sources, production and applications. John Wiley & Sons, Inc., Hoboken, NJ

Extremophilic Amylases: Microbial Production and Applications

7

Bhumi M. Bhatt, Ujjval B. Trivedi, and Kamlesh C. Patel

Abstract

Amylases are widely distributed and have various industrial applications. Microbes play a major role when bulk production of enzymes is desired. Enzymes which can work in harsh conditions like high or low pH, high salt concentration, and high or low temperature are of special interest. As starch solution has a lower pH, conversion of starch needs to be adjusted in terms of pH or require acid-stable amylases for working at lower pH. Thermophilic amylases are prerequisite for industrial processes like liquefaction followed by saccharification of starch at higher temperatures. Psychrophilic amylases are considered to reduce energy requirements when utilized in wastewater treatments during winters. Halophilic amylases may reduce contamination risk and can contribute to low-cost amylase production. Use of thermostable and acid-stable or -tolerant amylases can enhance conversion of starch into sugar by manyfold. *Bacillus licheniformis* is notable for thermophilic amylase while *Alteromonas haloplanktis* for psychrophilic amylase. Amylases are used in various industries like sugar, paper and pulp, textile, detergent, bread making, and brewing; biofuel production; and treatment of starch processing wastewater. In this chapter, we have discussed microbial sources for the production of distinct amylases working in extreme conditions.

B. M. Bhatt · U. B. Trivedi · K. C. Patel (✉)
P. G. Department of Biosciences, Sardar Patel University, Bakrol, Gujarat, India

N. K. Arora et al. (eds.), *Microbial Enzymes: Roles and Applications in Industries*, Microorganisms for Sustainability 11,
https://doi.org/10.1007/978-981-15-1710-5_7

7.1 Introduction

Amylases are one of the most valuable enzymes used in a broad range of industrial processes. These enzymes also have a wide potential in biotechnological applications and account for nearly 30% of the world's enzyme market (Rajagopalan and Krishnan 2008). The global market of the industrial enzyme was evaluated to about $4.2 billion in 2014 which is expected to reach nearly $6.2 billion till 2020 (Singh et al. 2016). Amylases break down starch and release glucose, maltose, and other oligosaccharides. Starch is the second major polysaccharide food reserve in nature after cellulose. Starch is ubiquitously distributed on the earth and synthesized by plants through photosynthesis in the presence of water and sunlight inside the plastids. Starch granules are produced in a variety of plant tissues including pollen, leaves, stems, roots, tubers, bulbs, rhizomes, fruits, and seeds.

Starch is a polymer of glucose subunits linked with glycosidic bonds. The starch polymer has two main components: amylose and amylopectin (Fig. 7.1). Amylose is a polymer of glucose subunits linked by α-1,4 glycosidic bond while amylopectin consists of a branching point of α-1,6 glycosidic bonds. Commercial starch is extracted from different sources in which maize is predominant while wheat, rice, potato, and sago also share significant contribution. Size and shape of granules are diverse depending on its source of origin. Amylases are produced by a variety of plants, animals, and microorganisms. Use of microorganisms for production of

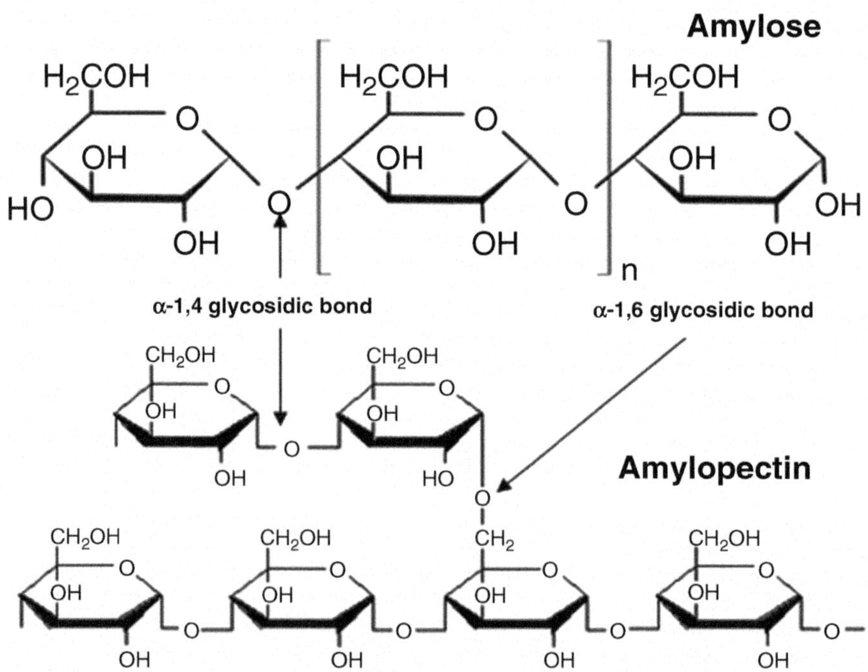

Fig. 7.1 Structure of amylose and amylopectin

amylases has many advantages like short life cycle, ease in handling, economical bulk production, and possibility of manipulation for desired characteristics and applications (Sundarram and Murthy 2014). Different microorganisms particularly bacteria, fungi, and yeast are mainly used for industrial production of amylase. Conventionally submerged fermentation (SmF) is used for the industrial processes but now by having known advantage over submerged fermentation, solid-state fermentation (SSF) is more preferred (Anto et al. 2006; Soccol et al. 2017). As amylases are extensively used in a variety of applications, they require some special characteristics. Industrial processes of starch liquefaction and gelatinization are performed at very high temperature and hence thermophilic property of amylase is a prerequisite. At the same time native starch has very low pH and so amylase which shows stability in acidic environments is also preferable. This chapter is focused on amylases with specialized characteristics to tolerate extreme conditions and their possible roles in industrial processes.

7.2 Classification of Amylases

Amylases are mainly classified into three different subtypes as α-amylase, β-amylase, and γ-amylase or glucoamylase. β-Amylases are produced by plants and microorganisms while α- and γ-amylase are found abundantly in animals and are also present in microorganisms (Benjamin et al. 2013). Starch-converting enzymes are further classified into four groups as endo, exo, debranching, and transferase (Fig. 7.2) (Van der Maarel et al. 2002).

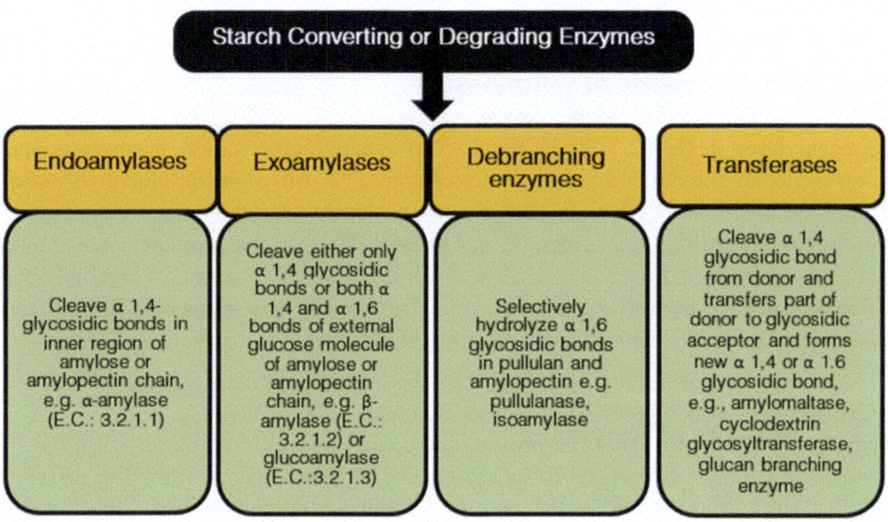

Fig. 7.2 Classification of starch-degrading enzymes

7.2.1 α-Amylase

α-Amylase is extensively studied among all amylases. α-Amylase (E.C. 3.2.1.1) is classified in the GH13 family. It is also known as 1,4-α-D-glucan glucanohydrolase (Hiteshi and Gupta 2014). Its systematic name is 4-α-D-glucan glucanohydrolase. α-Amylase belongs to an endoamylase class. It is an extracellular enzyme that randomly hydrolyzes α-1,4 glycosidic bond of the amylose chain. It is secreted as a primary metabolite of microorganisms (Kammoun et al. 2008). α-Amylase acts neither on terminal glucose residue nor on α-1,6 linkages of amylopectin (Whitcomb and Lowe 2007). Microbial α-amylases are the most stable and produced more economically compared to plant and animal α-amylases (Prajapati et al. 2015). Amylolytic enzymes contribute to the process of starch breakdown, but α-amylase is most important for initiation of this process (Tangphatsornruang et al. 2005). α-Amylase is an inducible enzyme and it is induced in the presence of starch or its hydrolytic product like maltose (Naidu and Saranraj 2013).

7.2.2 β-Amylase

β-Amylase is mainly of plant origin, but few microbial β-amylases are also known. β-Amylase (E.C. 3.2.1.2) is known as 1,4-α-D-glucan maltohydrolase and saccharogen amylase (Hiteshi and Gupta 2014). Its systematic name is 4-α-D-glucan maltohydrolase. Bacterial strains belonging to genera *Bacillus, Pseudomonas* and *Clostridium* (Rani et al. 2007) and fungal strains belonging to *Rhizopus* (Olufunke and Azeez 2012) and *Volvariella volvacea* (Olaniyi et al. 2010) have been reported to synthesize β-amylase. β-Amylase is distributed in higher plants such as soybean, sweet potato, and barley (Oudjeriouat et al. 2003). The properties of the β-amylase vary from source to source (Olufunke and Azeez 2012).

7.2.3 Glucoamylase or γ-Amylase

Glucoamylases (E.C. 3.2.1.3) are also known as amyloglucosidase, γ-amylase, and 1,4-α-D-glucan glucohydrolase. Its systematic name is 4-α-D-glucan glucohydrolase. They are exo amylases which release glucose molecule from the nonreducing end of starch and other oligosaccharides (Prajapati et al. 2013). Glucoamylases are extensively produced from fungal species. The enzyme cleaves α-1,4 linkages favorably but α-1,6 linkages are hydrolyzed at a low rate. This slower hydrolysis of α-1,6 linkage affects enzyme kinetics and saccharification effectiveness (Norouzian et al. 2006).

7.3 Sources of Amylases

Commercially, amylases are obtained from the plant, animal, and microbial sources. Microbial amylases are in demand because of various advantages which they offer like cost-effective industrial production, short life cycle, and ease of manipulation.

There are well-established tools available for genetic engineering which may be useful in the improvement of strain to achieve a higher yield of protein. Even protein engineering techniques can also be applied to get the desired enzyme property as per the need of application (Gupta et al. 2003). Bacterial and fungal amylases are extensively studied, along with actinomycetes and yeasts (Gupta et al. 2003). Recently, Archaea are also in focus due to the production of enzymes with special characteristics. Archaea were less studied as they are difficult to culture and observe. Bacterial amylases are more preferred than fungal because of the rapid growth rate of the organisms and easy handling and manipulation of the enzyme for desired characteristics (Benjamin et al. 2013). Bacterial amylases are mostly obtained from *Bacillus* species. Several species of *Bacillus* such as *B. subtilis*, *B. amyloliquefaciens*, *B. licheniformis*, and *B. stearothermophilus* are reported to produce amylase by submerged fermentation at industrial level (Benjamin et al. 2013).

Different types of fungi are known to secrete glucoamylase. *Aspergillus niger*, *Aspergillus awamori*, *Aspergillus foetidus*, *Aspergillus oryzae*, *Aspergillus terreus*, *Mucor rouxians*, *Mucor javanicus*, *Neurospora crassa*, *Rhizopus delemar*, *Rhizopus oryzae*, and *Arthrobotrys amerospora* are few examples of glucoamylase-producing fungi (Norouzian et al. 2006).

Enzymes from common bacterial and fungal source are mostly mesophilic and neutrophilic in nature. Many industrial processes need enzymes which may remain stable under harsh conditions, and so there is a need for extremophilic microbes and their hydrolytic enzymes.

7.4 Structural Characteristics of Amylases

Enzyme structure, including various domains and their biological role, is important for the preparation of bioengineered enzyme. The α-amylase structure is extensively studied among all amylases. Three-dimensional structure of α-amylase is known to have three main domains A, B, and C (Fig. 7.3). The domain A is a catalytic domain that has (β/α)8-barrel-shaped structure. $(\beta/\alpha)_8$ Barrel was first observed in chicken muscle triose phosphate isomerase (TIM); hence it is known as TIM barrel structure which is present in all members of the α-amylase family. The B domain is a small loop which overhangs between third β-strand and third α-helix of TIM barrel. It forms large substrate-binding cleft which varies and is supposed to play a vital role in substrate specificity differences in α-amylase (Van der Maarel et al. 2002). Domain C is a C-terminal antiparallel β-sheet composed of 5–10 strands subsequent to the catalytic $(\beta/\alpha)_8$ barrel. It is responsible for stability/folding of the protein in substrate binding (Robert et al. 2003). All known α-amylases contain calcium ion found at the interface between domains A and B (Linden et al. 2003). Calcium ion-binding site is positioned far away from the active site and contributes to the stabilization of the enzyme. Enzyme stability and catalytic efficiency are determined by the non-covalent interactions such as hydrogen bonds, ionic interactions, hydrophobic interactions, and van der Waals interaction and any factors negatively influencing these will result in unfolding or denaturation of the protein. Highly stable α-amylase is found

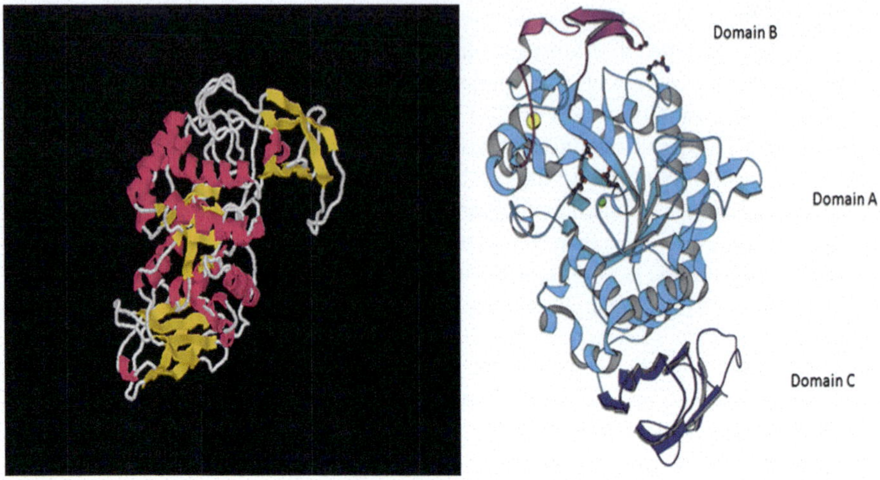

Fig. 7.3 3D structure of α-amylase from *Bacillus licheniformis* (left) and *Alteromonas haloplanktis* (right) (Hiteshi and Gupta 2014)

naturally from extremophiles while stability can be increased by the engineering of other naturally available α-amylase. Metagenomic approaches can play an important role in novel amylase from uncultivable microbes. Thermostability is a most prospecting feature among amylases from extremophiles (Sindhu et al. 2017).

7.5 Amylase from Extremophiles

Acidophiles, alkalophiles, thermophiles, psychrophiles, halophiles, and piezophiles are examples of extremophiles which exhibit stability under low pH, high pH, high temperature, low temperature, high salt concentration, and high osmotic pressure, respectively. Recently Chen and Jiang (2018) suggested that renewable energy sources cannot replace petroleum-based products. Hence the development of new technology known as "next-generation industrial biotechnology (NGIB)" is required. This technology will reduce energy and freshwater requirements, whereas the use of a low-cost substrate or waste material with great efficiency for the substrate to product conversion may offer low capital investment. Current practices require specific environments such as pH and temperature for growth of microorganisms for enzyme production. Most of the microbes used in industrial processes are neutrophiles and mesophiles, which cannot tolerate harsh environmental conditions and require specialized environments. In this context, extremophiles may be useful. Bacteria and Archaea both are known for their ability to survive in extreme conditions; however, in experimental conditions, the difficulty arises due to slower growth rate of Archaea. In the bioprocesses maintenance of sterility is also a major concern. If extremophiles are used, they may reduce the risk of contamination. It can also lower the cost of water and energy, which is required for maintaining sterility (Chen and Jiang 2018). Industrial production of bio-products involves upstream and downstream processes. Upstream processing involves substrate, its pretreatments, sterilization, aeration, agitation, temperature, and pH maintenance cost while downstream processes involve equipment and energy cost to separate microbial cells from recovered products. The cutdown in energy requirement at any point can reduce product cost. A resistant strain is needed for the development of NGIB. Major characteristics required for NGIB strains are fast growth at high or low temperature, pH, and/or osmotic pressure. It should be able to grow maximally on unusual substrates like long-chain fatty acids and solvents and tolerate toxic compounds including heavy metals, short-chain alcohols, and fatty acids with lower water and energy requirements. Microbial strain having these qualities may escalate amylase production to manyfold. Some archaea possess some of these traits but they are slow growers and application of genetic engineering tools has not done so well. Some bacteria and fungi can satisfy the properties like high or low temperature, pH, and/or osmotic pressure with lower water and energy requirements and can be further improved by using available engineering methods and tools especially for prokaryotic bacteria which are extensively studied (Chen and Jiang 2018). Organic solvent-tolerant amylases are known from *Haloarcula* sp. strain S-1, *Nesterenkonia* sp., and *Salimicrobium halophilum* strain LY20 (Dumorné et al. 2017). Thermostability for amylase enzyme plays an important role because starch starts solubilizing at high temperature (100 °C) and in acidic condition (pH 4.5–5.5). Several microbes have been reported to produce thermostable enzymes, including amylases, which can also show activity at high and low pH (Table 7.1).

Table 7.1 α-Amylases produced by microbes showing activity at high pH and temperature with their molecular weight

Microbes	Optimum temperature (°C)	Optimum pH	Molecular mass (kDa)	Reference
Alicyclobacillus acidocaldarius	75	3	160	Schwermann et al. (1994)
Bacillus licheniformis NH1	90	9	58	Hmidet et al. (2008)
B. stearothermophilus	80	5.6	59	Ali et al. (2001)
B. subtilis KCC103	65–70	5–7	53	Nagarajan et al. (2006)
Geobacillus sp. LH8	80	5–7	52	Mollania et al. (2010)
B. circulans	48	4.9	48	Dey et al. (2002)
B. amyloliquefaciens KCP2	65	8		Prajapati et al. (2015)
Cryptococcus flavus	50	5.5	84.5	Wanderley et al. (2004)
Cryptococcus sp. S2	37	6	66	Iefuji et al. (1996)
Thermococcus profundus	80	5.5	43	Kwak et al. (1998)
Staphylothermus marinus	100	5	82.5	Li et al. (2010)

7.5.1 Thermophilic Amylase

Thermophilic enzymes are stable above 55 °C, thermophilic extremozymes are stable above 75 °C, and hyperthermophilic enzymes are stable above 90 °C. Thermostability is an important characteristic of most amylases which are used in industrial processes (Sindhu et al. 2017). Among the species of genus *Bacillus*, *B. licheniformis* and *B. amyloliquefaciens* are commonly used for commercial thermostable amylases. Though both are mesophilic microorganisms, their amylases are thermophilic and have structure similarities (Vengadaramana 2013). Commonly, thermophilic proteins contain a higher concentration of charged residues on the surface. There is a higher ratio of Arg, Glu, Lys, and Val as compared to Asn, Gln, Ser, and Thr in thermophiles (Cambillau and Claverie 2000). Surface residues of proteins are flexible and show free intraprotein interactions which increases the thermotolerance of the proteins (Loladze et al. 1999). Thermophilic proteins contain a varied distribution of amino acids in comparison to mesophilic protein. Thermophilic proteins have less His and Asp among all charged amino acids at their exposed sites compared to buried sites. Hydrophobic amino acids like Val, Pro, Tyr, and Trp are increased at exposed sites; however, at buried sites, Leu is decreased and Val is increased (Chakravarty and Varadarajan 2000; Hiteshi and Gupta 2014). Thermostable enzyme provides various advantages like reduced cooling cost, higher solubility of the substrate, and fewer contamination risks.

7.5.2 Cold-Adapted or Psychrophilic Amylase

Limited research is carried out on psychrophilic α-amylase production. Psychrophiles are evolved in nature with several structural and functional adaptations (Russell 2000). Cold-active enzymes with high catalytic efficiency have integration of unsaturated fatty acids in cell membranes to sustain membrane fluidity and production of cold-shock proteins at low temperatures (Nam and Ahn 2011). Psychrophilic α-amylases are active under lower temperature with lower activation energies and lower melting temperatures. Protein's flexibility helps for easy accommodation of substrates at lower temperatures. The flexibility of protein structure is because of the reduction of salt bridges and fewer proline residues in loops as the cyclic structure of proline gives rigidity to the proteins (Feller and Gerday 1997). Psychrophiles can save energy and reduce the chances of undesirable chemical reactions which take place at higher temperatures. In the detergent industry, it can provide color protection to fabrics (Sindhu et al. 2017). Psychrophilic microbes can play an important role in bioremediation of solids and wastewater during winters (Hiteshi and Gupta 2014).

The structure of cold-active α-amylase from *Alteromonas haloplanktis* is similar to other mesophilic amylases (Fig. 7.3). The 3D structure of different amylases shows that only minor structural alterations are needed to adapt to lower temperatures. Usually, all changes related to cold adaptation occur outside the catalytic cleft, and active-site residues are highly conserved, which shows that fundamental pathway is not modified (Hiteshi and Gupta 2014).

7.5.3 Acid-Stable Amylase

Industrially amylases are used in liquefaction and saccharification processes. The native starch solution has lower pH (3.2–4.5) which needs to be adjusted for enzymatic conversion of starch to simple maltose and other oligosaccharides. Various bacterial amylases from *B. licheniformis* NH1 (4–9) (Hmidet et al. 2008), *Alicyclobacillus* sp. A4 (4.2), *A. acidocaldarius* (3), *B. caldolyticus* (5.5), *B. circulans* (4.9) (Dey et al. 2002), *Geobacillus* sp. LH8 (5–7) (Mollania et al. 2010), *Lactobacillus manihotivorans* (5.5), and *Pyrococcus furiosus* (5.5) are known as thermostable and acid-stable amylases (Sharma and Satyanarayana 2013). Acid-stable α-amylase from *A. acidocaldarius* shows charged residues replaced by neutral polar residues on the surface of the protein. An acidic environment is adapted by reducing the density of both positive and negative charges and avoiding electrostatic repulsion of charged groups at lower pH and contributing to acid stability (Schwermann et al. 1994).

7.5.4 Halophilic Amylase

Halophiles mainly belong to archaea. Halophilic enzymes have a reduction in water activity. These proteins have a high amount of negatively charged residues on their surface with increasing intramolecular salt bridges (Jaenicke and Böhm 1998). It promotes interaction with the salty environment without precipitation. This protein can be denaturing irreversibly if salt concentration decreases (Marhuenda-Egea and Bonete 2002). Halophilic enzymes do not exhibit specific structural properties. Repulsion of surface residue may be responsible for the instability of halophilic proteins in lower salt concentration (Jaenicke and Böhm 1998). *Halomonas meridiana* produced extracellular halophilic amylase with 5% salt in starch during exponential growth phase at pH 7. The optimum temperature for activity was 37 °C while optimum salinity was 10% NaCl and showed activity up to 30% salt concentration (Coronado et al. 2000).

7.5.5 Piezophilic Amylase

Although piezophilic microbial amylases are not reported yet, the effect of high pressure on commercial amylase has been studied. Abe and Horikoshi (2001) demonstrated that porcine pancreatic α-amylase at high pressure produces trisaccharide in place of maltobiose and tetrasaccharide, with maltooligosaccharide as a substrate, at great pressure and little energy. This reaction offers great industrial and biotechnological potential, particularly in the food industry (Dumorné et al. 2017). Vahidi et al. (2018) studied the effect of pressure on Taka-amylase. When they applied 1000–4000 bar pressure on the enzyme, barrel-shaped β-sheets were not modified but loops and helix were changed. Effect of high pressure is still not clear at the molecular level as the only change of structure is studied.

7.6 Production and Purification of Amylases

In nature, no bacteria work as pure culture system as in a lab environment. In the environmental conditions, mixed systems work and utilize sources available for their growth. Various microbial interactions are going on in natural environmental conditions. α-Amylase-producing two strains of *B. amyloliquefaciens* 04BBA15 and *Lactobacillus fermentum* 04BBA19 were analyzed for their interaction with *Saccharomyces cerevisiae*. They showed that commensalism occurs between *S. cerevisiae* and *B. amyloliquefaciens* 04BBA15 during fermentation while mutualism occurs between *S. cerevisiae* and *L. fermentum* 04BBA19 (Tatsinkou et al. 2014).

Traditionally industrial amylases are produced by submerged fermentation (SmF) processes using the pure culture of bacteria or fungi. During submerged fermentation, oxygen transfer rate is a critical point for maximum enzyme yield. Large-scale batch fermentation is carried out in stirred tank bioreactors. Aeration rate needs to be maintained for aerobic fermentation processes. Agitation can maintain dissolved oxygen in the medium. Agitation also plays a key role in viscous fermentation medium (Gangadharan et al. 2011).

Thermomonospora viridis, an actinomycete, produced amylase by SmF process. Maximum amylase yield was attained with 1.5% corn starch and 0.5% mycological peptone at pH 7 after 48 h when pH turned into alkaline (8.2) (Upton and Fogarty 1977). *Thermoactinomyces thalpophilus* produced amylase by submerged fermentation. Fermentation medium contained 2% sorghum as inexpensive carbon source (Uguru et al. 1997). In their study Alrumman et al. (2014) show that thermoalkaliphilic α-amylase was produced by *Bacillus axarquiensis* using potato wastewater.

Nowadays, for industrial production of enzymes, various types of agro-residues are utilized. If solid residues are used it solves waste disposal problem and also produces less wastewater (Gangadharan et al. 2011). Among various agro-residues, the most commonly used is wheat bran in industrial fermentation. *Thermomyces lanuginosus* has been studied for extracellular amylase production using wheat bran with solid-state fermentation process (Kunamneni et al. 2005). *Penicillium expansum* was used for α-amylase production by solid-state fermentation on waste loquat kernel (Specka et al. 1991).

B. amyloliquefaciens has been used for amylase production on various agro-residues; maximum amylase production achieved on wheat bran and groundnut oil cake with 1:1 mass ratio and 85% of initial moisture after 72 h (Gangadharan et al. 2006).

Akassou and Groleau (2018) have reported a high level of extracellular, thermostable amylolytic enzyme from *Thermus thermophilus* HB8. Nonionic detergent Triton X-100 has been used to extract cell surface-bound amylolytic enzyme. Amylolytic enzyme production was optimized using central composite design as 40 g/L of yeast extract, 41 g/L of peptone, 2.0 g/L of sodium chloride, and 25 g/L of starch with 20% inoculum giving maximum production. Thermophilic *Rhodothermus marinus* ITI90 strain showed maximum amylase and pullulanase production in optimized medium containing maltose and yeast extract. Soluble maize starch, glycogen, and pullulan showed lower enzyme production as compared to maltose (Gomes et al. 2003).

Table 7.2 Purification strategies employed for α-amylase produced by some microbes

Microbial source	Purification strategy	References
Geobacillus sp. LH8	$(NH_4)_2SO_4$ extraction, precipitation, Q-sepharose, mono Q-sepharose	Mollania et al. (2010)
B. licheniformis NH1	$(NH_4)_2SO_4$ precipitation, sephadex G-100 gel filtration, sepharose mono Q anion-exchange chromatography	Hmidet et al. (2008)
B. subtilis KCC103	Acetone precipitation, DEAE sephadex A-50	Nagarajan et al. (2006)
B. circulans	Acetone precipitation, sephadex G-100, CM-cellulose	Dey et al. (2002)
Cryptococcus flavus	Sephacryl S-100	Wanderley et al. (2004)

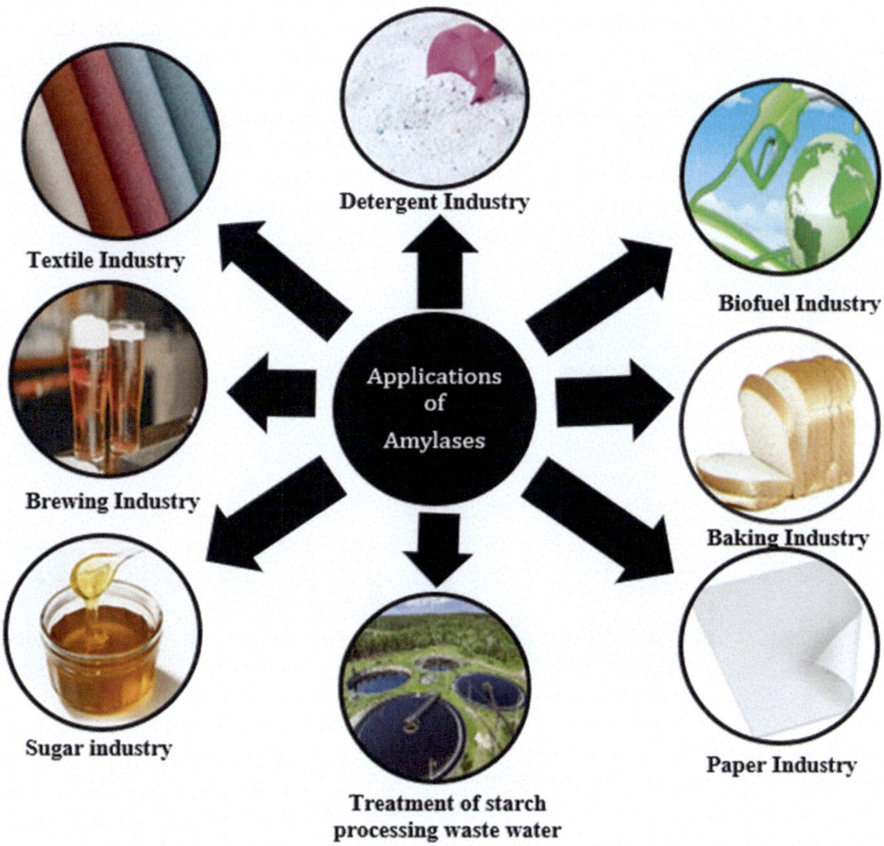

Fig. 7.4 Applications of amylases

Various purification technologies have been employed for purification of amylases in different studies. Ammonium sulfate precipitation and solvent precipitation are conventionally employed methods as the first step of purification strategy. Different chromatographic methods have exhibited key roles for purification of the enzyme. Some strategies used are shown in Table 7.2.

7.7 Applications of Amylases

Amylase was first produced by *A. niger*, which was used as a digestive aid for the pharmaceutical industry. In all types of amylases, α-amylase is a prominent enzyme having multiple roles in all industrial applications. Figure 7.4 shows various applications of amylases while Fig. 7.5 shows required characteristics of α-amylase for specific applications with bar graph plotted using the range of temperature and pH requirements of α-amylases to be used for different industrial applications. The temperature/pH range of α-amylases to be used in detergent industry, feed, baking, desizing, brewing, paper industry, and starch saccharification are 30–45 °C/10.0–11.5, 30–45 °C/4.5–7.0, 0–50 °C/4.5–5.5, 60–80 °C/5.5–6.5, 60–70 °C/5.5–6.0, 60–70 °C/4.5–5.5, and 95–100 °C/4.5–7.0, respectively (Mehta and Satyanarayana 2016). Some commercially available amylases published by the association of manufacturer and formulators of enzyme products (AMFEP) are provided in Table 7.3.

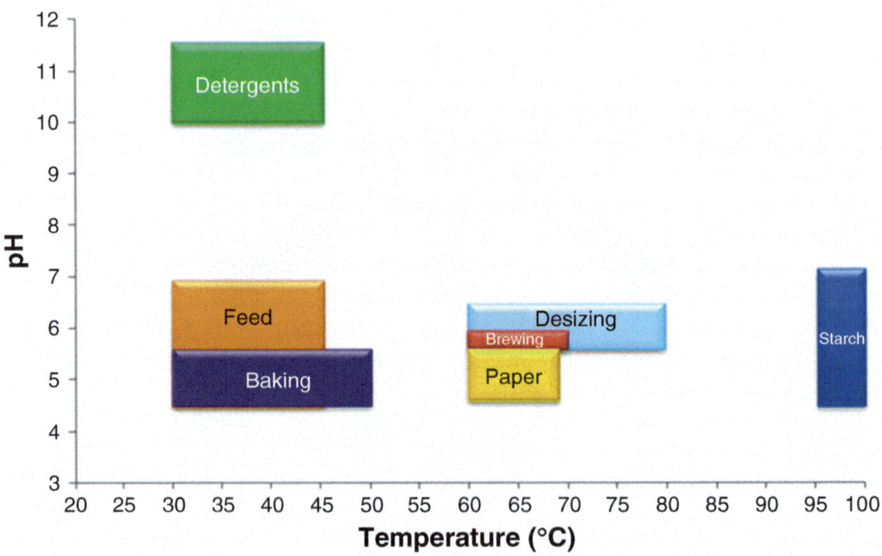

Fig. 7.5 Characteristic requirement of α-amylases for specific applications (Mehta and Satyanarayana 2016)

7.7.1 Sugar Industry

During starch hydrolysis, starch liquefaction and saccharification are two major processes needed for sugar production. Figure 7.6 shows the processes of liquefying and saccharifying starch using amylases for sugar syrup production. Conventionally, starch was hydrolyzed by chemical processes using acid or alkali treatments which are taken over by an enzymatic process. Enzymatic conversion gives many advantages over chemical method. Limitations of chemical methods are non-specificity, lack of controlling saccharide composition, less environment friendliness, and higher refining cost. Use of enzymes for this process avoids these limitations (Crabb and Shetty 1999). Conversion of starch into sugar, syrups, and dextrin forms the major part of the starch processing industry (Mojsov 2012). For liquefaction of starch, amylase is added when the slurry has a lower pH (6) and high temperature (more than 100 °C). Primarily, α-amylase from *B. amyloliquefaciens* was used for this process, which is replaced by *B. stearothermophilus* and *B. licheniformis* amylases due to their greater thermostability (Souza 2010). Thermophilic and acid-stable amylases are preferred for this kind of processes.

7.7.2 Baking Industry

Microbial amylases have been widely used in the baking industry (Hamer 1995). These enzymes can be added to the dough of bread to degrade the starch in the

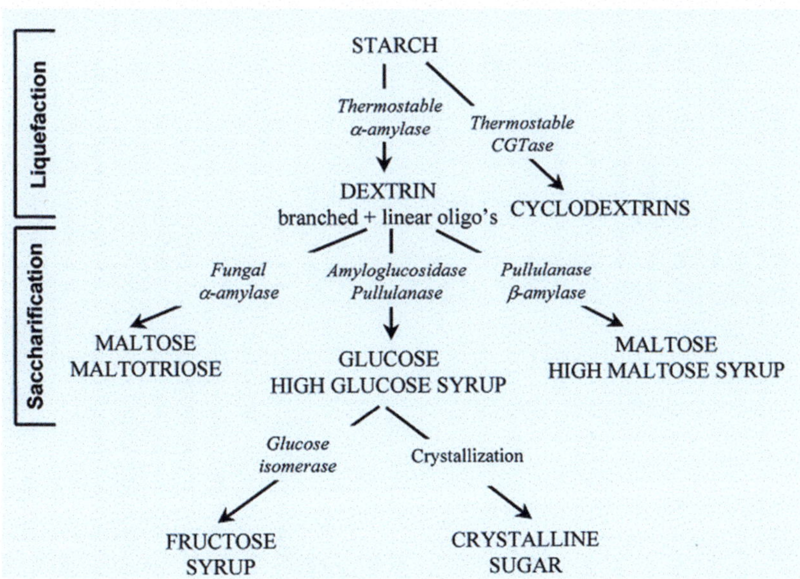

Fig. 7.6 Overview of the industrial processing of starch into cyclodextrins, maltodextrins, glucose or fructose syrups, and crystalline sugar (Van der Maarel et al. 2002)

Table 7.3 AMFEP list of commercial amylases

Enzyme name	Production organism	Donor organism	IUBMB number	Application field		
				Food	Feed	Technical
α-Amylase	*Aspergillus niger*	None	3.2.1.1	Yes		
α-Amylase	*A. oryzae*	None	3.2.1.1	Yes		Yes
α-Amylase	*B. amyloliquefaciens* or *subtilis*	*Bacillus* sp.	3.2.1.1	Yes	Yes	Yes
α-Amylase	*B. amyloliquefaciens* or *subtilis*	None	3.2.1.1	Yes	Yes	Yes
α-Amylase	*B. amyloliquefaciens* or *subtilis*	*Thermoactinomyces* sp.	3.2.1.1	Yes	Yes	
α-Amylase	*B. licheniformis*	*Bacillus* sp.	3.2.1.1	Yes		Yes
α-Amylase	*B. licheniformis*	None	3.2.1.1	Yes	Yes	Yes
α-Amylase	*B. stearothermophilus*	None	3.2.1.1	Yes		Yes
α-Amylase	*Microbacterium imperiale*	None	3.2.1.1	Yes		
α-Amylase	*S. violaceoruber*	*Streptomyces* sp.	3.2.1.1	Yes		
α-Amylase	*T. reesei* or *longibrachiatum*	*Aspergillus* sp.	3.2.1.1	Yes		Yes
β-Amylase	*Bacillus flexus*	None	3.2.1.2	Yes		
β-Amylase	*Barley*	None	3.2.1.2	Yes		
β-Amylase	*Soybean*	None	3.2.1.2	Yes		
β-Amylase	*Sweet potato*	None	3.2.1.2	Yes		
Glucoamylase or amyloglucosidase	*A. niger*	*Aspergillus* sp.	3.2.1.3	Yes		
Glucoamylase or amyloglucosidase	*A. niger*	None	3.2.1.3	Yes		Yes
Glucoamylase or amyloglucosidase	*A. niger*	*Rhizomucor* sp.	3.2.1.3	Yes		
Glucoamylase or amyloglucosidase	*A. niger*	*Talaromyces* sp.	3.2.1.3	Yes		Yes
Glucoamylase or amyloglucosidase	*R. delemar*	None	3.2.1.3	Yes		
Glucoamylase or amyloglucosidase	*R. oryzae*	None	3.2.1.3	Yes		
Glucoamylase or amyloglucosidase	*T. reesei* or *longibrachiatum*	*Trichoderma* sp.	3.2.1.3	Yes		Yes
Pullulanase	*B. acidopullulyticus*	None	3.2.1.41	Yes		
Pullulanase	*B. licheniformis*	*Bacillus* sp.	3.2.1.41	Yes		Yes
Pullulanase	*B. licheniformis*	None	3.2.1.41	Yes		

Pullulanase	*B. amyloliquefaciens* or *subtilis*	*Bacillus* sp.	3.2.1.41	Yes		
Pullulanase	*K. pneumoniae*	None	3.2.1.41	Yes		
Pullulanase	*Pullulanibacillus naganoensis*	None	3.2.1.41	Yes		

Source: https://amfep.org/_library/_files/Amfep_List_of_Enzymes_update_May_2015.pdf

flour to smaller dextrins, which are further fermented by the yeast. Amylases are also used as an anti-staling agent in bread baking, and they improve the softness retention of baked goods which increases the shelf life of these products (Gupta et al. 2003; Van der Maarel et al. 2002; Sahlström and Bråthen 1997). Addition of α-amylase to the dough enhances fermentation and reduces the viscosity of dough to increase the volume and texture of the product (Mojsov 2012). The high or low dose of α-amylase gives an adverse effect on bread quality, so branching, exo, or other maltogenic amylases can be used in the bakery industry for increasing the shelf life of the baked product. Commercially, *B. stearothermophilus* amylase is used in the bakery because it performs exo acting activity and little endo activity in the process which is important for preparing bakery items (Van der Maarel et al. 2002).

7.7.3 Detergent Industry

α-Amylases are being used in laundry and automatic dishwashing. Enzymes have superior stain removal ability compared to detergents. Amylases in detergent and dishwashing formulations degrade starchy foods like potatoes, gravies, custard, and chocolates to dextrins and other small oligosaccharides (Olsen and Falholt 1998; Mukherjee et al. 2009). Amylase removes starch from surfaces and also provides whiteness benefit since starch can be an attractant for many types of particulate soils. About 90% of all liquid detergents contain these enzymes (Gupta et al. 2003).

The oxidative stability of amylases is also needed because washing environments are highly oxidized (Kirk et al. 2002). Alkaline α-amylase is used as a component of detergents; the chelating agents in detergents remove calcium ions, which is essential for its stability. Thus calcium-independent α-amylase is needed (Mojsov 2012). *Alteromonas haloplanktis* amylase can be added in detergent for washing clothes at a lower temperature.

7.7.4 Textile Industry

In textile industries, starch is used as the sizing agent. Amylases are used for the desizing process. Sizing agents like starch are applied to the yarn before fabric production for fast and secure weaving. During weaving, the starch paste is applied for warping, which gives strength to the weaving process. It prevents loss of string by friction. The starch is removed, and the cloth goes to scouring and dying. Starch is usually removed by application of α-amylase (Hendriksen et al. 1999).

Starch is usually used for sizing due to advantages like its low cost, easy availability in most regions of the world, and easy removal. Starch is removed from woven fabric in a wet process in the textile finishing industry. The amylose is bioconverted by the α-amylase into glucose up to 100% whereas the amylopectin is converted 50% into glucose and maltose (Mojsov 2012). Bio-desizing using enzyme is preferred because of their high efficiency and specific action. Amylases result in complete removal of the size without any harmful effects on the fabric because of its eco-friendly behavior. The α-amylases selectively remove the starch and do not attack the fibers (Ahlawat et al. 2009; Feitkenhauer 2003; Gupta et al. 2003). Amylases from *Bacillus* strains have been employed in textile industries for over a long time.

7.7.5 Paper Industry

α-Amylase has a wide application in the pulp and paper industry. It is used for modification of starches in coated paper. In the paper industry, sizing of paper with starch is performed to protect the paper against mechanical damage during processing (Bruinenberg et al. 1996; Gupta et al. 2003; Van der Maarel et al. 2002). The coating treatment improves the quality of the finished product, and enhances stiffness and elasticity of paper (Gupta et al. 2003; Bruinenberg et al. 1996). Starch is added to paper at a temperature range of 45–60 °C. Since the viscosity of the natural starch is too high for paper sizing, partial degradation of this polymer is needed. α-Amylase is employed for this purpose (Gupta et al. 2003; Mojsov 2012).

7.7.6 Brewing Industry

In beer industries, microbial amylases are used to aid cereal amylase in the production of fermentable sugar (Mojsov 2012). Amylases may be used to hydrolyze barley and starchy additive for lowering the cost of beer (Singh et al. 2016). Enzymes such as α-amylase, glucoamylase, and cellulases are essential in generating fermentable sugars by hydrolyzing starch for the production of distilled alcoholic beverages (Kirk et al. 2002; Singh et al. 2016).

7.7.7 Biofuel Industry

Over the past decades, the demand for ethanol as a biofuel has increased. Biofuels are becoming popular at the global level as a sustainable means of energy production. Ethanol-based biofuels can be easily produced from agricultural waste materials. Starch is the most used substrate due to its low price and ease of availability as a raw material in most regions of the world (Chi et al. 2009). The bioconversion of starch into ethanol involves liquefaction and saccharification, where it is converted into sugar using an amylolytic microorganism or enzymes such as α-amylase, followed by fermentation, where sugar is converted into ethanol using an ethanol-fermenting microorganism such as *S. cerevisiae* (De Moraes et al. 1999; Mojsov 2012).

7.7.8 Treatment of Starch-Processing Wastewater

Food processing plant waste containing excess starch may cause pollution of water bodies. Biological treatment of food processing wastewater can purify the effluent and also produce valuable products like microbial biomass and protein (Friedrich et al. 1987; Klingspohn et al. 1993; Mojsov 2012). During winters, cold-adapted or psychrophilic amylase can also be used for wastewater treatment.

7.8 Conclusion

Various types of amylases are found in nature; however, α- and γ-amylases from microorganisms are widely used in various industries. Mesophilic and neutrophilic amylases are commonly found in microorganisms. Amylases from extremophiles possessing specialized characteristics would become more useful. These amylases can work in different conditions like high pH and temperature. Thermophilic amylases have more hydrophobic residues on the surface while acid-stable amylases have neutral polar residues on the surface of the protein. However, in the case of psychrophilic amylases, minor structural alterations far from the active site are possible; hence, the ultimate pathway cannot be modified. Halophilic amylases have negatively charged residues on the surface with higher intramolecular salt bridges. Thermophilic and acid-stable amylases are important for starch conversion into simple sugars. Psychrophilic amylases can be used for wastewater treatment during winters. Halophiles and thermophiles reduce the risk of contaminations during enzyme production. Alkaline and calcium-independent amylases are already playing a key role in detergents. In the enzyme industry, production of specialized amylase active in multiple extreme conditions is always required and microbes are the only way to fulfil the need. Research priorities on understanding microbial systems and pathways for producing such extremozymes will boom the enzyme industry.

References

Abe F, Horikoshi K (2001) The biotechnological potential of piezophiles. Trends Biotechnol 19:102–108

Ahlawat S, Dhiman SS, Battan B, Mandhan RP, Sharma J (2009) Pectinase production by *Bacillus subtilis* and its potential application in biopreparation of cotton and micropoly fabric. Process Biochem 44:521–526

Akassou M, Groleau D (2018) Optimization of the production of an extracellular and thermostable amylolytic enzyme by *Thermus thermophilus* HB8 and basic characterization. Extremophiles 22:189–202

Ali MB, Mhiri S, Mezghani M, Bejar S (2001) Purification and sequence analysis of the atypical maltohexaose-forming α-amylase of the *B. stearothermophilus* US100. Enzyme Microb Technol 28:537–542

Alrumman SA, Mostafa YS, Eifan SA, Alamri SA, Hesham AE (2014) Isolation of thermoalkalophilic-α-amylase producing bacteria and optimization of potato waste water medium for enhancement of α-amylase production. Adv Life Sci Technol 20:41–51

Anto H, Trivedi UB, Patel KC (2006) Glucoamylase production by solid-state fermentation using rice flake manufacturing waste products as substrate. Bioresour Technol 97(10):1161–1166

Benjamin S, Smitha RB, Jisha VN, Pradeep S, Sajith S, Sreedevi S, Prakasan P, Unni KN, Josh MS (2013) A monograph on amylases from *Bacillus spp.* Adv Biosci Biotechnol 4:227–241

Bruinenberg PM, Hulst AC, Faber A, Voogd RH (1996) A process for surface sizing or coating of paper. European Patent Application, 690, A1

Cambillau C, Claverie JM (2000) Structural and genomic correlates of hyperthermostability. J Biol Chem 275:32383–32386

Chakravarty S, Varadarajan R (2000) Elucidation of determinants of protein stability through genome sequence analysis. FEBS Lett 470:65–69

Chen GQ, Jiang XR (2018) Next generation industrial biotechnology based on extremophilic bacteria. Curr Opin Biotechnol 50:94–100

Chi Z, Chi Z, Liu G, Wang F, Ju L, Zhang T (2009) *Saccharomycopsis fibuligera* and its applications in biotechnology. Biotechnol Adv 27:423–431

Coronado MJ, Vargas C, Hofemeister J, Ventosa A, Nieto JJ (2000) Production and biochemical characterization of an α-amylase from the moderate halophile *Halomonas meridiana*. FEMS Microbiol Lett 183:67–71

Crabb WD, Shetty JK (1999) Commodity scale production of sugars from starches. Curr Opin Microbiol 2:252–256

De Moraes LM, Astolfi Filho S, Ulhoa CJ (1999) Purification and some properties of an α-amylase glucoamylase fusion protein from *Saccharomyces cerevisiae*. World J Microbiol Biotechnol 15:561–564

Dey G, Palit S, Banerjee R, Maiti BR (2002) Purification and characterization of maltooligosaccharide-forming amylase from *Bacillus circulans* GRS 313. J Ind Microbiol Biotechnol 28:193–200

Dumorné K, Córdova DC, Astorga-Eló M, Renganathan P (2017) Extremozymes: a potential source for industrial applications. J Microbiol Biotechnol 27:649–659

Feitkenhauer H (2003) Anaerobic digestion of desizing wastewater: influence of pretreatment and anionic surfactant on degradation and intermediate accumulation. Enzyme Microb Technol 33:250–258

Feller G, Gerday C (1997) Psychrophilic enzymes: molecular basis of cold adaptation. Cell Mol Life Sci 53:830–841

Friedrich J, Cimerman A, Perdih A (1987) Mixed culture of *Aspergillus awamori* and *Trichoderma reesei* for bioconversion of apple distillery waste. Appl Microbiol Biotechnol 26:299–303

Gangadharan D, Sivaramakrishnan S, Nampoothiri KM, Pandey A (2006) Solid culturing of *Bacillus amyloliquefaciens* for α-amylase production. Food Technol Biotechnol 44:269–274

Gangadharan D, Nampoothiri KM, Pandey A (2011) α-Amylase production by *Bacillus amyloliquefaciens* using agro wastes as feed stock. Food Technol Biotechnol 49:336–340

Gomes I, Gomes J, Steiner W (2003) Highly thermostable amylase and pullulanase of the extreme thermophilic eubacterium *Rhodothermus marinus*: production and partial characterization. Bioresour Technol 90:207–214

Gupta R, Gigras P, Mohapatra H, Goswami VK, Chauhan B (2003) Microbial α-amylases: a biotechnological perspective. Process Biochem 38:1599–1616

Hamer RJ (1995) Enzymes in the baking industry. Enzymes in food processing. Springer, Boston, MA, pp 190–222

Hendriksen HV, Pedersen S, Bisgard-Frantzen H (1999) A process for textile warp sizing using enzymatically modified starches. Patent Application WO, 99:35325

Hiteshi K, Gupta R (2014) Thermal adaptation of α-amylases: a review. Extremophiles 18:937–944

Hmidet N, Bayoudh A, Berrin JG, Kanoun S, Juge N, Nasri M (2008) Purification and biochemical characterization of a novel α-amylase from *Bacillus licheniformis* NH1: cloning, nucleotide sequence and expression of amyN gene in *Escherichia coli*. Process Biochem 43:499–510

Iefuji H, Chino M, Kato M, Iimura Y (1996) Raw-starch-digesting and thermostable alpha-amylase from the yeast *Cryptococcus sp.* S-2: purification, characterization, cloning and sequencing. Biochem J 318:989

Jaenicke R, Böhm G (1998) The stability of proteins in extreme environments. Curr Opin Struct Biol 8:738–748

Kammoun R, Naili B, Bejar S (2008) Application of a statistical design to the optimization of parameters and culture medium for α-amylase production by *Aspergillus oryzae* CBS 819.72 grown on gruel (wheat grinding by-product). Bioresour Technol 99:5602–5609

Kirk O, Borchert TV, Fuglsang CC (2002) Industrial enzyme applications. Curr Opin Biotechnol 13:345–351

Klingspohn U, Bader J, Kruse B, Kishore PV, Schügerl K, Kracke-Helm HA, Likidis Z (1993) Utilization of potato pulp from potato starch processing. Process Biochem 28:91–98

Kunamneni A, Permaul K, Singh S (2005) Amylase production in solid state fermentation by the thermophilic fungus *Thermomyces lanuginosus*. J Biosci Bioeng 100:168–171

Kwak YS, Akiba T, Kudo T (1998) Purification and characterization of α-amylase from hyperthermophilic archaeon *Thermococcus profundus*, which hydrolyzes both α-1, 4 and α-1, 6 glucosidic linkages. J Ferment Bioeng 86:363–367

Li D, Park JT, Li X, Kim S, Lee S, Shim JH, Park SH, Cha J, Lee BH, Kim JW, Park KH (2010) Overexpression and characterization of an extremely thermostable maltogenic amylase, with an optimal temperature of 100°C, from the hyperthermophilic archaeon *Staphylothermus marinus*. N Biotechnol 27:300–307

Linden A, Mayans O, Meyer-Klaucke W, Antranikian G, Wilmanns M (2003) Differential regulation of a hyperthermophilic α-amylase with a novel (Ca, Zn) two-metal center by zinc. J Biol Chem 278:9875–9884

Loladze VV, Ibarra-Molero B, Sanchez-Ruiz JM, Makhatadze GI (1999) Engineering a thermostable protein via optimization of charge– charge interactions on the protein surface. Biochemistry 38:16419–16423

Marhuenda-Egea FC, Bonete MJ (2002) Extreme halophilic enzymes in organic solvents. Curr Opin Biotechnol 13:385–389

Mehta D, Satyanarayana T (2016) Bacterial and archaeal α-amylases: diversity and amelioration of the desirable characteristics for industrial applications. Front Microbiol 7:1129

Mojsov K (2012) Microbial alpha-amylases and their industrial applications: a review. Int J Manage IT Eng 2:583–609

Mollania N, Khajeh K, Hosseinkhani S, Dabirmanesh B (2010) Purification and characterization of a thermostable phytate resistant α-amylase from *Geobacillus sp.* LH8. Int J Biol Macromol 46:27–36

Mukherjee AK, Borah M, Rai SK (2009) To study the influence of different components of fermentable substrates on induction of extracellular α-amylase synthesis by *Bacillus subtilis* DM-03 in solid-state fermentation and exploration of feasibility for inclusion of α-amylase in laundry detergent formulations. Biochem Eng J 43:149–156

Nagarajan DR, Rajagopalan G, Krishnan C (2006) Purification and characterization of a maltooligosaccharide-forming α-amylase from a new *Bacillus subtilis* KCC103. Appl Microbiol Biotechnol 73:591–597

Naidu MA, Saranraj P (2013) Bacterial amylase: a review. Int J Pharm Biol Arch 4:274–287

Nam ES, Ahn JK (2011) Isolation and characterization of cold adapted bacteria producing lactose hydrolyzing enzyme isolated from soils of Nome area in Alaska. Int Res J Microbiol 2:348–355

Norouzian D, Akbarzadeh A, Scharer JM, Young MM (2006) Fungal glucoamylases. Biotechnol Adv 24:80–85

Olaniyi OO, Akinyele BJ, Arotupin DJ (2010) Purification and cold-adapted characterization of beta amylase from *Volvariella volvacea*. Nigerian J Microbiol 24:76–82

Olsen HS, Falholt P (1998) The role of enzymes in modern detergency. J Surfactant Deterg 1:555–567

Olufunke FOT, Azeez II (2012) Purification and characterization of beta-amylase of *Bacillus subtilis* isolated from Kolanut Weevil. J Biol Life Sci 4:68–78

Oudjeriouat N, Moreau Y, Santimone M, Svensson B, Marchis-Mouren G, Desseaux V (2003) On the mechanism of α-amylase. FEBS J 270:3871–3879

Prajapati VS, Trivedi UB, Patel KC (2013) Optimization of glucoamylase production by *Colletotrichum sp.* KCP1 using statistical methodology. Food Sci Biotechnol 22:31–38

Prajapati VS, Trivedi UB, Patel KC (2015) A statistical approach for the production of thermostable and alkalophilic alpha-amylase from *Bacillus amyloliquefaciens* KCP2 under solid-state fermentation. 3 Biotech 5:211–220

Rajagopalan G, Krishnan C (2008) α-Amylase production from catabolite derepressed *Bacillus subtilis* KCC103 utilizing sugarcane bagasse hydrolysate. Bioresour Technol 99:3044–3050

Rani T, Takagi T, Toda H (2007) Bacterial and mold amylases. The enzymes. Academic Press Inc., New York, pp 235–271

Robert X, Haser R, Gottschalk TE, Ratajczak F, Driguez H, Svensson B, Aghajari N (2003) The structure of barley α-amylase Isozyme 1 reveals a novel role of domain C in substrate recognition and binding. Structure 11(8):973–984

Russell NJ (2000) Toward a molecular understanding of cold activity of enzymes from psychrophiles. Extremophiles 4(2):83–90

Sahlström S, Bråthen E (1997) Effects of enzyme preparations for baking, mixing time and resting time on bread quality and bread staling. Food Chem 58:75–80

Schwermann B, Pfau K, Liliensiek B, Schleyer M, Fischer T, Bakker EP (1994) Purification, properties and structural aspects of a thermoacidophilic α-amylase from *Alicyclobacillus Acidocaldarius* ATCC 27009. FEBS J 226:981–991

Sharma A, Satyanarayana T (2013) Microbial acid-stable α-amylases: characteristics, genetic engineering and applications. Process Biochem 48:201–211

Sindhu R, Binod P, Madhavan A, Beevi US, Mathew AK, Abraham A, Pandey A, Kumar V (2017) Molecular improvements in microbial α-amylases for enhanced stability and catalytic efficiency. Bioresour Technol 245:1740–1748

Singh R, Kumar M, Mittal A, Mehta PK (2016) Microbial enzymes: industrial progress in 21st century. 3 Biotech 6:174

Soccol CR, da Costa ESF, Letti LAJ, Karp SG, Woiciechowski AL, de Souza Vandenberghe LP (2017) Recent developments and innovations in solid state fermentation. Biotechnol Res Innov 1(1):52–71

Souza PMD (2010) Application of microbial α-amylase in industry-a review. Braz J Microbiol 41:850–861

Specka U, Mayer F, Antranikian G (1991) Purification and properties of a thermoactive glucoamylase from *Clostridium thermosaccharolyticum*. Appl Environ Microbiol 57:2317–2323

Sundarram A, Murthy TPK (2014) α-amylase production and applications: a review. J Appl Environ Microbiol 2:166–175

Tangphatsornruang S, Naconsie M, Thammarongtham C, Narangajavana J (2005) Isolation and characterization of an α-amylase gene in cassava (*Manihot esculenta*). Plant Physiol Biochem 43:821–827

Tatsinkou BF, Tavea F, Ayeoffe LF, Ndjouenkeud R, Wanji S (2014) Microbial interactions for enhancement of α-amylase production by *Bacillus amyloliquefaciens* 04BBA15 and *Lactobacillus fermentum* 04BBA19. Biotechnol Rep 4:99–106

Uguru GC, Akinyanju JA, Sani A (1997) The use of sorghum for thermostable amylase production from *Thermoactinomyces thalpophilus*. Lett Appl Microbiol 25:13–16

Upton ME, Fogarty WM (1977) Production and purification of thermostable amylase and protease of *Thermomonospora viridis*. Appl Environ Microbiol 33:59–64

Vahidi SH, Bozorgmehr MR, Morsali A, Beyramabadi SA (2018) Does high pressure have any effect on the structure of alpha amylase and its ability to binding to the oligosaccharides having 3 to 7 residues? Molecular dynamics study. J Mol Graph Model 80:85–94

Van Der Maarel MJ, Van Der Veen B, Uitdehaag JC, Leemhuis H, Dijkhuizen L (2002) Properties and applications of starch-converting enzymes of the α-amylase family. J Biotechnol 94:137–155

Vengadaramana A (2013) Industrial important microbial alpha-amylase on starch-converting process. Sch Acad J Pharm 2:209–221

Wanderley KJ, Torres FA, Moraes LM, Ulhoa CJ (2004) Biochemical characterization of α-amylase from the yeast *Cryptococcus flavus*. FEMS Microbiol Lett 231:165–169

Whitcomb DC, Lowe ME (2007) Human pancreatic digestive enzymes. Dig Dis Sci 52:1–17

Microbial Lipases and Their Versatile Applications

8

Neha Vishnoi, Sonal Dixit, and Jitendra Mishra

Abstract

Lipases are important biocatalysts and known for their versatile biotechnological applications. Lipases are well known as triacylglycerol hydrolases converting fats into simpler compounds as glycerol and fatty acids at the water-lipid interface and vice versa in nonaqueous medium. Amongst, microorganisms mainly bacteria, fungi, and yeasts are the sources of lipase. Important lipase-producing bacterial genera are *Bacillus*, *Pseudomonas*, *Botryococcus*, and *Burkholderia*. Microbial lipases have received much attention due to their high substrate specificity, lesser processing time, low energy need, high stability, and inexpensive industrial production by using easily available raw materials. The high stability of lipases has extended their application to different industries, and they are taking the lead in the industrial synthesis of oleochemicals, surfactants, drugs, and bioactive compounds. Microbial lipases have wide commercial applicability and are used for the manufacture of several products like oil, food, soaps and detergents, cosmetics, paper, leather, fabrics, and biodiesel. In the fat and oil industries various trans-esterification and inter-esterification chemical reactions are catalyzed for the alteration of fats/oils in the production of nutritionally important triacylglycerols and polyunsaturated fatty acids (PUFA), substitutes for cocoa butter, fatty acid-enriched oils, etc. At the global level, vigorous research for finding novel microbial lipases with industrial value is being carried out.

N. Vishnoi
Department of Environmental Science, Babasaheb Bhimrao Ambedkar University, Lucknow, India

S. Dixit (✉)
Department of Botany, University of Lucknow, Lucknow, India

J. Mishra (✉)
DST-CPR, Babasaheb Bhimrao Ambedkar (Central) University, Lucknow, India

N. K. Arora et al. (eds.), *Microbial Enzymes: Roles and Applications in Industries*, Microorganisms for Sustainability 11,
https://doi.org/10.1007/978-981-15-1710-5_8

This chapter emphasizes the microbial synthesis of lipases and applications in varying sectors of industries.

8.1 Introduction

Since the last few decades, enzymes emerged as potential biocatalysts being used in different industrial sectors (Pliego et al. 2015). Currently, some of the microbial enzymes are playing a significant role in the industrially driven process such as textile, detergent, food, and pharmaceuticals (Cheetham 1995). While talking about lipases, they occupy third place after proteases and amylases having multiple applications in several industries (Snellman et al. 2002; Ulker et al. 2011). Claude Bernard first reported lipolytic action of pancreatic lipase in the year 1856 (Rodriguez de Romo and Borgstein 1999). Since then, other sources of lipases which include plants, bacteria, fungi, and archaea are well known. Lipases at water-lipid interface catalyze the hydrolysis of fats and convert them to simpler compounds (fatty acids and glycerol) whereas in the nonaqueous medium they reverse the chemical process (Lee et al. 2015; Ramos-Sanchez et al. 2015; Ullah et al. 2015). Lipases (EC 3.1.1.3) are glycerol ester hydrolases and during water-oil medium hydrolyse ester linkages of glycerides. Lipases select acyl faction from glycerides to form a lipase-acyl complex during hydrolysis and transfer its acyl group to hydroxyl (OH) group in aqueous condition whereas in the absence of water, they transfer acyl factions of carboxylic acids to nucleophiles (Martinelle and Hult 1995). Thus, these enzymes can acylate sugars, alcohols, amines, and thiols and synthesize a range of stereo-specific esters, thioesters, sugar esters, and amides (Dellamora-ortiz et al. 1997; Singh et al. 2003). Lipases are also known to have enantioselective qualities which can be utilized in biotransformation reactions to catalyze esterification, inter-esterification, trans-esterification, aminolysis, and acidolysis (Hasan et al. 2009). Organic solvents might denature lipases and can cause alteration in the configuration, and hence affect both functional and catalytic properties (Guo et al. 2015).

Lipases have a three-dimensional structure (Fig. 8.1) with the characteristic α/β-- hydrolase fold in core structure (Ollis et al. 1992). The catalytic center contains Ser-Asp-His (Glu is replaced in some lipases by Asp) and holds a consensus sequence G-X1-S-X2-G as the catalytic component, where G = glycine, X1 = histidine, S = serine, and X2 = aspartic or glutamic acid (Kapoor and Gupta 2012; Farrokh et al. 2014; Faouzi et al. 2015; Priji et al. 2015). The three-dimensional arrangement of enzyme plays a significant role in the engineering and designing of lipase-related functions. The Lipase Engineering Database (LED) (http://www.led.uni-stuttgart.de/) has been created to provide a systematic analysis of the relationship of sequence, structure, and function of lipases from various sources.

The investigation of chemo-specific, region-specific, and enantio-specific trait of lipases is likely to be a frontier research area for the workers. As in the past few decades, the use and application of enzyme-based products have increased significantly.

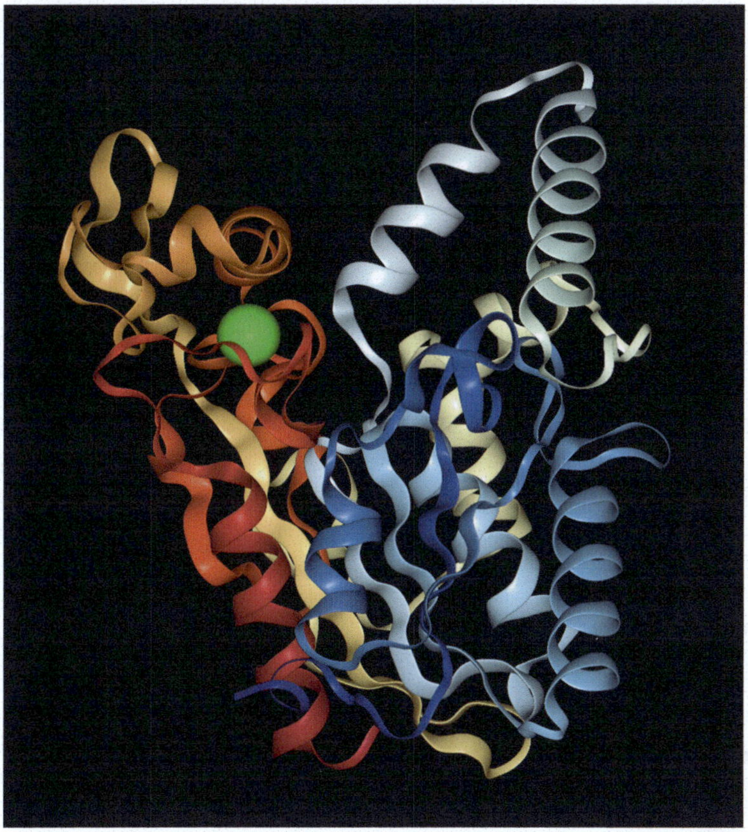

Fig. 8.1 Lipase 3D structure (source: https://www.rcsb.org/3d-view/1OIL/1)

According to Markets and Markets™, a survey company, the market for microbial lipases was estimated to be USD 425.0 million in 2018, which is further projected to reach USD 590.2 million by 2023, at a calculated annual growth rate (CAGR) of 6.8%. According to Guerrand (2017), the percentage share of lipase is also very high, i.e., 90% of the microbial enzyme market. High catalytic activity, lesser requirement of expensive growth media, and ease in genetic alteration are the preferred traits essential for bulk production of microbial lipases (Dey et al. 2014; Lee et al. 2015; Ullah et al. 2015). Apart from this, numerous efforts for improving catalytic and physicochemical properties have also been reported. Lipases are extensively used in bioprocess technology. However, trans-esterification by lipase can be used in biodiesel production in an efficient, energy-saving, and eco-friendly manner compared to chemical catalysis process (Fjerbaek et al. 2009). Based on regiospecificity or positional specificity, lipases can be categorized into three different classes.

1. **Nonspecific lipases:** Catalyze triglycerides and convert them into glycerol and free fatty acids with some intermediates (monoglycerides and diglycerides) and can eliminate fatty acids from any place of the substrate. Hydrolysis of triglycerides occurs slowly as compared to monoglycerides and diglycerides (Ribeiro et al. 2011; Kapoor and Gupta 2012).
2. **1,3-Specific lipases:** Liberate fatty acid only from positions 1 and 3 of the triglycerides and are not able to hydrolyze ester bonds at secondary positions. Hydrolysis of triglycerides by 1,3-specific lipases to diglycerides is a much quicker process than that into monoglycerides (Kapoor and Gupta 2012).
3. **Fatty acid-specific lipases:** Show selectivity towards fatty acid and catalyze the process of hydrolysis of esters having long-chain fatty acids with double bonds in cis position between carbons 9 and 10 (Ribeiro et al. 2011). Lipases remain active in organic solvents and do not require cofactor for their activity (Lee et al. 2015). Trans-esterification reaction can use glycerides (mono, di, and tri) and free fatty acids; reaction time is shorter, and it shows high yield in nonaqueous media and resistance to acidic pH (Ashfaq 2015).

8.2 Types of Lipases

8.2.1 Extra- and Intracellular Lipases

Lipases of microbial origin can be extracellular, intracellular, or bound to the membrane, but the extracellular form is more widely used (Tan et al. 2003). Extracellular lipases are produced outside and require further purification by suitable techniques (Robles-Medina et al. 2009). However, obtaining a pure form of extracellular lipase is a relatively complex process and depends on the source and composition of the lipase enzyme (Saxena et al. 2003). They can be commercially produced by solid-state fermentation (SSF) or by submerged fermentation (SF) process. According to Balaji and Ebenezer (2008), mostly in fermentation process, mixed proteins are formed, and further purification is essential for obtaining pure form of enzyme with the selective catalytic property. The relative cost of purification of lipases with suitable techniques can be comparatively higher than the production (Joseph et al. 2008). The application of intracellular and cell-bound lipases can reduce the purification cost in industrial processes facilitating the use of this enzyme. Ramakrishnan et al. (2016) reported an intracellular alkaline lipase from *Enterococcus faecium* MTCC5695 (MTCC5695), with optimal activity at pH 10.8 and 40 °C. Nunes et al. (2014) reported intracellular lipase by a yeast *Yarrowia lipolytica* strain IMUFRJ 50682 using different carbon sources including glucose, glycerol, crude glycerol, olive oil, and frying oil. Matsumoto et al. (2001) reported overproduction of intracellular lipase by *Rhizopus oryzae* applicable to biodiesel production.

8.2.2 Immobilized Lipases

As mentioned above, the refinement of extracellular lipases is a costlier process; therefore, production of intracellular lipases immobilized on solid supports has gained much attention (Jegannathan et al. 2008). In this process, the whole cell is used for the synthesis of intracellular lipases and immobilization of the cell on solid support enhances the biocatalytic potential by manyfold (Iftikhar et al. 2012). Moreover, the immobilized enzyme also offers additional benefits that include reusability, easier control of the reaction, choice in the use of reactor configurations, etc. (Silva et al. 2012; Gomes et al. 2004). Adamczak and Bednarski (2004) showed the enhanced activity of intracellular lipases from *Rhizomucor miehei* and *Y. lipolytica* when immobilized on biomass support particles. According to Fukuda et al. (2001), immobilization of microbial lipase can reduce purification cost, thereby making the product cheaper compared with the extracellular lipases. For more information on immobilized lipases one can see review by Facin et al. (2019). Three major immobilized lipases which are commercially being used are Novozym 435 (from *Candida antarctica*), Lipozyme TL IM, and Lipozyme RM (from *Thermomyces lanuginosus* and *R. miehei*) (Robles-Medina et al. 2009).

8.2.3 Bacterial, Fungal, and Yeast Lipase

Amongst all microbes, bacteria, fungi, and yeast are the major lipase producers. Although bacterial lipases are less diverse and their use is also limited in industries, they are still in demand due to high yield and ability to work at alkaline pH (Gupta et al. 2004).

Bacterial lipases show substantial heterogeneity, as observed in molecular and catalytic characteristics (Javed et al. 2018). Usually, lipases produced from bacteria are glycoproteins, but some extracellular lipases are lipo-proteinaceous in nature. The extracellular production of bacterial lipases mainly depends on factors like carbon and nitrogen sources, lipids, oxygen, and inorganic salts supplemented during fermentation (Winkler et al. 1979; Jaeger et al. 1994). Many bacterial genera are known to produce lipases; however, *Bacillus, Burkholderia, Pseudomonas*, and *Staphylococcus* are well defined. There are reports on the production of bacterial lipases showing optimum activity at extreme conditions. For example, Eddehech et al. (2019) reported a thermo-active, alkaline lipase from a newly isolated *Serratia* sp. W3 strain. Musa et al. (2019) reported halophilic lipase from a newly isolated moderate halophilic bacterium *Marinobacter litoralis* SW-45 which showed activity and stability at high pH range (7.0–9.5), temperature (30–50 °C), and NaCl concentration (0–21%).

Fungal lipases have wide commercial applications. Their characteristics, like thermal stability, substrate specificity, pH stability, and tolerance in organic solvents, make them an excellent choice in bioprocess applications. Use of fungal lipases is preferable than bacteria because they are extracellularly produced in medium and can be easily separated, which also reduces the cost of production. In the fungal kingdom, zygomycetes, hyphomycetes, and ascomycetes have been well studied to produce extracellular lipase. The fungal genera *Aspergillus* spp., *Penicillium* spp., and *Rhizopus* spp. are the main sources of lipase (Singh and Mukhopadhyay 2012). The LipozymeTM IM (from *R. miehei*) and Lipozyme TL IM (from a thermophilic fungus *T. lanuginosus*) are the two main immobilized forms of fungal lipase known for enormous applications in several industries.

Yeast lipases have attracted various industries due to their efficiency, stability, and high yield in production efficiency. The most important genus of yeast in lipase production belongs to *Candida* and its species include *C. rugosa*, *C. curvata*, *C. tropicalis*, *C. antarctica*, *C. parapsilosis*, *C. deformans*, *C. cylindracea*, and *C. valida* (Pandey et al. 1999). However, amongst them *C. rugosa* and *C. antarctica* are versatile and recognized for their extracellular and immobilized form of lipase, used in several industries. Lipases of *C. rugosa* are used in biotransformation, biolubricants, biodiesel, omega-3 polyunsaturated fatty acids, and plasticizer production (Cavalcanti et al. 2018; Morais Júnior et al. 2017; Marzuki et al. 2015). *C. antarctica* has been used in the production of many industrially relevant compounds such as nitrogenated organic compounds, biodiesel production from vegetable oil, and biodiesel from waste cooking oil (Gotor-Fernández et al. 2006; Shimada et al. 1999; Mehrasbi et al. 2017). Apart from abovementioned yeasts, *Y. lipolytica, Rhodotorula* spp., *Pichia* spp., *Saccharomycopsis crataegenesis, Trichosporon asteroides, Torulaspora globosa,* and *Geotrichum candidum* are also used in several industrial processes (Holmquist 1998; Vakhlu 2006).

8.3 Applications of Lipases

Microbial lipases are an essential tool used in the biotechnological processes and gained much attention of workers around the globe (Hasan et al. 2006; Kapoor and Gupta 2012). Microbial lipases are generally more stable as compared to those of plant or animal origin. Microbial lipases remain active under ambient environmental conditions and can withstand higher temperature and pH (Ashfaq 2015). They also exhibit good organic solvent tolerance needed in many industrial applications (Javed et al. 2018). In the case of microbial lipases, unwanted reaction product generated is also not a big issue. They also offer cost-effective downstream processing. Broadly, lipases are applied as biocatalysts to synthesize supplementary products in the food, fat, and oil processing industry, as well as for the manufacturing of fine chemicals including pharmaceuticals. However, recently detergents, degreasing, paper manufacture, cosmetics, and biofuels have been some other applications in which microbial lipases are taking the lead over other enzymes (Rubin and Dennis 1997; Masse et al. 2001; Takamoto et al. 2001; Hasan et al. 2006; Rajendran et al. 2009; Parra et al. 2015). Some of the significant applications of microbial lipases are depicted in Fig. 8.2, and the detailed applications are discussed in this section.

Fig. 8.2 Application of lipases in different industries

8.3.1 Food Industry

Microbial enzymes are being used in food preparation for a very long time. Amylases, proteases, and lipases are the three major enzymes used in a variety of food applications (Raveendran et al. 2018). Lipases are majorly used in dairy, bakery and confectioneries, fruit juices, and inter-esterification product of fats and oils

(Ghorai et al. 2011; David 2017). *R. miehei* lipases are utilized in milk fat hydrolysis, ripening of cheese, lipolysis of butter fat, and flavor enhancement (Seitz 1974; Sirisha et al. 2010; Boonmahome and Mongkolthanaruk 2013; Ferreira-Dias et al. 2013; Adrio and Demain 2014; Farrokh et al. 2014; Ullah et al. 2015). Nowadays, sn-1,3 lipases are also used to produce human milk fat substitutes (David 2017). Lipases hydrolyze fat triglycerides into free fatty acids which are widely used in flavor enhancement of butter, cheese, margarine, and chocolate milk (Joseph et al. 2007; Xiao et al. 2017). Processing of egg yolk for the synthesis of mayonnaise is also catalyzed by phospholipases (Ray 2012).

Food industries are now using lipase to generate short-chain aliphatic ester flavor for developing fragrance and flavor in food commodities (Macedo et al. 2003). Lipases catalyze trans-esterification reaction for the synthesis of ethyl caprylate, methyl butyrate, octyl acetate, and isoamyl acetate to impart fruity notes such as apple, peach, pineapple, strawberry, orange, and banana (Torres et al. 2009; Ahmed et al. 2010; Garlapati and Banerjee 2013; SÁ et al. 2017). Lipases from immobilized *Staphylococcus simulans* have been used for the ethyl valerate (green apple flavor) and hexyl acetate (pear flavor) in a solvent-free system (Karra-Châabouni et al. 2006). An alkaline lipase from organic solvent-tolerant *Acinetobacter* sp. EH28 is reported for synthesis of ethyl caprylate (Ahmed et al. 2010). Recently, Cong et al. (2019) reported synthesis of flavor esters by a novel lipase from *Aspergillus niger* in a soybean-solvent system. Fermented food variants like sour dough, olives, vegetable sausages, and cheese have been synthesized by halotolerant lipase obtained from *Lactobacillus plantarum* (Esteban-Torres et al. 2015). Lipases from coagulase-negative catalase-positive cocci (*Staphylococcu*s spp.) are known to release free short-chain fatty acids responsible for the aroma development in fermented sausage. Lipases are employed to change the characteristic of lipids by modifying the position of fatty acid chains in glycerol by replacing them with other fatty acids (Undurraga et al. 2001). Immobilized lipases from *C. antarctica* (CAL-B), *C. cylindracea* AY30, *Humicola lanuginosa*, *Pseudomonas* sp., and *G. candidum* were reported for the esterification of functional phenols for producing lipophilic antioxidants in sunflower oil (Buisman et al. 1998). Lipases have been used to remove fat and acidity reduction in fish products (Kazlauskas and Bornscheuer 1998; Mata et al. 2017).

Currently, there is a high demand for novel economic and green technologies for modification of vegetable oils with customized physicochemical and nutritional properties. Fatty acid-specific and regiospecific microbial lipases have gained significance and may be used for modification of vegetable oils. Upgradation of cheap oils can also be achieved through lipases to produce nutritionally vital triacylglycerols such as oleic acid-enriched oils, cocoa butter alternate, and low-calorie triacylglycerols. Lipase-mediated amendments are expected to occupy an important position in the oil industries for tailoring structured lipids (Gupta et al. 2003). Castor oil and distinctive vegetable oils have large amounts (about 90%) of ricinoleic acid, which is a hydroxy monounsaturated fatty acid of commercial importance. Lipase-catalyzed process has been utilized for the synthesis of this acid by castor oil hydrolysis (Goswami et al. 2012).

8.3.2 Fat and Oleochemical Industry

Combination of enzyme technology with fat chemistry has emerged as a novel way to produce oleochemicals from oils and fats which are of plant or animal origin. In this context, the use of microbial lipase is of much concern. Microbial lipase catalyzes hydrolysis, esterification, trans-esterification, and intra-esterification reactions, which are directly involved in the inexpensive synthesis of fat and oleochemicals in industries (Baumann et al. 1988). Use of nonspecific triacylglycerol lipases to produce polyunsaturated fatty acids (PUFA) and other dietary chemicals has been reported for a long time (McNeill et al. 1996; Jaeger and Reetz 1998; Nakajima et al. 2000). However 1,3-specific lipases can be used for hydrolysis of cruciferous oils to produce very-long-chain monounsaturated fatty acids such as gadoleic, erucic, and nervonic in the oleochemical industry. Okada and Morrissey (2008) used immobilized *C. rugosa* lipase for concentrating n-3 polyunsaturated fatty acids (n-3 PUFAs) from sardine oil. Free PUFAs and their mono- and diglycerides are employed in the manufacture of many pharmaceutical products including thrombolytics, anticholesterolemics, and anti-inflammatories (Belarbi et al. 2000). Köse et al. (2002) reported trans-esterification of refined cottonseed oil by immobilized *C. antarctica* lipase in a solvent-free medium for producing valuable intermediates in oleochemistry. Lipase-mediated hydrolysis and synthesis reactions are also commonly used for the upgradation of few of the less wanted fats into cocoa butter alternates (Undurraga et al. 2001). Lipase-catalyzed inter-esterification of butter fat is reported (Pabai et al. 1995). Application of microbial lipase in the synthesis of bio-plasticizer is also evaluated. In a study, Kim et al. (2019) showed the use of *C. antarctica*-immobilized lipase (Novozym 435) in 100% conversion of adipic acid and ethylhexanol used as a substrate into diethylhexyl adipate (DEHA) plasticizer. Earlier, epoxidation of soybean lecithin, a by-product obtained during vegetable oil, was also reported for the synthesis of epoxy lecithin plasticizer by Novozym 435 (Reddy et al. 2013).

8.3.3 Paper and Pulp Industry

For making various stuff in the paper and pulp industry wood processing is required. As wood also contains lipophilic compounds (also called as wood resin) and wood-extractive compounds soluble in nonpolar and polar solvents massive amount of organic solvents are needed during wood pulping and refining of paper (Gutiérrez et al. 2001). Organic solvents are not very useful because of two reasons: (1) lipophilic extracts generate colloidal pitch (constituents of wood mainly resin, waxes, and triglycerides) which may persist for a long time in waters and cause halt of the process (Hillis and Sumimoto 1989; Jaeger and Reetz 1998) and (2) chemical reaction of solvents with wood component also releases some recalcitrant compounds in the environment (Thompson et al. 2001). Hence, the enzymatic method for pitch control by using lipases was tried in the papermaking process at a larger scale (Bajpai 1999). Lipases from *C. rugosa* hydrolyze triglycerides by about 90% level in the pitch into glycerol/monoglycerides and fatty acid which are less sticky and hydrophilic and hence can be washed easily (Jaeger and Reetz 1998). Lipase can also be used for the deinking process in wastepaper for increased brightness and intensity (Bajpai 2014). Fukuda et al. (1990) showed that the lipase produced from *Pseudomonas* sp. (KWI-56) when added to a deinking composition enhanced paleness of paper and minimized residual ink spots. This approach also lessens chemical usage and saves energy, time, and manufacturing cost.

8.3.4 Detergents

Lipases are being used as potential additives in detergent formulations. These are also recognized as an ideal enzyme used in the detergent industry. Earlier, Godfrey and West (1996) estimated that about 1000 tons of lipases were yearly sold in the detergent market. In the detergent industry, particularly those enzymes which can act at high pH and temperature, withstand oxidizing and chelating agents, contain broad substrate specificity, and perform very well at low concentration are desired and lipases fulfill all of them (Anon 1983; Fuji et al. 1986; Ito et al. 1998). *Aspergillus oryzae*- and *Acinetobacter radioresistens*-derived lipases were found to remain active under alkaline conditions and showed their potential use in the laundry (Minoguchi and Muneyuki 1989; Gerhartz 1990; Nishioka et al. 1990; Satsuki and Watanabe 1990; Umehara et al. 1990). In a study, Cherif et al. (2011) reported a novel lipase from *Staphylococcus* sp. which showed extreme stability towards nonionic and anionic surfactants and relative stability towards oxidizing agents and hence excellent stability and compatibility in commercial production of solid and liquid detergents. Lipases from *Pseudomonas aeruginosa* have also been evaluated as detergent additives (Grbavčić et al. 2011; Dey et al. 2014). Role of *Bacillus sonorensis* lipase in corn oil stain removal in cotton fabric is reported by Nerurkar et al. (2013). Cold-active lipases are also being used as a functional additive and in synthesis of organic chiral compounds in detergent production at commercial level (Aboualizadeh et al. 2011; Zheng et al. 2011). Li et al. (2014) isolated cold-adapted lipase from *Pseudomonas stutzeri* PS59 and confirmed its broad substrate specificity and compatibility in the presence of surfactants, oxidizing agents, and other detergent additives required in the laundry industry. Alkaline and thermotolerant lipase from *P. aeruginosa* strain BUP2 has also been employed in detergent industries (Unni et al. 2016). Rathi et al. (2001) also reported a novel alkaline lipase from *Burkholderia cepacia* and its use in detergent formulation. Niyonzima and More (2015) discussed the role of lipases obtained from several species of bacilli and other bacterial strains in the detergent industry.

8.3.5 Textile Industry

In the textile industry, lipases are used in the removal of the residue of machine lubricants and desizing of fabrics (Lange 1997). Microbial lipases have already been used in improving the moisture regain, wettability, and dyeability of polyethylene terephthalate (PET) fabrics (Kim and Song 2006). In hydrolytic action lipase creates more hydrophilic groups on the surface of PET fabrics, which results in greater absorbency to the fabric for enhanced levelness in dyeing (Kim and Song 2008). Nerurkar et al. (2015) found that lipase from a marine bacteria *B. sonorensis* can be used for bioscouring (enzymatic removal of impurities from cotton fabric) which is more effective than conventional alkaline treatment. However, microbial lipase in combination with pectinase in a one-step process was found to reduce the bioscouring time and provided excellent dyeing performance and fabrics properties (Kalantzi et al. 2010). Lipases, together with alpha-amylase, are employed for the desizing in denims and other cotton fabrics (Rowe 2001). Buchert et al. (2000) also showed combined use of pectinases, proteases, and lipases in bioscouring of cotton fabric. Lipase can even remove tough fatty stains from cloth (Shaikh 2010). Kantouch et al. (2005) found that lipase pretreatment facilitated wool dyeing under mild temperature, increased dye consumption, and enhanced the rate of dyeing which can have a major impact on energy saving and reduction of pollution impact.

8.3.6 Leather Industry

The high demand for organic solvents, chemicals, and surfactants in the leather industry has not only increased the commercial burden of industries but also upstretched environmental issues. In the tannery, before the tanning process, skins and hides of animals which mainly consist of proteins and fat are removed. For this, several hazardous chemicals are used on a routine basis. However, instead of using chemicals, microbial enzyme-based approaches have proven to be more sustainable (Kamini et al. 1999). Lipases remove fat in the degreasing stages without generating toxic by-products which are harmful to humans and environment. Earlier, lipase from *Rhizopus nodosus* was applied for the degreasing of suede clothing leathers from wooled sheep skins (Muthukumaran and Dhar 1982). Currently, microbial lipase, in combination with proteases, is being utilized in the leather industry for dehairing of animal hides and skin (Sanchez and Demain 2017). Alkaline lipases are exclusively used for the processing of high-quality leather (Wanyonyi and Mulaa 2019). These lipases impart an extra consistency in color and a cleaner look of leather. Acid-active lipases have been applied in the treatment of skins preserved in a pickled state. An acid lipase named NovoCor AD is used in stages of degreasing of hides and skins. Lipases have also been utilized in the synthesis of hydrophobic and waterproof leather (Hasan et al. 2006).

8.3.7 Cosmetics

Microbial lipase can be used as an emollient in the manufacture of cosmetics products such as suntan creams, bath oils, and moisturizing lotion skin. Lipases are potentially used in ingredient dermocosmetic formulation development possessing sensory or biological activities (Yvergnaux 2017). Ansorge-Schumacher and Thum (2013) reviewed the application of immobilized lipases in the cosmetics industry. They highlighted that process simplification, product quality, and environmental footprint are the selective advantages of lipases, which can replace traditional chemical processes. Unichem International (Spain) and Croda Universal Ltd. are using *C. cylindracea* lipase for production of quality products. Noh et al. (2019) review production of butyl butyrate as a fragrance and flavoring compound by microbial lipase. Recently, the trans-esterification process for the synthesis of geranyl acetate from geraniol and ethyl acetate was catalysed by *C. antarctica* lipase (Bhavsar and Yadav 2019). This greener solvent-free system has been found to be very useful for the perfume industry. The commercial application of immobilized and lyophilized *C. antarctica* lipase in the synthesis of an aromatic ester benzyl propionate used as a fragrance ingredient in several cosmetics is also well explored (de Meneses et al. 2019). Retinoids, vitamin A, and its water-soluble derivatives were made by the catalytic action of immobilized lipase and are being used in personal care skin products and pharmaceuticals (Maugard et al. 2002).

8.3.8 Biofuel

The limited availability of nonrenewable energy resources, high prices of crude oil, and environmental concerns raised the demand of using vegetable oils as an alternative to fossil fuels (Shah et al. 2004). Biodiesel also called as biofuel mainly consists of esters of long-chain fatty acids and short-chain alcohols. Currently, biodiesel catalysts are grouped as alkali, acid, or enzyme. Amongst them, the use of enzyme is an eco-friendly and cost-effective approach. Lipases catalyze trans-esterification process and convert fatty acids and short-chain alcohols into methanol (Vicente et al. 2004; Lotti et al. 2015). In this process of trans-esterification dislocation of alcohol from an ester takes place by another alcohol of high commercial value (Srivastava and Prasad 2000). The trans-esterification by lipase of *Acinetobacter venetianus* RAG-1 was used in the production of biodiesel (Boonmahome and Mongkolthanaruk 2013). Some workers reported the application of K_2SO_4-coated lipase microcrystals as a substitute for chemical catalysts for the synthesis of biodiesel (Sirisha et al. 2010; Zheng et al. 2012). According to Hegde et al. (2015), lipase-mediated trans-esterification reactions may reduce the cost of downstream processing in the production of biodiesel. Apart from this, the use of vegetable oil-based biofuel is a cleaner alternative because it does not emit oxides of sulfur whereas the concentration of particulate matter is also found to be about one-third as compared to petroleum. Iso et al. (2001) stated that due to environmental advantages, biodiesel fuel could be a proper substitute for conventional fuel. In a study, Karmee et al. (2015) reported that lipases convert methanol and waste oil generated from cooking waste into usable products like biodiesel and glycerol. Lipases stable at alkaline pH are more valuable in the biodiesel synthesis as alkaline pH upgrades oil solubility and homogeneity of the reaction mixture which makes trans-esterification rates higher (Christopher et al. 2015; Li et al. 2016). Liu et al. (2011)

found that lipases obtained from *B. cepacia* and then immobilized on NKA resin can act as a biocatalyst in the synthesis of biodiesel. Abdulla and Ravindra (2013) showed that lipase from *B. cepacia* immobilized on the matrix of alginate and k-carrageenan remained active after six consecutive cycles of re-usage in the production of biodiesel. Immobilized lipase obtained from *Botryococcus* sp. showed excellent trans-esterification reaction in biodiesel production (Sivaramakrishnan and Incharoensakdi 2016). In a study, Yoo et al. (2011) showed that when soya bean and palm oil were used, lipase of *Ralstonia* sp. catalyzed the synthesis of diesel at pH 8 with 5% methanol and 20% water content. Use of lipase as biocatalyst for the synthesis of biodegradable polymers like 1-butyl oleate, which reduces the level of viscosity in biodiesel in the cold season, is also reported (Linko et al. 1998). Lubricating properties of the oil can be enhanced by generating tri-methylolpropane (TMP) tri-ester by the action of immobilized lipase from thermophilic fungus *T. lanuginosus* (Ghamgui et al. 2004). A novel lipase SL-4 from *Burkholderia ubonensis* showed the production of biodiesel with a conversion ratio of 92.24% by catalyzing soybean oil in a solvent-free system. Yang et al. (2016) reported a novel thermostable lipase as a biocatalyst for the synthesis of biodiesel.

8.3.9 Therapeutic and Pharmaceuticals

Lipases have a wide role in the medicinal and pharmaceutical field. The enantioselective, trans-esterification, and inter-esterification reactions catalyzed by lipases have immense importance in the drug-based industry (Stinson 1995). Since past, lipases have been used in making specialized lipids, digestive aids, and alteration of monoglycerides (used as emulsifiers) in the pharmaceutical industry (Vulfson 1994; Sharma et al. 2001). Yang et al. (1997) found that a lipase from *C. rugosa* immobilized on nylon performed regioselective acylation reaction in the synthesis of lovastatin drug, used for lowering serum cholesterol levels. Uttatree et al. (2010) reported that a thermostable and organic solvent-tolerating lipase from *Acinetobacter baylyi* could act as a strong catalyst in generating bioenergy and trans-esterification of palm oil used for medicinal purpose. Earlier, Saphir (1967) got a patent for the role of lipases in hair waving process. Sangeetha (2011) discussed medicinal applications of lipase in the treatment of diseases related to skin of scalp and hair loss. Cold-adapted lipase B from *C. antarctica* has been utilized for solvent-free synthesis of citronellol laurate (Ganapati and Piyush 2005). Shimada et al. (2001) also showed solvent-free synthesis of a potentially useful pharmaceutical substance, ethyl docosahexaenoate (EtDHA), from lipases B of *C. antarctica.* Similar results were obtained for extracellular and immobilized lipases B of *C. antarctica* where these were used for the synthesis of amino sugar fatty acid esters including aryl aliphatic glycolipids (Otto et al. 2000; Pöhnlein et al. 2014). Lipases from *Bacillus* sp. having specificity towards fatty acid ester can be employed to synthesize enantiopure compounds in medical industries (Guncheva and Zhiryakova 2011). *Staphylococcus* lipase was used to produce antioxidant chemicals, including eugenol benzoate, tyrosol acetate, and propyl gallate (Horchani et al. 2012). Lipases are also used in the diagnosis of tuberculosis (TB) where *Mycobacterium tuberculosis* lipase is employed to detect the infection with high specificity and sensitivity (Brust et al. 2011). They are also utilized in the treatment of tumors, malignant in nature, by activating the tumor necrosis factor (Nagarajan 2012).

8.3.10 Biosensors

The combination of a thin layer of the insolubilized (immobilized) enzyme with an electrochemical probe can be collectively used as a sensing element or biosensor (Trojanowicz 2014). Along with esterases, and phospholipases, lipases also have a role in enzyme-based biosensor applications. These are being used in the food, chemical, polymer, and medical industries (Sandoval and Herrera-López 2018). Their application as a diagnostic tool to detect cholesterol and triglyceride levels in blood samples has revolutionized the role of microbe-based sensor in clinical field (Starodub 2006). Earlier work also confirmed utilization of lipase biosensors combined with glucose oxidase onto pH/oxygen electrodes in determinations of both triglycerides and cholesterol level in the blood (Karube and Sode 1988; Imamura et al. 1989). In the last few years, enzyme-labeled probes replaced the use of unstable and harmful isotopes in biosensor applications (Gurung et al. 2013). Oligonucleotide-labeled lipases have been employed to sense the presence of complementary nucleic acids by hybridization (Kynclova et al. 1995). In this context, fungal lipases have proven to be useful due to high sensitivity and turnover number, thermostability, and stability at room temperature under working conditions (Kynclova et al. 1995). Zehani et al. (2014) developed novel impedimetric biosensors by using two lipases, one from *C. rugosa* (microbial source) and another from porcine pancreas (animal source) immobilized on the gold electrode for highly sensitive and rapid quantitative detection of diazinon pesticide in an aqueous medium. Nowadays, several other lipase-based biosensors have been used in different sectors.

8.3.11 Waste Treatment

Since a long time, lipases are being used in activated sludge and other aerobic waste processes. Initially, Bailey and Ollis (1986) found that the thin layers of fat were repeatedly removed by lipases of *C. rugosa* from the surface of aerated tanks to permit oxygen availability. Utilization of lipases in effluent processing units such as poultry waste, abattoirs, food processing industry, leather industry, anaerobic digesters, and household sewage was reported by Godfrey and Reichelt (1983). Tschocke (1990) also showed that immobilized forms of lipases hydrolyze triglycerides in the fat present in wastewater treatment plants. Dharmsthiti and Kuhasuntisuk (1998) reported that *P. aeruginosa* LP602 and its lipase can be used for the treatment of wastewater rich in lipid content. Microbial enzyme-based products are commercially produced at industrial level for bioremediation of oil-contaminated sites. A product named WW07P manufactured by Oasis Environmental Ltd. company contains a specific microbe that is being used in the cleanup of greases, fats, and oils in wastewater treatment comprising elevated amounts of grease, oil, and fat (http://www.oasisenviro.co.uk/ww07pproductinfo.html). Immobilized form of lipase was also utilized as biocatalysts to form simple alkyl ester derivatives of restaurant grease. *Pseudomonas cepacia* was used to catalyze the degradation of hydrocarbons in grease (Hsu et al. 2002). In anaerobic digester, extracellular lipase converts sewage sludge and particulate organic matter into simpler compounds of relatively lower molecular weights (Whiteley et al. 2003). Parmar et al. (2001) reported that lipase, protease, and cellulase in equivalent proportions were commercially produced and employed in the reduction of 30–50% total suspended solids in sludge and hence improved settling ability of solid substances. Vasileva-Tonkova and Galabova (2003) reported the role of lipase of bacteria isolated from wastewater in the bioaugmentation process for cleaning of wastewater, heavily contaminated with hydrocarbons and organic polymers.

8.4 Conclusion

Lipases are extremely useful biomolecules, and along with other microbial enzymes they offer great biotechnological potential. There are various microbial sources of lipase which are used for relatively cheaper production of enzyme at the industry level. Lipases do not require cofactors and can be easily immobilized on support materials and hence researchers have also utilized this trait and developed techniques for fabricating these enzymes on cheap carrier materials for effective utilization of product recovery with reuse (of enzyme). Lipases show versatility in terms of substrate utilization and transform different substrates into products of high commercial value, which cannot be synthesized by the conventional chemical processes. In the last few years, vigorous research in the field of microbial enzyme technology opened up possibilities of low-cost production of microbial lipases. Utilization of microbial enzyme from extremophiles could provide add-on features to the existing microbial enzyme technology (Das et al. 2019). The catalytic properties, such as stability at high temperatures and pH, are the additional benefits of extremozymes, which are very useful to industries (Arora and Panosyan 2019). Although many microbial strains have been identified including cold-adapted, thermotolerant, alkaline, and acidic lipases, further research for harnessing their use in the broader context of industrial sustainability is needed. Similarly, engineered lipases possessing numerous characteristics amenable for bioprocess and environmental applications are yet to be explored. Thus, there is a need to delve all possible measures which could be used to enhance microbial lipase production, not only for replenishing the current demand but also for achieving the goal of industrial sustainability.

Acknowledgement The author Sonal Dixit is thankful to DSKPDF Cell, Pune, India, and University Grant Commission, New Delhi, India, for the award of the D.S. Kothari Postdoctoral Fellowship (F4-2/2006 (BSR)/BL/15-16/0156).

References

Abdulla R, Ravindra P (2013) Immobilized *Burkholderia cepacia* lipase for biodiesel production from crude *Jatropha curcas L.* oil. Biomass Bioenergy 56:8–13

Aboualizadeh F, Kaur J, Behzad-Behbahani A, Khalvati B (2011) Induction of mutation in *Bacillus subtilis* lipase gene using error-prone PCR. Jundishapur J Microbiol 4:153–158

Adamczak M, Bednarski W (2004) Enhanced activity of intracellular lipases from *Rhizomucor miehei* and *Yarrowia lipolytica* by immobilization on biomass support particles. Process Biochem 39:1347–1361

Adrio JL, Demain AL (2014) Microbial enzymes: tools for biotechnological processes. Biomolecules 4:117–139

Ahmed EH, Raghavendra T, Madamwar D (2010) An alkaline lipase from organic solvent tolerant *Acinetobacter sp.* EH28: application for ethyl caprylate synthesis. Bioresour Technol 101:3628–3634

Anon (1983) Enzymes in consumer products technology. Chem Eng Technol 55:589–590

Ansorge-Schumacher MB, Thum O (2013) Immobilised lipases in the cosmetics industry. Chem Soc Rev 42:6475–6490

Arora NK, Panosyan H (2019) Extremophiles: applications and roles in environmental sustainability. Environ Sustain 2(3):217–218

Ashfaq M (2015) Basmati rice a class apart (a review). J Rice Res 3:88–95

Bailey JE, Ollis DF (1986) Applied enzyme catalysis. In: Biochemical engineering fundamentals, 2nd edn. McGraw-Hill, New York, NY, pp 157–227

Bajpai P (1999) Application of enzymes in the pulp and paper industry. Biotechnol Prog 15:147–157

Bajpai P (2014) Deinking with enzymes. In: Bajpai P (ed) Recycling and deinking of recovered paper. Elsevier, Amsterdam, pp 139–153

Balaji V, Ebenezer P (2008) Optimization of extracellular lipase production in *Colletotrichum gloeosporioides* by solid state fermentation. Indian J Sci Technol 1:1–8

Baumann H, Bühler M, Fochem H, Hirsinger F, Zoebelein H, Falbe J (1988) Natural fats and oils—renewable raw materials for the chemical industry. Angew Chem Int Ed Engl 27:41–62

Belarbi EH, Molina E, Chisti Y (2000) RETRACTED: a process for high yield and scaleable recovery of high purity eicosapentaenoic acid esters from microalgae and fish oil

Bhavsar KV, Yadav GD (2019) Synthesis of geranyl acetate by transesterification of geraniol with ethyl acetate over *Candida antarctica* lipase as catalyst in solvent-free system. Flavour Fragr J 34:288–293

Boonmahome P, Mongkolthanaruk W (2013) Lipase-producing bacterium and its enzyme characterization. J Life Sci Technol 1:196–200

Brust B, Lecoufle M, Tuaillon E, Dedieu L, Canaan S, Valverde V, Kremer L (2011) *Mycobacterium tuberculosis* lipolytic enzymes as potential biomarkers for the diagnosis of active tuberculosis. PLoS One 6(9):e25078

Buchert J, Pere J, Puolakka A, Nousiainen P (2000) Scouring of cotton with pectinases, proteases, and lipases. Text Chemist Colorist Am Dyestuff Rep 32:4852

Buisman GJH, van Helteren CTW, Kramer GFH, Veldsink JW, Derksen JTP, Cuperus FP (1998) Enzymatic esterifications of functionalized phenols for the synthesis of lipophilic antioxidants. Biotechnol Lett 20:31–136

Cavalcanti EDC, Aguieiras ÉCG, da Silva PR, Duarte JG, Cipolatti EP, Fernandez-Lafuente R, da Silva JAC, Freire DMG (2018) Improved production of biolubricants from soybean oil and different polyols via esterification reaction catalyzed by immobilized lipase from *Candida rugosa*. Fuel 215:705–713

Cheetham PSJ (1995) The applications of enzymes in industry. In: Wiseman A (ed) Handbook of enzyme biotechnology. Ellis Horwood, Chichester, pp 419–522
Cherif S, Mnif S, Hadrich F, Abdelkafi S, Sayadi S (2011) A newly high alkaline lipase: an ideal choice for application in detergent formulations. Lipids Health Dis 10:221
Christopher LP, Zambare VP, Zambare A, Kumar H, Malek L (2015) A thermoalkaline lipase from a new thermophile *Geobacillus thermodenitrificans* AV-5 with potential application in biodiesel production. J Chem Technol Biotechnol 90:2007–2016
Cong S, Tian K, Zhang X, Lu F, Singh S, Prior B, Wang ZX (2019) Synthesis of flavor esters by a novel lipase from *Aspergillus niger* in a soybean-solvent system. 3 Biotech 9(6):244
Das D, Kalra I, Mani K, Salgaonkar BB, Braganca JM (2019) Characterization of extremely halophilic archaeal isolates from Indian salt pans and their screening for production of hydrolytic enzymes. Environ Sustain 2(3):227–239
David G (2017) Lipases industrial applications: focus on food and agroindustries. OCL 24(4):D403
de Meneses AC, Lerin LA, Araújo PHH, Sayer C, de Oliveira D (2019) Benzyl propionate synthesis by fed-batch esterification using commercial immobilized and lyophilized Cal B lipase. Bioprocess Biosyst Eng 42(10):1625–1634
Dellamora-ortiz GM, Martins RC, Rocha WL, Dias AP (1997) Activity and stability of a *Rhizomucor miehei* lipase in hydrophobic media. Biotechnol Appl Biochem 26:31–37
Dey A, Chattopadhyay A, Mukhopadhyay S (2014) An approach to the identification and characterisation of a psychrotrophic lipase producing *Pseudomonas sp.* ADT3 from Arctic Region. Adv Biosci Biotechnol 5:322–332
Dharmsthiti S, Kuhasuntisuk B (1998) Lipase from *Pseudomonas aeruginosa* LP602: biochemical properties and application for wastewater treatment. J Ind Microbiol Biotechnol 21:75–80
Eddehech A, Zied Z, Aloui F, Smichi N, Noiriel A, Abousalham A, Gargouri Y (2019) Production, purification and biochemical characterization of a thermoactive, alkaline lipase from a newly isolated Serratia sp. W3 Tunisian strain. Int J Biol Macromol 123:792–800
Esteban-Torres M, Mancheno JM, de las Rivas B, Munoz R (2015) Characterization of a halotolerant lipase from the lactic acid bacteria *Lactobacillus plantarum* useful in food fermentations. LWT Food Sci Technol 60:246–252
Facin BR, Melchiors MS, Valério A, Oliveira JV, Oliveira DD (2019) Driving immobilized lipases as biocatalysts: 10 years state of the art and future prospects. Ind Eng Chem Res 58(14):5358–5378
Faouzi L, Fatimazahra EB, Moulay S, Adel S, Wifak B, Soumya E, Iraqui M, Saad KI (2015) Higher tolerance of a novel lipase from *Aspergillus flavus* to the presence of free fatty acids at lipid/water interface. Afr J Biochem Res 9:9–17
Farrokh P, Yakhchali B, Karkhane AA (2014) Cloning and characterization of newly isolated lipase from *Enterobacter sp*. Bn12. Braz J Microbiol 45:677–687
Ferreira-Dias S, Sandoval G, Plou F, Valero F (2013) The potential use of lipases in the production of fatty acid derivatives for the food and nutraceutical industries. Electron J Biotechnol 16:1–28
Fjerbaek L, Christensen KV, Norddahl B (2009) A review of the current state of biodiesel production using enzymatic transesterification. Biotechnol Bioeng 102:1298–1315
Fuji T, Tatara T, Minagawa M (1986) Studies on application of lipolytic enzyme in detergent industries. J Am Oil Chem Soc 63:796–799
Fukuda S, Hayashi S, Ochiai H, Iizumi T, Nakamura K (1990) Improvers for deinking of wastepaper. Japanese Patent 2:229–290
Fukuda H, Kondo A, Noda H (2001) Biodiesel fuel production by transesterification of oils. J Biosci Bioeng 92:405–416
Ganapati DY, Piyush SL (2005) Lipase-catalyzed transesterification of methyl acetoacetate with n-butanol. J Mol Catal B: Enzym 32:107–113
Garlapati VK, Banerjee R (2013) Solvent-free synthesis of flavour esters through immobilized lipase mediated transesterification. Enzyme Res 2013:6
Gerhartz W (1990) Industrial uses of enzymes. In: Enzymes in industry production and application. VCH, Weinheim, pp 77–148

Ghamgui H, Karra-Chaâbouni M, Gargouri Y (2004) 1-butyl oleate synthesis by immobilized lipase from *Rhizopus oryzae*: a comparative study between n-hexane and solvent-free system. Enzyme Microb Technol 35(4):355–363

Ghorai S, Banik SP, Verma D, Chowdhury S, Mukherjee S, Khowala S (2011) Comprehensive biotechnology. In: Moo-Young M (ed) Fungal biotechnology in food and feed processing, 2nd edn. Academic Press, Burlington, MA, pp 603–615

Godfrey T, Reichelt J (1983) Industrial applications. In: Industrial enzymology—applications of enzymes in industry. The Nature Press, London, pp 170–465

Godfrey T, West S (1996) Industrial enzymology. Macmillan Press, London, p 3

Gomes FM, Pereira EB, de Castro HF (2004) Immobilization of lipase on chitin and its use in nonconventional biocatalysis. Biomacromolecules 5(1):17–23

Goswami D, Basu JK, De S (2012) Lipase applications in oil hydrolysis with a case study on castor oil: a review. Crit Rev Biotechnol 33:1–16

Gotor-Fernández V, Busto E, Gotor V (2006) Candida antarctica lipase B: an ideal biocatalyst for the preparation of nitrogenated organic compounds. Adv Synth Catal 348:797–812

Grbavčić S, Bezbradica D, Izrael-Zivkovic L, Avramovic N, Milosavic N, Karadzic I, Knezevic-Jugovic Z (2011) Production of lipase and protease from an indigenous *Pseudomonas aeruginosa* strain and their evaluation as detergent additives: compatibility study with detergent ingredients and washing performance. Bioresour Technol 102:11226–11233

Guerrand D (2017) Lipases industrial applications: focus on food and agroindustries. OCL 24:D403

Guncheva M, Zhiryakova D (2011) Catalytic properties and potential applications of *Bacillus* lipases. J Mol Catal B: Enzym 68:1–21

Guo J, Chen CP, Wang SG, Huang XJ (2015) A convenient test for lipase activity in aqueous-based solutions. Enzyme Microb Technol 71:8–12

Gupta R, Gupta N, Rathi P (2004) Bacterial lipases: an overview of production, purification and biochemical properties. Appl Microbiol Biotechnol 64(6):763–781

Gupta R, Rathi P, Bradoo S (2003) Lipase mediated upgradation of dietary fats and oils. Crit Rev Food Sci Nutr 43:635–644

Gurung N, Ray S, Bose S, Rai V (2013) A broader view: microbial enzymes and their relevance in industries, medicine, and beyond. Biomed Res Int 2013:329121

Gutiérrez A, del Río JC, Martínez MJ, Martínez AT (2001) The biotechnological control of pitch in paper pulp manufacturing. Trends Biotechnol 19:340–348

Hasan F, Shah AA, Hameed A (2006) Industrial applications of microbial lipases. Enzyme Microb Technol 39(2):235–251

Hasan F, Shah AA, Hameed A (2009) Methods for detection and characterization of lipases: a comprehensive review. Biotechnol Adv 27:782–798

Hegde K, Chandra N, Sarma SJ, Brar SK, Veeranki VD (2015) Genetic engineering strategies for enhanced biodiesel production. Mol Biotechnol 57:606–624

Hillis WE, Sumimoto M (1989) Effect of extractive on pulping. In: Rowe JW (ed) Natural Products of Woody Plants II. Springer, Berlin, pp 880–920

Holmquist M (1998) Insights into the molecular basis for fatty acyl specificities of lipases from *Geotrichum candidum* and *Candida rugosa*. Chem Phys Lipids 93:57–65

Horchani H, Aissa I, Ouertani S, Zarai Z, Gargouri Y, Sayari A (2012) Staphylococcal lipases: biotechnological applications. J Mol Catal B: Enzym 76:125–132

Hsu AF, Jones K, Foglia TA, Marmer WN (2002) Immobilized lipase catalysed production of alkyl esters of restaurant grease as biodiesel. Biotechnol Appl Biochem 36:181–186

Iftikhar T, Niaz M, Ali EA, Jabeen R, Abdullah R (2012) Production process of extracellular lipases by *Fusarium sp.* using agricultural by products. Pak J Bot 44:335–339

Imamura S, Takahashi M, Misaki H, Matsuura K (1989) Method and reagent containing lipases for enzymatic determination of triglycerides. West Germany Patent 3,912,226

Iso M, Chen B, Eguchi M, Kudo T, Shrestha S (2001) Production of biodiesel fuel from triglycerides and alcohol using immobilized lipase. J Mol Cat B Enzymat 16:53–58

Ito S, Kobayashi T, Ara K, Ozaki K, Kawai S, Hatada Y (1998) Alkaline detergent enzymes from alkaliphiles: enzymatic properties, genetics and structures. Extremophiles 2:185–190

Jaeger KE, Ransac S, Dijkstra BW, Colson C, van Heuvel M, Misset O (1994) Bacterial lipases. FEMS Microbiol Rev 15(1):29–63

Jaeger K-E, Reetz MT (1998) Microbial lipases form versatile tools for biotechnology. Trends Biotechnol 16:396–403

Javed S, Azeem F, Hussain S, Rasul I, Siddique MS, Riaz M, Afzal M, Kouser A, Nadeem H (2018) Bacterial lipases: a review on purification and characterization. Prog Biophys Mol Biol 132:23–34

Jegannathan KR, Abang S, Poncelet D, Chan ES, Ravindra P (2008) Production of biodiesel using immobilized lipase-A critical review. Crit Rev Biotechnol 28:253–264

Joseph B, Ramteke PW, Thomas G (2008) Cold active microbial lipases: some hot issues and recent developments. Biotechnol Adv 26:457–470

Joseph B, Ramteke PW, Thomas G, Shrivastava N (2007) Standard review cold active microbial lipases: a versatile tool for industrial applications. Biotechnol Mol Biol Rev 2:39–48

Kalantzi S, Mamma D, Kalogeris E, Kekos D (2010) Improved properties of cotton fabrics treated with lipase and its combination with pectinase. Fibres Text East Eur 18(5):86–92

Kamini NR, Hemachander C, Geraldine Sandana Mala J, Puvanakrishnan R (1999) Microbial enzyme technology as an alternative to conventional chemicals in leather industry. Curr Sci 77(1):80–86

Kantouch A, Raslan WM, El-Sayed H (2005) Effect of lipase pretreatment on the dyeability of wool fabric. J Nat Fibers 2(2):35–48

Kapoor M, Gupta MN (2012) Lipase promiscuity and its biochemical applications. Process Biochem 47:555–569

Karmee SK, Patria RD, Lin CSK (2015) Techno-economic evaluation of biodiesel production from waste cooking oil-A case study of Hong Kong. Int J Mol Sci 16:4362–4371

Karra-Châabouni M, Ghamgui H, Bezzine S, Rekik A, Gargouri Y (2006) Production of flavour esters by immobilized Staphylococcus simulans lipase in a solvent-free system. Process Biochem 41(7):1692–1698

Karube I, Sode K (1988) Biosensors for lipids. In: Applewhite TH (ed) Proceedings of the World Conference on Biotechnology in the Fats and Oils industry. Campaign, USA, American Oil Chemists Society, pp 215–218

Kazlauskas RJ, Bornscheuer UT (1998) In: Rehm HJ, Reed G, Pühler A, Stadler PJW, Kelly DR (eds) Biotechnology: biotransformations with lipases, vol 8. VCH, Weinheim, pp 37–191

Kim H, Kim T, Choi N, Kim BH, Oh SW, Kim IH (2019) Synthesis of diethylhexyl adipate by *Candida antarctica* lipase-catalyzed esterification. Process Biochem 78:58–62

Kim HR, Song WS (2006) Lipase treatment of polyester fabrics. Fibers Polymers 7(4):339–343

Kim H-R, Song W-S (2008) Effects of triton X-100 and calcium chloride on the porcine pancreas lipase treatment of PET fabrics. J Korean Soc Cloth Text 32:911–917

Köse Ö, Tüter M, Aksoy HA (2002) Immobilized *Candida antarctica* lipase-catalyzed alcoholysis of cotton seed oil in a solvent-free medium. Bioresour Technol 83(2):125–129

Kynclova E, Hartig A, Schalkhammer T (1995) Oligonucleotide labelled lipase as a new sensitive hybridization probe and its use in bio-assays and biosensors. J Mol Recognit 8:139–145

Lange NK (1997) Lipase-assisted desizing of woven cotton fabrics. Text Chemist Colorists 29(6):23–26

Lee LP, Karbul HM, Citartan M, Gopinath SCB, Lakshmipriya T, Tang TH (2015) Lipase secreting *Bacillus species* in an oil-contaminated habitat: promising strains to alleviate oil pollution. Biomed Res Int 2015:1–9

Li XL, Zhang WH, Wang WD, Dai YJ, Zhang HT, Wang Y, Wang HK, Lu FP (2014) A high-detergent-performance, cold-adapted lipase from *Pseudomonas stutzeri* PS59 suitable for detergent formulation. J Mol Catal B Enzym 102:16–24

Li N, Li DD, Zhang YZ, Yuan YZ, Geng H, Xiong L, Liu DL (2016) Genome sequencing and systems biology analysis of a lipase-producing bacterial strain. Genet Mol Res 15:1–12

Linko YY, Lamsa M, Wu X, Uosukainen E, Seppala J, Linko P (1998) Biodegradable products by lipase biocatalysis. J Biotechnol 66(1):41–50

Liu Y, Chen D, Yan Y, Peng C, Xu L (2011) Biodiesel synthesis and conformation of lipase from *Burkholderia cepacia* in room temperature ionic liquids and organic solvents. Bioresour Technol 102:10414–10418

Lotti M, Pleiss J, Valero F, Ferrer P (2015) Effects of methanol on lipases: molecular, kinetic and process issues in the production of biodiesel. Biotechnol J 10:22–30

Macedo GA, Lozano MMS, Pastore GM (2003) Enzymatic synthesis of short chain citronellyl esters by a new lipase from *Rhizopus sp.* J Biotechnol 6:72–75

Martinelle M, Hult K (1995) Kinetics of acyl transfer reactions in organic media catalysed by *Candida antarctica* lipase B. Biochim Biophys Acta 1251:191–197

Marzuki NHC, Mahat NA, Huyop F, Aboul-Enein HY, Wahab RA (2015) Sustainable production of the emulsifier methyl oleate by *Candida rugosa* lipase nanoconjugates. Food Bioprod Process 96:211–220

Masse L, Kennedy KJ, Chou SP (2001) The effect of an enzymatic pretreatment on the hydrolysis and size reduction of fat particles in slaughterhouse wastewater. J Chem Technol Biotechnol 76:629–635

Mata TM, Correia D, Pinto A, Andrade S, Trovisco I, Matos E, Caetano NS (2017) Fish oil acidity reduction by enzymatic esterification. Energy Procedia 136:474–480

Matsumoto T, Takahashi S, Kaieda M, Ueda M, Tanaka A, Fukuda H, Kondo A (2001) Yeast whole-cell biocatalyst constructed by intracellular overproduction of *Rhizopus oryzae* lipase is applicable to biodiesel fuel production. Appl Microbiol Biotechnol 57:515–520

Maugard T, Rejasse B, Legoy MD (2002) Synthesis of water-soluble retinol derivatives by enzymatic method. Biotechnol Prog 18:424–428

McNeill GP, Ackman RG, Moore SR (1996) Lipase-catalyzed enrichment of long-chain polyunsaturated fatty acids. J Am Oil Chem Soc 73(11):1403–1407

Mehrasbi MR, Mohammadi J, Peyda M, Mohammadi M (2017) Covalent immobilization of *Candida antarctica* lipase on core-shell magnetic nanoparticles for production of biodiesel from waste cooking oil. Renew Energy 101:593–602

Minoguchi M, Muneyuki T (1989) Immobilization of lipase on polyacrylamide and its use in detergents. Japanese Patent 1:285,188

Morais Júnior WG, Fernández-Lorente G, Guisán JM, Ribeiro EJ, De Resende MM, Costa Pessela B (2017) Production of omega-3 polyunsaturated fatty acids through hydrolysis of fish oil by *Candida rugosa* lipase immobilized and stabilized on different supports. Biocatal Biotransformation 35:63–73

Musa H, Hafiz Kasim F, Nagoor Gunny AA, Gopinath SCB, Azmier Ahmad M (2019) Enhanced halophilic lipase secretion by Marinobacter litoralis SW-45 and its potential fatty acid esters release. J Basic Microbiol 59(1):87–100

Muthukumaran N, Dhar SC (1982) Comparative studies on the degreasing of skins using acid lipase and solvent with reference to the quality of finished leathers. Leather Sci 29:417–424

Nagarajan S (2012) New tools for exploring old friends-microbial lipases. Appl Biochem Biotechnol 168:1163–1196

Nakajima M, Snape J, Khare SK (2000) Method in non-aqueous enzymology. In: Gupta MN (ed) Biochemistry. Birkhauser Verlag, Basel, Switzerland, pp 52–69

Nerurkar M, Joshi M, Adivarekar R (2015) Bioscouring of cotton using lipase from marine bacteria Bacillus sonorensis. Appl Biochem Biotechnol 175:253–265

Nerurkar M, Joshi M, Pariti S, Adivarekar R (2013) Application of lipase from marine bacteria *Bacillus sonorensis* as an additive in detergent formulation. J Surfact Deterg 16:435–443

Nishioka M, Joko K, Takama M (1990) Lipase manufacture with *Candida* for use in detergents. Japanese Patent 292,281

Niyonzima FN, More SS (2015) Coproduction of detergent compatible bacterial enzymes and stain removal evaluation. J Basic Microbiol 55:1149–1158

Noh HJ, Lee SY, Jang YS (2019) Microbial production of butyl butyrate, a flavor and fragrance compound. Appl Microbiol Biotechnol 103:2079

Nunes P, Martins AB, Santa Brigida AI, Miguez Da Rocha Leao MH, Amaral P (2014) Intracellular lipase production by *Yarrowia lipolytica* using different carbon sources. Chem Eng Trans 38:421–426
Okada T, Morrissey MT (2008) Production of n-3 polyunsaturated fatty acid concentrate from sardine oil by immobilized Candida rugosa lipase. J Food Sci 73(3):C146–C150
Ollis DL, Cheah E, Cygler M, Dijkstra B, Frolow F, Franken SM, Sussman JL (1992) The α/β hydrolase fold. Protein Eng Des Sel 5(3):197–211
Otto RT, Scheib H, Bornscheuer UT, Pleiss J, Syldatk C, Schmid RD (2000) Substrate specificity of lipase B from *Candida antarctica* in the synthesis of arylaliphatic glycolipids. J Mol Catal B: Enzym 8(4):201–211
Pabai F, Kermasha S, Morin A (1995) Lipase from Pseudomonas fragi CRDA 323: partial purification, characterization and interesterification of butter fat. Appl Microbiol Biotechnol 43:42–51
Pandey A, Benjamin S, Soccol CR, Nigam P, Krieger N, Soccol VT (1999) The realm of microbial lipases in biotechnology. Biotechnol Appl Biochem 29:119–131
Parmar N, Singh A, Ward OP (2001) Enzyme treatment to reduce solids and improve settling of sewage sludge. J Ind Microbiol Biotechnol 26:383–386
Parra LP, Espina G, Devia J, Salazar O, Andrews B, Asenjo JA (2015) Identification of lipase encoding genes from Antarctic seawater bacteria using degenerate primers: expression of a cold-active lipase with high specific activity. Enzyme Microb Technol 68:56–61
Pliego J, Mateos JC, Rodriguez J, Valero F, Baeza M, Femat R, Camacho R, Sandoval G, Herrera-Lopez EJ (2015) Monitoring lipase/esterase activity by stopped flow in a sequential injection analysis system using p-nitrophenyl butyrate. Sensors Switz 15:2798–2811
Pöhnlein M, Slomka C, Kukharenko O, Gärtner T, Wiemann LO, Sieber V, Syldatk C, Hausmann R (2014) Enzymatic synthesis of amino sugar fatty acid esters. Eur J Lipid Sci Technol 116:423–428
Priji P, Unni KN, Sajith S, Binod P, Benjamin S (2015) Production, optimization and partial purification of lipase from *Pseudomonas sp.* strain BUP6, a novel rumen bacterium characterized from Malabari goat. Biotechnol Appl Biochem 62:71–78
Rajendran A, Palanisamy A, Thangavelu V (2009) Lipase catalyzed ester synthesis for food processing industries. Braz Arch Biol Technol 52:207–219
Ramakrishnan V, Goveas LC, Suralikerimath N, Jampani C, Halami PM, Narayan B (2016) Extraction and purification of lipase from *Enterococcus faecium* MTCC5695 by PEG/phosphate aqueous-two phase system (ATPS) and its biochemical characterization. Biocatal Agric Biotechnol 6:19–27
Ramos-Sanchez LB, Cujilema-Quitio MC, Julian-Ricardo MC, Cordova J, Fickers P (2015) Fungal lipase production by solid-state fermentation. J Bioprocess Biotech 5:203
Rathi P, Saxena RK, Gupta R (2001) A novel alkaline lipase from *Burkholderia cepacia* for detergent formulation. Process Biochem 37(2):187–192
Raveendran S, Parameswaran B, Ummalyma SB, Abraham A, Mathew AK, Madhavan A, Rebello S, Pandey A (2018) Applications of microbial enzymes in food industry. Food Technol Biotechnol 56:16–30
Ray A (2012) Application of lipase in industry. Asian J Pharm Technol 2:33–37
Reddy JR, Rao BV, Karuna MS, Raju KV, Krishna AV, Prasad RB (2013) Lipase-mediated preparation of epoxy lecithin and its evaluation as plasticizer in polyester laminates. J Appl Polym Sci 127:2945–2951
Ribeiro BD, Castro AM, Coelho MAZ, Freire DMG (2011) Production and use of lipases in bioenergy: a review from the feedstocks to biodiesel production. Enzyme Res 2011:1–16
Robles-Medina A, Gonzalez-Moreno PA, Esteban-Cerdan L, Molina-Grima E (2009) Biocatalysis: towards ever greener biodiesel production. Biotechnol Adv 27:398–408
Rodriguez de Romo AC, Borgstein J (1999) Claude Bernard and pancreatic function revisited after 150 years. Vesalius 5(1):18–24
Rowe HD (2001) Biotechnology in the textile/clothing industry: a review. J Consumer Stud Home Econ 23:53–61

Rubin B, Dennis EA (1997) Lipases: part A biotechnology methods in enzymology, vol 284. Academic Press, New York, pp 1–408

SÁ AGA, de Meneses AC, Araújo PH, de Oliveira D (2017) A review on enzymatic synthesis of aromatic esters used as flavor ingredients for food, cosmetics and pharmaceuticals industries. Trends Food Sci Technol 69:95–105

Sanchez S, Demain AL (2017) Useful microbial enzymes – an introduction. Biotechnology of microbial enzymes. Elsevier, Amsterdam, pp 1–11

Sandoval G, Herrera-López EJ (2018) Lipase, phospholipase, and esterase biosensors (review). In: Sandoval G (ed) Lipases and phospholipases. Methods in molecular biology, vol 1835. Humana Press, New York, NY

Sangeetha R (2011) Bacterial lipases as potential industrial biocatalysts: an overview. Res J Microbiol 6:1–24

Saphir J (1967) Permanent hair waving. West Germany Patent 1,242,794

Satsuki T, Watanabe T (1990) Application of lipase to laundry detergents. Bio Ind 7:501–507

Saxena RK, Sheoran A, Giri B, Davidson WS (2003) Purification strategies for microbial lipases. J Microbiol Methods 52:1–18

Seitz EW (1974) Industrial applications of microbial lipases-a review. J Am Oil Chem Soc 51:12–16

Shah S, Sharma S, Gupta MN (2004) Biodiesel preparation by lipase catalysed transesterification of Jatropha oil. Energy Fuel 18:154–159

Shaikh MA (2010) Enzymes: a revaluation in textile processing. Pak Text J:48–51

Sharma R, Chisti Y, Banerjee UC (2001) Production, purification, characterization, and applications of lipases. Biotechnol Adv 19:627–662

Shimada Y, Watanabe Y, Samukawa T, Sugihara A, Noda H, Fukuda H, Tominaga Y (1999) Conversion of vegetable oil to biodiesel using immobilized *Candida antarctica* lipase. J Am Oil Chem Soc 76:789–793

Shimada Y, Watanabe Y, Sugihara A, Baba T, Ooguri T, Moriyama S, Terai T, Tominaga Y (2001) Ethyl esterification of docosahexaenoic acid in an organic solvent-free system with immobilized *Candida antarctica* lipase. J Biosci Bioeng 92:19–23

Silva JA, Macedo GP, Rodrigues DS, Giordano RLC, Gonçalves LRB (2012) Immobilization of *Candida antarctica* lipase B by covalent attachment on chitosan-based hydrogels using different support activation strategies. Biochem Eng J 60:16–24

Singh AK, Mukhopadhyay M (2012) Overview of fungal lipase: a review. Appl Biochem Biotechnol 166(2):486–520

Singh SK, Felse AP, Nunez A, Foglia TA, Gross RA (2003) Regioselective enzyme-catalyzed synthesis of sophorolipid esters, amides, and multifunctional monomers. J Org Chem 68:5466–5477

Sirisha E, Rajasekar N, Narasu ML (2010) Isolation and optimization of lipase producing bacteria from oil contaminated soils. Advan Biol Res 4:249–252

Sivaramakrishnan R, Incharoensakdi A (2016) Purification and characterization of solvent tolerant lipase from *Bacillus sp.* for methyl ester production from algal oil. J Biosci Bioeng 121:517–522

Snellman EA, Sullivan E, Colwell RR (2002) Purification and properties of the extracellular lipase, LipA, of *Acinetobacter sp*. RAG-1. FEBS J 269:5771–5779

Srivastava A, Prasad R (2000) Triglycerides-based diesel fuels. Renew Sustain Energy Rev 4:111–133

Starodub NF (2006) Biosensors for the evaluation of lipase activity. J Mol Catal B: Enzym 40(3):155–160

Stinson SC (1995) Fine and intermediate chemical markers emphasize new products and process. Chem Eng News 73:10–26

Su F, Peng C, Li G-L, Xu L, Yan Y-J (2016) Biodiesel production from woody oil catalyzed by *Candida rugosa* lipase in ionic liquid. Renew Energy 90:329–335

Takamoto T, Shirasaka H, Uyama H, Kobayashi S (2001) Lipase-catalyzed hydrolytic degradation of polyurethane in organic solvent. Chem Lett 6:492–493

Tan T, Zhang M, Wang B, Ying C, Deng I (2003) Screening of high lipase producing *Candida* sp. and production of lipase by fermentation. Process Biochem 39:459–465
Thompson G, Swain J, Kay M, Forster CF (2001) The treatment of pulp and paper mill effluent: a review. Bioresour Technol 77(3):275–286
Torres S, Baigorí MD, Swathy S, Pandey A, Castro GR (2009) Enzymatic synthesis of banana flavour (isoamyl acetate) by *Bacillus licheniformis* S-86 esterase. Food Res Int 42(4):454–460
Trojanowicz M (2014) Enantioselective electrochemical sensors and biosensors: a mini-review. Electrochem Comm 8:47–52
Tschocke C (1990) Enzymatic treatment of fats in wastewater treatment plants. Eau Ind Nuisances 138:63–64
Ulker S, Ozel A, Çolak A, Karaoglu SA (2011) Isolation, production, and characterization of an extracellular lipase from *Trichoderma harzianum* isolated from soil. Turk J Biol 35:543–550
Ullah N, Daud M, Shabir M, Ozkan T, Qasim M (2015) Screening identification and characterization of lipase producing soil bacteria from upper dir and Mardan Khyber Pakhtunkhwa, Pakistan. Int J Biosci 6:49–55
Umehara K, Masago Y, Mukaiyama T, Okumura O (1990) Behaviour of alkaline lipase on detergency. Yukagaku 39:321–326
Undurraga D, Markovits A, Erazo S (2001) Cocoa butter equivalent through enzymic interesterification of palm oil mid fraction. Process Biochem 36:933–939
Unni KN, Priji P, Sajith S, Faisal PA, Benjamin S (2016) *Pseudomonas aeruginosa* strain BUP2, a novel bacterium inhabiting the rumen of Malabari goat, produces an efficient lipase. Biol Bratisl 71:378–387
Uttatree S, Winayanuwattikun P, Charoenpanich J (2010) Isolation and characterization of a novel thermophilic-organic solvent stable lipase from *Acinetobacter baylyi*. Appl Biochem Biotechnol 162:1362–1376
Vakhlu J (2006) Yeast lipases: enzyme purification, biochemical properties and gene cloning. Electron J Biotechnol 9(1)
Vasileva-Tonkova E, Galabova D (2003) Hydrolytic enzymes and surfactants of bacterial isolates from lubricant-contaminated wastewater. Z Naturforsch C 58:87–92
Vicente G, Martinez M, Aracil J (2004) Integrated biodiesel production: a comparison of different homogeneous catalysts systems. Bioresour Technol 92:297–305
Vulfson EN (1994) Industrial applications of lipases. In: Woolley P, Peterson SB (eds) Lipases-their structure, biochemistry and application. Cambridge University Press, Cambridge, pp 271–288
Wanyonyi WC, Mulaa FJ (2019) Alkaliphilic enzymes and their application in novel leather processing technology for next-generation tanneries. Adv Biochem Eng Biotechnol. Springer, Berlin, Heidelberg
Whiteley CG, Enongene G, Pletschke BI, Rose P, Whittington-Jones K (2003) Co-digestion of primary sewage sludge and industrial wastewater under anaerobic sulphate reducing conditions: enzymatic profiles in a recycling sludge bed reactor. Water Sci Technol 48:29–138
Winkler KW, Ulrich K, Stuckmann M (1979) Glycogen, hyaluronate and some other polysaccharides greatly enhance the formation of exolipase by *Serratia marcescens*. J Bacteriol 138:663–670
Xiao F, Li Z, Pan L (2017) Application of microbial lipase and its research progress. Progress Appl Microbiol:8–14
Yang F, Weber TW, Gainer JL, Carta G (1997) Synthesis of lovastatin with immobilized *Candida rugosa* lipase in organic solvents: effects of reaction conditions on initial rates. Biotechnol Bioeng 56:671–680
Yang W, He Y, Xu L, Zhang H, Yan Y (2016) A new extracellular thermo-solvent stable lipase from *Burkholderia ubonensis* SL-4: identification, characterization and application for biodiesel production. J Mol Catal B: Enzym 126:76–89
Yoo HY, Simkhada JR, Cho SS, Park DH, Kim SW, Seong CN, Yoo JC (2011) A novel alkaline lipase from *Ralstonia* with potential application in biodiesel production. Bioresour Technol 102:6104–6111

Yvergnaux F (2017) Lipases: particularly effective biocatalysts for cosmetic active ingredients. OCL 24(4):D408

Zehani N, Dzyadevych SV, Kherrat R, Jaffrezic-Renault NJ (2014) Sensitive impedimetric biosensor for direct detection of diazinon based on lipases. Front Chem 2:44

Zheng J, Xu L, Liu Y, Zhang X, Yan Y (2012) Lipase-coated K_2SO_4 micro-crystals: preparation, characterization, and application in biodiesel production using various oil feedstocks. Bioresour Technol 110:224–231

Zheng X, Chu X, Zhang W, Wu N, Fan Y (2011) A novel cold-adapted lipase from *Acinetobacter sp*. XMZ-26: gene cloning and characterisation. Appl Microbiol Biotechnol 90:971–998

Role of Fungal Enzymes in the Removal of Azo Dyes

9

Garima Singh, S. K. Dwivedi, and Jitendra Mishra

Abstract

Azo dyes are very commonly used for coloring purposes. At the global level, due to their strong retention behavior, a huge amount of azo dyes are used in the carpet, textile, and paper industries. A major drawback associated with the use of azo dyes is related to their excessive flush off in the surroundings, which pollutes the environment. Therefore, effective mitigation methods are required for the removal of azo dyes from environmental sources. Microbes have the capability to decolorize azo dyes, which is an ecologically sound and cost-effective strategy of dye remediation. Fungi degrade or decolorize azo dyes by two broad mechanisms: (1) through enzymatic activities and (2) adsorption. Fungal enzymes also play an important role in the breakdown of toxic by-products formed by azo dyes. Fungal enzymes such as lignin peroxides, manganese peroxides, and laccase are useful in the biodegradation of azo dyes. Several fungi are known to produce dye-decolorizing enzymes responsible for break down of azo dyes into lesser toxic forms. Fungal enzymes such as lignin peroxides, manganese peroxides, and laccase are useful in the biodegradation of azo dyes. This chapter describes the essential roles and mechanisms of fungi involved in removal of azo dye from the environment. Apart from this, how fungal enzyme can be utilized for the decolorization process at the industrial level for making the environment cleaner and sustainable is also discussed.

G. Singh · S. K. Dwivedi (✉)
Department of Environmental Science, Babasaheb Bhimrao Ambedkar (Central) University, Lucknow, India

J. Mishra
DST-CPR, Babasaheb Bhimrao Ambedkar (Central) University, Lucknow, India

N. K. Arora et al. (eds.), *Microbial Enzymes: Roles and Applications in Industries*, Microorganisms for Sustainability 11,
https://doi.org/10.1007/978-981-15-1710-5_9

9.1 Introduction

Color is the most important constituent that made human life more aesthetic and fascinating in the real world. The process of dyeing is as old as human civilization and has existed since humans started to use natural colors obtained from different sources for their creative works. Even in "Atharva Veda" use of natural dyes is depicted. Wall paintings of "Ajanta and Ellora" demonstrate the effectiveness of dyeing craft which had been native from ancient times in India (Grierson et al. 1989). Natural dyes are made from natural sources such as plant leaves, flowers, vegetables, insects, and many others. Earlier textile stuff was dyed with natural dyes of dull color range and less retention. Later, the industrialization process equipped with modern machinery facilitated the synthesis of bright color with high retention capacity, which subsequently attenuated the use of natural dyes. Due to lack of versatility and less retention on the fabric, natural dyes were soon replaced by synthetic dyes. Now coloring industries have also started to favor synthetic dyes instead of natural (Kaushik and Malik 2009).

Perkin discovered the first synthetic dye "mauvine" during the study of coal and tar (Kant 2012). Now a days synthetic dyes with diverse chemical nature are available and widely being used in the coloring of leather, textiles, paper, cosmetics, printing materials, plastics, and food industries (Prasad and Rao 2013). Synthetic dye molecule contains two important components: (1) chromophores (–C=C–,–C=O,–C=N–,–NO_2,–N=N– and quinonoid ring) and (2) auxochromes (–NH_2,–OH,–OCH_3, CH_3CO–, halones, etc.). Chromophores absorb the visible range of light and are responsible for generating different dye colors (Forgacs et al. 2004). Auxochromes play a role in water solubility and also enhance the similarity towards natural colors (Christie 2001). The widely used synthetic azo dyes are anthraquinone, triphenylmethane, and indigo dyes. Among them, azo dyes are mostly used in the coloring of the textile and carpet industries.

About 10–15% of azo dyes are discharged in water during the dyeing process (Asad et al. 2007; Sen et al. 2016). Data indicates that till 1999, about 50,000 tons of azo dyes were discarded into the environment, which severely polluted the surrounding ecosystem (Singh and Shukla 2015). Recently, it has been estimated that dye concentration drastically increased by 6.9 million tons in 2017 in wastewater (Rawat et al. 2018). There are many deleterious effects associated with the discharge of azo dyes in the environment (Fig. 9.1). In the aquatic system, azo dyes are related to the reduction of photosynthesis, exhaustion of dissolved oxygen, and generation of toxic effect on flora and fauna (Aksu et al. 2007; Saratale et al. 2009; Mugdha and Usha 2012). Apart from this, penetration of azo dyes in the freshwater system makes water unpleasant whereas breakdown of dye products (colorless aromatic amines) is found to be carcinogenic, toxic, and mutagenic (Xu et al. 2005; Tan et al. 2014; Mahmoud et al. 2017). When untreated effluent-containing azo dyes reach the marine ecosystem it affects chemical oxygen demand and total organic carbon, which results in the death of aquatic organisms (Saratale et al. 2009).

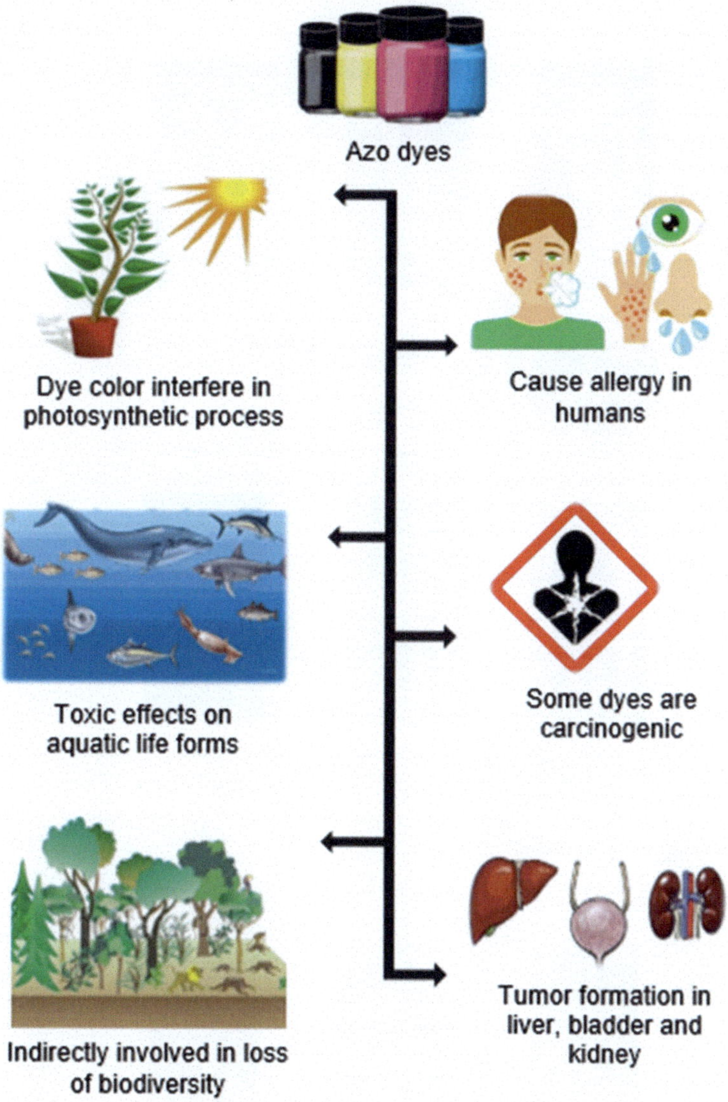

Fig. 9.1 Harmful effects of azo dyes on humans and environment

Decolorization and degradation of azo dyes have emerged as a serious issue globally. The physical and chemical methods have already been developed and are used for the removal of dye from wastewater (Fig. 9.2). The methods used are advanced oxidation processes (such as application of hydrogen peroxide (H_2O_2), ozone (O_3), and ultraviolet light), membrane filtration, ion exchange, adsorption, coagulation, flocculation, and precipitation (Mondal et al. 2006). There are evidences which suggest that most of the methods used are expensive, time consuming,

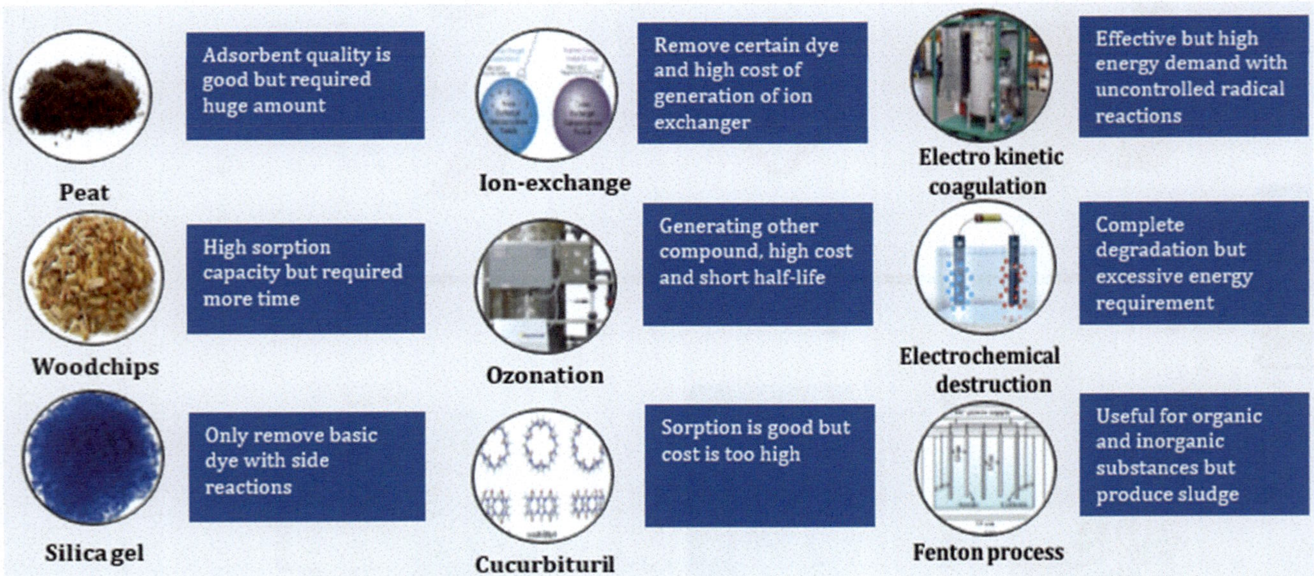

Fig. 9.2 Different physiochemical methods used for azo dye removal

and of limited application, i.e., treating only certain type of dye (Hai et al. 2007; Ali 2010; Kurade et al. 2012; Khan et al. 2013). Further, with the perspective of the environment, these methods are found to be unsuitable as during the processes of dye removal high amount of sludge is generated (secondary pollutant) and incomplete degradation of azo dyes occurs (Kurade et al. 2016).

Over the last two decades, biological methods have emerged as eco-friendly and economically viable tools for removing the azo dyes from different environmental sources (Kaushik and Malik 2009). Several workers have documented the role of microorganisms in dye decolorization (Bankole et al. 2018; He et al. 2018). Research indicates that microbe-mediated aerobic and anaerobic process of the degradation/decolorization of azo dyes is more effective in comparison to the chemical methods (Jadhav et al. 2016). Due to the fast growth and high reactivity towards dye molecule, bacterial strains are more frequently used in dye decolorization. However, in some cases, decolorization product of dyes such as aromatic amines also inhibits bacterial growth and activity (Qu et al. 2010). The biosorbent characteristic of algae is also useful in dye removal; however, demerit associated with algal biosorbent is the early diffusion of dye. Over last few decades, research confirmed that fungi have the unique quality of dye degradation and removal of several contaminants (Asses et al. 2018). Fungi adsorb dyes in their biomass (live/dead) and also release several extracellular enzymes, i.e., laccase, manganese peroxidase, and lignin peroxidase, which play a vital role in the degradation of dyes (Gowri et al. 2014). Fungal enzymes are known for their capability to convert toxic azo dye products into lesser toxic forms. The main objective of this chapter is to elucidate the role and status of fungi in the removal of azo dyes. Apart from this, the enzymatic mechanisms involved in the remediation of azo dyes are also discussed. Information on the mass production of fungal enzymes and constraints associated with their lower applicability at the industrial level is also discussed.

9.2 Fungi as Azo Dye Decolorizers

Among microbes, fungi play a very important role in the degradation and decolorization of azo dyes. The fungi remove dye by two processes: (1) adsorbing dyes in their cellular biomass and (2) releasing extracellular enzymes. Mainly oxidative and nonspecific ligninolytic enzymes perform a major role in the degradation and decolorization of azo dyes. In the early 1990s, researchers revealed the definitive role of white-rot fungi in the degradation and adsorption of recalcitrant compounds, including azo dyes (Aust et al. 2003). Genera of white-rot fungi, such as *Trametes hirsute, Irpex lacteus,* and *Trametes versicolor* belonging to the family of basidiomycetes, are reported for their ability to degrade or decolorize dyes from industrial waste effluent (Robinson et al. 2001; Mishra et al. 2011). Fungal enzymes such as laccase (Lac), lignin peroxidase (Lip), and manganese peroxidase (MnP) are called as "arsenal enzymes" whereas glutamate oxidase, polyphenol oxidase, aryl alcohol oxidase, etc. are known as "auxiliary enzymes" used in the treatment of azo dyes. Few non-basidiomycetes fungi also use demethylation and oxidation process to metabolize a wide range of compounds including azo dyes. Some species of *Penicillium* (*P. geastrivorus, P. ochrochloron*, etc.) and *Aspergillus* (*A. terreus, A. niger*) have been reported to remove azo dyes from liquid medium (Abadulla et al. 2000; Almeida and Corso 2014). In a study, Kasinath et al. (2003) showed that MnP and Lac enzymes of *Phanerochaete chrysosporium* and *I. lacteus* removed 100% of brilliant blue azo dyes. Similarly, a white-rot fungus *T. hirsute* is also reported to remove 100% of indigo carmine by the release of Lac (Couto and Sanromán 2006). Several workers also confirmed the role of fungal MnP in the degradation of different azo dyes. For example, Pazarlioglu et al. (2005) showed 95% degradation of direct blue dye by *P. chrysosporium*; Nilsson et al. (2006) showed 70% of reactive blue by *T. versicolor*; and Couto and Sanromán (2006) reported 84% decolorization of methyl orange by MnP released from *T. hirsute.* However, in the case of fungal Lac, similar results were obtained by several workers showing 70–88% removal of dye (Mishra et al. 2011; Bankole et al. 2018). In few instances, all three enzymes (Lac, LiP, and MnP) were collectively used to remove dye molecule (Hanapi et al. 2018).

9.3 Mechanism of Azo Dye Decolorization/Degradation

9.3.1 Adsorption

The term "adsorption" means uptake of a molecule or ions on the surface of a solid and the association between molecule and surface can be chemically mediated (chemisorption), or it may be purely physical (physical adsorption or physicosorption) (Hai et al. 2007; Kurade et al. 2012). The material which is taken up is called adsorbate, and the solid surface on which uptake occurs is called as adsorbent (Dabrowski et al. 2005). The process of adsorption mainly depends on the properties of adsorbate, which include molecular weight, structure and size, polarity, and solution concentration (Malakootian and Heidar 2018). The use of adsorption methods in the remediation of azo dye is considered very effective due to its simple process design, ease of operation, and cost-effectiveness (Rafatullah et al. 2010). The fungus may adsorb dyes in their growing/living cellular biomass or in dead cell (Table 9.1). The process is also known as biosorption (Ali 2010). In the case of fungus, increased cell-to-surface ratio entails higher physical adsorption of dye (Fu and Viraraghavan 2001). Fungal biosorption involves complexation, surface ionization, precipitation, and entrapment of dye in inner spaces of fungal mycelium (Yeddou-Mezenner 2010). Studies confirmed that in comparison to living biomass, dead biomass is more useful in dye removal and the problem of toxicity due to the generation of secondary compounds and nutritional requirements is also not a matter of concern (Lavanya et al. 2014). Although biosorption is the primary means of dye decolorization, in many incidences subsequent decolorization of dye occurs by the release of intra- or extracellular fungal oxidative enzymes (Rani et al. 2014). This is why biosorption has been suggested as an initial step, before enzymatic treatment, to concentrate the dye (Aretxage et al. 2001). It has been found that the pH value of the

Table 9.1 Live or dead fungal biomass for adsorption of synthetic dyes

Fungal culture	Dyes	% of removal	Contact time (h)	Reference
Trichoderma sp.	Acid brilliant red	100.0	24.0	Xin et al. (2012)
Ganoderma lucidum, Irpex lactus	Black dycem	90.0	72.0	Baccar et al. (2011)
Trametes versicolor	Sirius blue K	62.62	96.0	Erden et al. (2009)
Aspergillus niger	Direct blue 199	44.90	4.00	Xiong et al. (2010)
Agaricus bisporus	Reactive blue 49	72.86	90.0	Akar et al. (2009)
Cunninghamella elegans	Direct red 80, reactive blue 19	100 and 84–98	24.0	Prigione et al. (2008)
Penicillium chrysogenum	Acid orange 8, reactive orange 16	70.4 and 67.6	14.0	Low et al. (2008)
Aspergillus niger	Synazol red	88.00	18.0	Khalaf (2008)
Thuja orientalis	Acid blue 40	48.50	1.20	Akar et al. (2008)
Aspergillus fumigates	Reactive brilliant red	94.70	120	Wang et al. (2012)
Trametes versicolor	Direct blue-1	63.20	6.00	Bayramoglu and Arica (2007)
Trametes versicolor, Aspergillus niger	Reactive green, reactive blue	86 and 83	1.00	Kumari and Abraham (2007)
Rhizopus stolonifer	Bromophenol blue	88.00	20.0	Zeroual et al. (2006)
P. chrysosporium	Astrazone blue	60.00	2.00	Asma et al. (2006)
Neurospora crassa	Acid red 57	98.78	0.40	Akar et al. (2006)
Cephalosporium aphidicola	Acid red 57	29.20	2 0.00	Kiran et al. (2006)

dye/pollutants solution has a profound effect on the adsorption capacity of fungus (Kaushik and Malik 2009). The adsorption time and efficiency of the dye removal can be greatly dependent on the pH. For more details on biosorption of dye by fungus, one could see Aksu and Cagatay (2006), Bayramoglu and Arica (2007), and Khalaf (2008).

9.3.2 Enzymatic Mechanisms

The use of fungal enzymes is a greener technology of mitigation of azo dyes from the environment (Ezeronye and Okerentugba 1999). The fungal decolorization of azo dyes mainly occurs by biodegradation and biosorption mechanism (Vanhulle et al. 2008). In biodegradation process extracellular, nonselective, and nonspecific enzyme system or cocktails are involved. However, dye decolorization capability of the fungal enzyme is directly linked with the availability of nutrient and growth conditions in the production process (Kaushik and Malik 2009). White-rot fungi have the highest capability to produce nonspecific enzymes such as Lac, MnP, and Lip. These enzymes are involved in generating complex reaction systems enabling the degradation of dye structure into simpler and nontoxic forms (Pointing 2001; Knapp et al. 2001). However, the action of fungal enzymes seldom leads to the mineralization of dyes and very much depends on the chemical structure. A high degree of mineralization occurs when dye contains substituted aromatic ring structure compared to the unsubstituted rings (Singh 2006). Certain bonds in the dye molecules are cleaved and utilized as carbon source, although chromophore is not affected. This mechanism preferably occurs when a consortium of microorganisms are involved (Knapp et al. 2001; Singh 2006;). Fungi release many oxidative enzymes such as Lac, Lip, MnP, polyphenol oxidases, N-demethylase, tyrosinase, dye-decolorizing peroxidase, and cellobiose dehydrogenase and all these take part in decolorization of dye (Oturkar et al. 2011; Telke et al. 2011; Martorell et al. 2012).

Since many years ligninolytic enzymes, i.e., MnP, LiP, and Lac, are known for their bio-oxidation capability of harmful chemical contaminants released in the environment (Gold and Alic 1993; Torres et al. 2003; He et al. 2018; Ghobadi et al. 2019). Similarly, peroxidases from plants, microbes, and animals have also been studied for bio-oxidation of azo dyes from industry effluents (Martinez 2002; Ferreira-Leitao et al. 2007; Dos Santos et al. 2007). Table 9.2 provides detail of some fungal enzymes involved in decolorization of different azo dyes.

9.3.2.1 Laccases (Lac)

Laccases are copper (Cu^{2+} state)-containing oxygen oxidoreductases and are a member of polyphenol oxidase group of enzymes (Bourbonnais and Paice 1992). These are one of the oldest known enzymes and were first reported in a Japanese tree *Toxicodendron verniciﬂuum* (Janusz et al. 2015). Later their presence in several other plants, fungi, bacteria, and insects was also confirmed (Dwivedi et al. 2011; Zeinab et al. 2013; Qin et al. 2018). Among fungi, white-rot fungi are the major Lac producers (Zavarzina et al. 2018). Lac enzymes are found in various forms like monomeric, dimeric, and tetrameric glycol proteins having four Cu atom monomers situated at its catalytic sites (Murugesan et al. 2006). As Lac comes under the polyphenol oxidases, it has redox ability of copper ion to catalyze at least one electron for the oxidation of the substrate with immediate reduction of molecular oxygen into water molecule (Abadulla et al. 2000). Although Lac can only oxidize compounds with lower redox potential (means lower than Lac itself) and hence only phenols and aromatic or aliphatic amines can be catalyzed, however, in the presence of redox mediator non-phenolic compounds can also be catalyzed (Galhaup et al. 2002). Degradation of azo dyes by Lac begins with the asymmetrical cleavage of azo bonds which is further followed by the process of desulfonation, oxidative cleavage, dihydroxylation, and demethylation and depends on the structure of dyes to be degraded (Yang et al. 2015; Zheng et al. 2016). Azo dyes can also be degraded without the process of cleavage (Chen 2006; Pereira et al. 2009). This mechanism of dye degradation includes the formation of phenolic compounds that have resulted from the extremely nonspecific free radicals (Chen 2006). Two white-rot fungi *Pleurotus* spp. and *Trametes* spp. have been reported to be extensively involved in industrial dye decolorization (Couto and Sanromán 2006). Their Lac has been found

Table 9.2 Fungal enzymes involved in azo dye decolorization/degradation

Fungal species	Name of dye	Decolorization (%)	Enzyme used	Reference
Peroneutypa scoparia	Acid red 97	75.0	Lac	Pandi et al. (2018)
Cerrena sp., *Phanerochaete* sp.	Reactive black 5 and methylene blue	86.0 and 88.0	Lac and MnP	Hanapi et al. (2018)
Phoma tropica, Dichotomomyces cejpii	Congo red, methyl red	90.0 and 92.0	Lac	Krishnamoorthy et al. (2018)
Achaetomium strumarium	Acid red 88	99.55	NADH-DCIP reductase and Lac	Bankole et al. (2018)
Trichoderma tomentosum	Acid red 3	94.9	MnP	He et al. (2018)
Peyronellaea prosopidis	Scarlet RR	85.0	LiP, Lac and MnP	Bankole et al. (2018)
Phanerochaete chrysosporium	Congo red	93.0	LiP and MnP	Bosco et al. (2016)
Pleurotus ostreatus and *Fusarium oxysporum*	Brilliant green and Evans blue	79.20	LiP	Przystas et al. (2015)
Trametes gibbosa	Reactive black 5	87.70	Lac	Adnan et al., 2014
Coprinus plicatilis	Reactive blue 19 (RB19)	99.0	Lac	Akdogan et al. (2014)
Schizophyllum commune IBL-06	Solar brilliant red 80	100.0	Lac, LiP, and MnP	Asgher et al. (2013)
Pleurotus eryngii	Reactive black 5	93.56	Lac, LiP, and MnP	Hadibarata et al. (2013)
Pycnoporus sanguineus	Reactive blue 4 and Orange G	81.0 and 97.0	Lac and MnP	Christiane and Steeve (2013)
Armillaria sp. F022	Reactive black 5	80.0	Lac	Hadibarata et al. (2012)
Phanerochaete chrysosporium	Direct red 180	100.0	LiP	Sen et al. (2012)
Ganoderma sp. En3	Methyl orange, crystal violet, bromophenol blue, and malachite green	96.70, 75.0, 90.0, and 91.0	Lac	Zhuo et al. (2011)
Trametes sp.	Brilliant blue R250 and Orange II	100.0	Lac and MnP	Grinhut et al. (2011)
Trametes trogii	Brilliant blue R, indigo carmine, bromophenol blue, and direct red	82.0, 84.5, 75.0, and 80.0	Lac and MnP	Grassi et al. (2011)
Phanerochaete chrysosporium	Reactive black 5	90.0	LiP and MnP	Enayatizamir et al. (2011)

(continued)

Table 9.2 (continued)

Fungal species	Name of dye	Decolorization (%)	Enzyme used	Reference
Phanerochaete chrysosporium	Direct red 88	100.0	LiP	Singh et al. (2010)
Trametes hirsute, Phanerochaete chrysosporium	Remazol procion blue, remazol golden yellow	87.0 and 80.0	Lac	Jadhav et al. (2010)
Datronia sp.	Reactive blue and reactive black	95.0 and 90.0	Lac and MnP	Vaithanomsat et al. (2010)
Phanerochaete chrysosporium	Orange II	85.0	MnP	Sharma et al. (2009)
Trametes versicolor	Reactive blue 4	90.0	Lac	Yemendzhiev et al. (2014)

to degrade several dye molecules including indigoid, brilliant blue R, anthraquinone, remazol triarylmethane, and triphenylmethane (Wong and Yu 1999; Abadulla et al. 2000; Zheng et al. 2016). In a study, Mendoza et al. (2011) developed a Lac/mediator system in a membrane reactor for evaluating reusability of *T. versicolor* Lac and mediators (2,6,6-tetramethylpiperidine 1-oxyl (TEMPO) and syringaldehyde) for the removal of some azo dye. They found that Lac/syringaldehyde provided the best system for decolorization red FN-2BL, red BWS, remazol blue R, and blue 4BL and decolorization yields were 98%, 88%, 80%, and 78%, respectively. The potential of a novel extracellular form of Lac from a fungal strain, *Lentinus* sp., was explored in the degradation of anthraquinone and azo dyes by Hsueh and Chen (2008). Lac enzyme from a macro-fungus, *Podoscypha elegans* (G. Mey.), was found to decolorize five azo dyes, viz. Orange G, Congo Red, direct blue 15, rose bengal, and direct yellow.

9.3.2.2 Peroxidases

Peroxidases are ubiquitous heme-containing oxidative enzymes found in several life forms, including bacteria, fungi, animals, and plants (Sen et al. 2016). Peroxidases catalyze the oxidation of a wide variety of organic compounds using hydrogen peroxide (H_2O_2) as a donor (Carmona-Ribeiro et al. 2015; Ali et al. 2016; Pandey et al. 2017). It helps in the degradation of lignin and other aromatic compounds. Among all peroxidases, fungal peroxidases are best known for their pivotal role in lignin degradation (Conesa et al. 2002; Wesenberg et al. 2003; Hammel and Cullen 2008). In contrary to Lac, fungal peroxidase can also catalyze non-phenolic aromatic compounds with very high ionization potentials (Sen et al. 2016). Here in this section, the role and mechanism of ligninolytic peroxidases, i.e., LiPs, MnPs, and VPs, are being described.

Lignin Peroxidases (LiPs)

The enzyme was reported by Tien and Kirk (1983) as an extracellular oxygenase released by a white-rot fungus *P. chrysosporium*. Later other white-rot fungal genera were also identified to release isoenzyme of LiPs in the extracellular medium (Mukherjee and Kumar 2018). LiPs are extracellular monomeric heme-protein belonging to the family of oxidoreductases (Erden et al. 2009). LiPs are one among three fungal peroxidases (LiPs, MnPs, and VPs) which take part in hydrogen peroxide-dependent oxidative degradation of lignin. The oxidative properties of LiPs include generation of radical cation through one-electron oxidation and then side-chain cleavage, demethylation, intramolecular addition, and rearrangements (Wong 2009; Zeinab et al. 2013). Hydroxylation of benzylic methylene groups and oxidation of benzyl alcohols to their corresponding aldehydes or ketones are other ways to act (Furukawa et al. 2014). As LiP has a higher redox potential than Lac and MnP they catalyze both phenolic and non-phenolic compounds including azo dye. Research confirmed the role of LiPs produced by *P. chrysosporium* in decolorization of several azo dyes (Spadaro et al. 1992; Podgornik et al. 1999). Apart from *P. chrysosporium*, *Phanerochaete ostreatus*, *Polyporus* sp., and *Trametes* sp. are also associated with the production of LiPs as the major reason to decolorize a wide range of structurally different azo dyes.

Manganese Peroxidases (MnPs)

MnPs are glycosylated heme-oxidoreductases produced extracellularly only in certain basidiomycetes families (Agaricales, Corticiales, Polyporales, Hymenochaetales). MnPs oxidize Mn^{2+} ions to highly reactive Mn^{3+} ions which function as diffusible oxidant and get released from the enzyme surface to degrade phenolic moieties in lignin and other similar compounds (Zhao et al. 2005). There are several isozyme forms of MnPs which have been identified. The oxidation of Mn^{2+} to Mn^{3+} by MnPs requires chelating agents, usually organic acids which stimulate the catalytic activity. Chelation of Mn with organic acids facilitates dissociation of the enzyme–manganese complex (Sen et al. 2016). A wide variety of white-rot fungi are known to produce extracellular MnPs (Hofrichter 2002). Studies confirmed that the molecular weight of MnPs ranges from 38 to 62.5 kDa, and 43% of similarity with LiP sequences (Mart ́inez et al. 1996). MnPs act in pH ranges from 4.0 to 7.0 with a peak at pH 5.5 whereas their activity at 50–70 °C temperature makes them more versatile than other peroxidases (Bosco et al. 2016). In a study, Zheng et al. (2016) reported high stability of MnP from a white-rot fungus *Cerrena unicolor* BBP6 towards many metal ions. This MnP showed optimum activity at 60 °C temperature and 4.5 pH, respectively, and effectively decolorized many types of dyes. Recently a new MnP, TP55, isolated from *T. pubescens* strain i8 showed a very high catalytic efficiency for textile dye decolorization (Rekik et al. 2019). This MnP TP55 was also able to tolerate organic solvent and detergent compatibility than other best-known MnPs such as horseradish peroxidase (HRP), MnP from *Bjerkandera adusta* strain CX-9 (MnP BA30), and MnP from *P. chrysosporium* (MnP PC). Similarly, solvent- and metal-tolerant MnP-Tra-48424 was also purified and characterized from a white-rot fungus *Trametes* sp. 48424 (Zheng et al. 2016). This MnP-Tra-48424 was strongly resistant to metal ions such as Ni^{2+}, Li^{+}, Ca^{2+}, K+, and Mn^{2+} and organic solvents such as propanediol, glycerol, and glycol. There are also reports where non-white-rot fungi showed their potential in decolorization of azo dye. For example, a fast-growing *Trichoderma tomentosum* was spotted with remarkable ability to degrade several azo dyes (94.9–99.2% within 72 h).

This fungus showed high level of MnP and low level of LiP activities with no aromatic amine production during the process of decolorization (He et al. 2018).

Versatile Peroxidases (VPs)

Versatile peroxidases were discovered in 1999 in members of genus *Pleurotus* (Qin et al. 2018). VPs are heme glycoproteins of class II fungal peroxidases and oxidize low-, medium-, and high-redox-potential compounds in both Mn^{2+}-mediated and Mn^{2+}-independent modes of action (Knapp et al. 2001). The high redox potential of VPs enables oxidation of several azo dyes as well as other phenolic and non-phenolic aromatic compounds that are the substrates of other related peroxidases (Gomez-Toribio et al. 2001). In a strict sense, VPs have catalytic features of both MnPs (for the oxidation of Mn^{2+}) and LiPs (oxidizing the redox mediator veratryl alcohol at the catalytic Trp) (Fernández-Fueyo et al. 2012). The ability of VPs oxidation of high-redox-potential substrates is mediated through three long-range electron transfer (LRET) pathways to the heme at acid pH (Fernández-Fueyo et al. 2012). In recent years, abundant information on VPs structure–function relationships generated a specific interest of this enzyme in the degradation of a variety of recalcitrant compounds that other peroxidases are not able to oxidize directly (Ruiz-Duenas et al. 2011). VPs from *P. ostreatus* were used for removal of reactive black-5 (Knapp et al. 2001). VPs from *I. lacteus* decolorized direct blue-1 dye (Qin et al. 2018).

9.4 Factors Affecting Dye Decolorization/Degradation

There are several external factors which play an important role in decolorization/degradation of azo dyes. Some of them are discussed below:

9.4.1 pH

It is a well-established fact that pH plays an important role in fungal growth and development. It has also been reported that low pH (4–5) favors the growth of most fungi. Even for better decolorization of azo dye by fungi, low pH is favorable but not considered as an essential condition. As in solution, dyes have ionic form and fungal biomass also retains electrical charge; hence the pH of the solution plays a major role in dye binding (Fu and Viraraghavan 2001). In several studies, it has been found that fungi show better dye decolorization rate at the pH range from 6 to 10 (Chen et al. 2003; Guo et al. 2007; Kilic et al. 2007). According to Dhanjal et al. (2013), higher amount of dye decolorization occurred in slightly acidic condition (at pH 6). Hadibarata et al. (2013) reported that *Pleurotus eryngii* can exceptionally decolorize 93% of reactive black 5 azo dye at pH 3. In a study, Bankole et al. (2018) showed that acid red 88 dye decolorized at different pH, i.e., 3, 4, 5, 7, 9, and 10. However, maximum decolorization up to 99% was observed at pH 4 with the incubation period of 96 h. Earlier, Murugesan et al. (2006) showed maximum Lac activity at pH 5 involved in dye degradation in *Pleurotus sajor-caju*, which further decreases by increasing the temperature.

There are also some reports which also showed the decolorization of dye in alkaline conditions. For example, in fungal genera of *P. sajor-caju* and *P. chrysosporium* maximum amount of azo dye decolorization occurred at pH of 7 and 8 (Zeinab et al. 2013). Singh and Chen (2008) also showed the production of fungal Lac in the pH range of 4–10 and their role in the degradation of indigo carmine dye.

9.4.2 Temperature

Decolorization or degradation of azo dyes is very much affected by variation in temperature. Change in temperature also changes the degradation or decolorization rate of synthetic dyes and interferes with enzymatic mechanism (Chakraborty et al. 2013; Almeida and Corso 2014). The optimum temperature required for the fungal growth is found to vary in the range of 25–35 °C (Fu and Viraraghavan 2001; Pietikainen et al. 2005; Lasram et al. 2010). According to Saratale et al. (2009) higher temperature decreases decolorization rate by the deactivation of azo reductase enzyme. Most of the studies confirmed that 25 and 30 °C are the best-suited temperatures for degradation of azo dyes. Shedbalkar and Jadhav (2011) reported that *P. ochrochloron* MTCC 517 decolorized maximum amount of cotton blue dyes at the optimum temperature of 25 °C. In a study, Zeinab et al. (2013) showed that both *P. chrysosporium* and *P. ostreatus* fungal isolates decolorize azo dyes at 30 °C. In *P. sajor-caju*, maximum Lac activity was observed at 30 °C, which drastically decreases with further increase in temperature (Murugesan et al. 2006). There are also reports which confirmed high temperature variability in dye decolorization in few fungal isolates. For example, in a yeast *Galactomyces geotrichum*, decolorization of methyl red was reported at 5 °C, 30 °C, and 50 °C, but the maximum decolorization occurred at 30 °C (Jadhav et al. 2011). Similarly, in another study fungus *Achaetomium strumarium* decolorized acid red 88 dye at 30 °C, 40 °C, 50 °C, and 60 °C but here maximum decolorization (99%) was noticed at 40 °C (Bankole et al. 2018). These findings indicate that temperature is very important parameter which governs the behavior of fungal strains towards dye decolorization.

9.4.3 Concentration and Structure of Dyes

Primary factor which decide degradability of any dye is its concentration and structure (Levin et al. 2004; Wells et al. 2006). The structure of some azo dyes cannot be easily breakable by the microbial attack and further presence of free radicals' structures also created an obstacle in the binding of several azo-reductase to dye molecule (Sen et al. 2016). Presence of a lower number of carbon and nitrogen atoms in azo dyes increases the recalcitrant nature of dye molecule (Lavanya et al. 2014). That is why supplement of external sources of carbon and nitrogen basically in the form of complex organic sources like yeast extract powder, potato dextrose powder, peptone, or mixture of carbohydrates is essential for the removal of azo dyes (Kilic et al. 2007). Synthetic dyes with low molecular weight demonstrate the maximum amount of decolorization. In few cases, dye removal rate becomes very low with the exchange of electron-withdrawing assemblage. When the electron donors such as acetate ions and glucose, which induce the reductive cleavage of azo bond in dye structure, are used up, degradation becomes more tougher (Bankole et al. 2018). In the anaerobic condition of a bioreactor, types and accessibility of electron donors vary significantly and create hindrance in attaining higher concentration of dye decolorization (Sen et al. 2016). He et al. (2018) showed that when concentration of methyl red dye increases from 750 to 1000 ppm decolorization rate also decreases in the same manner by *Sphingomonas paucimobilis*. In a study, Spadaro et al. (1992) found that *P. chrysosporium* has the capability to degrade azo dyes by converting aromatic rings with few substituents like hydroxyl, amino, or nitro function which mineralized the unsubstituted rings. As compared to carboxyl group, the presence of sulfo group with stronger electron withdrawing effect provides more decolorization of azo dyes (Hsueh and Chen 2008).

9.4.4 Redox Mediators (RMs)

The oxidative or reductive process influences degradation and decolorization of azo dyes (Cho et al. 2007). The RMs such as veratryl alcohol, 1-hydroxybenzotriazole, 2-methoxyphenothiazine, violuric acid, and many others are frequently used for enhancing the degradation potential of azo dyes (Husain and Husain 2011). Natural compounds such as acetosyringone and syringaldehyde have also been used as eco-friendly Lac mediators (Cho et al. 2007). The ubiquitous sources of RMs including reduced forms of nicotinamide adenine dinucleotide (phosphate) (NAD(P)H) were found to reduce azo dyes, even in the absence of microbes or their enzymes (Sen et al. 2016). In the presence of RMs substrate can go through a one-electron oxidation process and convert into radical cation (Fabbrini et al. 2002). 1-Hydroxybenzotriazole, a redox mediator, is used to degrade several dyes like direct red 23, direct blue 80, direct red 239, and direct yellow. Maximum degradation/decolorization of malachite green dye is observed in vanillin mediator (Husain et al. 2010). Furthermore, Bibi et al. (2011) showed in a study that Lac/mediator system produces fewer toxic metabolites compared to original compound. Recently, new findings on natural RMs have extended our understanding towards reductive transformation of pollutants, including azo dyes (Khlifi et al. 2010). Lavanya et al. (2014) found that redox-active humin isolated from a paddy soil stimulated azo dye reduction. Additionally, high-end techniques revealed the existence of redox-active quinone moieties and other oxygen-containing groups in humin. The humin was found to reduce by sulfide, which stimulated the abiotic reduction of acid red 27 (AR27) and four other azo dyes. For more information on the application of RMs in azo dye biotransformation, one can see Martinez (2002), Cho et al. (2007), and Sen et al. (2016).

9.4.5 Carbon and Nitrogen Sources

As the least amount of carbon is present in azo dyes, external addition of carbon and nitrogen sources are required for their proper degradation (Sani and Banerjee 1999). There are different carbon sources such as glucose, starch, acetate, and many others used in enhancing the decolorization rate of azo dyes under anaerobic conditions (Van der Zee and Villaverde 2005). Although, azo dye decolorization increases in the presence of carbon and nitrogen sources but in some cases, addition of carbon sources was found to be less effective (Saratale et al. 2009). Nitrogen sources such as peptone, urea, beef extract, and yeast extract can also be used in the deactivation of NADH which works as an electron donor in microbial azo dye reduction or degradation (Chang et al. 2001a). In synthetic media, nitrogen sources like rice husk and rice straw can be used in the form of supplements to enhance the azo dye decolorization rate (Saratale et al. 2009).

9.4.6 Aeration and Agitation

Oxygen plays a significant role in dye decolorization process due to its essential role in the physiological features, including cell growth and dye reduction (Chen et al. 2003). Proper availability of oxygen in azo dye reduction reaction is required to break down, ring opening and hydroxylation reactions (Pandey et al. 2007). It has also been noticed that toxic intermediate aromatic amines which are formed during cleavage of azo bond can be further metabolized in aerobic microbial systems. According to Buitrón et al. (2004) in a batch reactor, up to 99% decolorization of acid red 151 dye may occur whereas in anaerobic conditions only 14–16% was noticed. A very small quantity of oxygen is essential for rejuvenation of reducing cofactors NADH and NADPH, because under anaerobic process, activities of reductive enzymes are higher while oxidative enzymes are mainly involved in azo dye degradation. Consequently, aerobic conditions are also preferable for complete mineralization of dye molecule (You and Teng 2009). Tan et al. (2014) found that a yeast *Candida tropicalis* TL-F1 showed efficient decolorization and degradation (up to 20 mg/L) of various azo dyes in aerobic conditions.

Agitation or shaking conditions too have a crucial role in deciding the rate of decolorization of azo dyes. Agitation rate (0–50 rpm) provides more uptake of dissolved oxygen and nutrient to free cells (Chang et al. 2001a). In a study, Mahmoud et al. (2017) showed effect of agitation speed (in case of *A. niger*) on decolorization of azo dye (direct red). They found maximum decolorization of 80% at 350 rpm while decreasing agitation speed to 250, 150 and 50 rpm, decreased dye degradation by 75, 70 and 50% respectively. Similar, results were also obtained in case of *Alternaria alternata* by Chakraborty et al. (2013) and in *Ganoderma lucidum* by Selvakumar et al. (2013).

9.5 Mass Production of Enzymes

Despite the enormous knowledge of fungal biology, at the commercial levels, large-scale production of dye-decolorizing enzymes has proven to be very challenging. In particular, during mass production of enzyme, native fungal strains do not grow effectively and stringent parameters are required for better production of enzyme. The variability in optimization conditions such as substrate concentration, temperature, pH, and presence of metal ions have been identified as the major bottleneck in fungal enzyme production. Submerged fermentations (SmF) and solid-state fermentation (SSF) are the two methods widely used for the production of fungal enzymes. However, as SSF provides a natural environment for the growth of fungi, it is preferably used in the production of dye-decolorizing enzyme. In SSF system, fungal mycelium can grow either on the moist surface of solid particles or inside/in between two porous particles (Pandey 2003). Post-fermentation, the crude enzyme may be directly recovered in SSF. There are evidences which show that optimization in the fermentation process may provide better yield of enzymes. For example, Kocyigit and Ugurlu (2015) reported that adding metal compound (copper sulfate 300 μM) increased the production of Lac two times as compared with control (in SSF). Similarly, Vantamuri and Kaliwal (2016) reported maximum Lac activity (2000 U/L) at pH 6 and 40 °C temperature when *Marasmius* sps. was used in SSF. Nowadays, industries are more concerned towards cutting down the cost of microbial enzyme and in this context utilization of cheap and readily available raw material in the fermentation process is recommended (Singh et al. 2016). In a study, Ruiz et al. (2012) reported that addition of lemon peel pomace as substrate in SSF process enhanced enzyme activity (2181 U/L) when *Aspergillus* and *Penicillium* sp. were used. Use of corn stover as substrate in SSF showed better production of fungal enzyme (Lac 97.47 U/mL, MnP 614.23 U/mL, and LiP 1007.39 U/mL) from *Schizophyllum commune* (Yasmeen et al. 2013). Gomes et al. (2009) reported that utilization of wheat bran can enhance production of Lac (200 U/mL) from *Coriolopsis byrsina*. However, the same substrate was found to increase the activity of MnP (7 U/mL) and LiP (8 U/mL) when *Lentinus* sp. was used. The use of agro-industrial residues as a low-cost substrate for the production of fungal enzyme and their application in rapid decolorization of azo dyes is also evaluated by Singh et al. (2016). They found that pine sawdust, wheat bran, rice straw, wheat stack, peanut shell, and soybean powder when used as substrate, showed maximum activity of the enzymes MnP (1200 U/L), Lac (586 U/L), and LiP (109 U/L), respectively, on day 5 of SSF. However, optimization of the dye decolorization system with crude MnP enhanced the rates of decolorization of Orange IV and Orange G dye to 76% and 57%, respectively.

In SmF system, microorganisms utilize liquid substrates, and end products are liberated into the fermentation broth. In SmF, process parameters such as pH, temperature, moisture, oxygen transfer, and aeration can be easily controlled (Kunamneni et al. 2005). However, due to high moisture content, SmF is mostly used for bacterial enzyme production. It has been reported that when fungi are used in SmF, low recovery of enzyme occurs due to slower release of the enzyme bound to mycelium (Sabu et al. 2006). There are some selected studies which indicate that in fungal SmF system, optimization of carbon and nitrogen concentration in liquid substrate may improve the yield of dye-decolorizing enzymes. For example, Songulashvili et al. (2011) found that in SmF, supplementation of culture medium with KNO_3 increased Lac and MnP activities of *G. lucidum* 447 strain by 75% and 27%, respectively. Similarly, the substitution of synthetic carbon source in liquid substrate with different agro-waste residues enhanced Lac production in SmF (Songulashvili et al. 2011, 2015).

9.6 Future Perspectives

For effective and maximum utilization of fungal enzymes in the removal of azo dye, vigorous research in the field of fungal biotechnology is needed. However, for commercial production, improvement in fermentation technology can play an additional role. It has also been realized that the laboratory level research does not meet the criterion required for industrial enzyme production. Hence for the microbial production of azo dye-decolorizing enzymes, these hurdles need to be addressed. Several workers tried to improve the production and efficacy of fungal azo dye-removing enzymes. According to the Bankole et al. (2018) search and identification of potential fungal strains at genomic level could be useful in the selection of natural azo dye-degrading fungi whereas selection of method applied or fermentation technology can play essential role in enhancing the yield of fungal azo dye-removing enzymes. Generally, use of packed-bed reactors (PBR), rotating biological contactor (RBC), and stirred-tank reactors (STR) is preferred in microbial enzyme production; however, their application may vary by the types of enzyme and strain selected (Longo et al. 2008). The PBR are more frequently used in the removal of azo dyes from the industrial effluents. The PBR can be used to remove 90% of orange II dye from textile effluents (Almeida and Corso 2014). However, in STR, use of *T. versicolor* provided 99% removal of reactive black-5 and reactive red-198 dye (Sen et al. 2016). For a long time, SSF and SmF processes have been used in the production of fungal dye-decolorizing enzymes such as Lac, LiP, and MnP. Researches in recent past have shown that a slight modification in media and optimization in SSF or SmF processes might be useful for enhancing enzyme yield (Songulashvili et al. 2015; Vendruscolo et al. 2016). For example, *T. versicolor* fungus produced 229 U/L of laccase when nylon sponge was used in SSF while least amount (126 U/L) was produced when it (nylon sponge) was not used in the cultivation of fungi (Singh et al. 2016). Similarly, *I. lacteus* produced higher amount of MnP enzyme (950 U/L) when temperature and pH level were changed in SSF bioreactor (Zhao et al. 2005). However, it is essential to know that for fungi, SSF system is always better than SmF (Sabu et al. 2006). Recently, immobilization of fungal enzymes on nanofiber has found to enhance the rate of decolorization. For example, when LiP enzymatic extracts of *P. ostreatus* (PLO9) and *G. lucidum* (GRM117) were immobilized on carbon nanotubes, an increase of 18 and 27 fold in LiP-specific activity was observed compared to the free enzyme (Oliveira et al. 2018).

Sometimes, an exhaustive search for novel fungal strains showing the potential of dye decolorization does not provide additional benefit to the enzyme industry as after repeated use in the fermentation process, the strain may not perform well. In such conditions, a slight modification in culture conditions or growth parameters by the addition of some additives, or optimization of the substrate may lead to an increase in enzyme production (also known as "one strain–many compounds approach") (Romano 2018).

In recent past, genetic engineering-based techniques have also proven useful. Gene cloning, heterologous expression, and recombination of the gene have been used to develop superior strain with the capability to produce dye-decolorizing enzymes (Singh et al. 2016). Genetically modified fungi (GMF) have also been tried for the removal of azo dyes from waste. For example by transferring laccase gene from *T. versicolor* in a yeast *Yarrowia lipolytica*, a recombinant laccase was constructed and it is widely used for the treatment of synthetic dyes from textile effluent (Theerachat et al. 2012). There are several other examples such as laccase released by *T. versicolor* was used in engineered *P. chrysosporium* for the degradation of phenolic dyes (Coconi-Linares et al. 2015). In another study, MnP gene was used to construct bioengineered *S. cerevisiae* by which production of MnP increased manyfold which also enhanced the commercial applicability of fungal azo dye-decolorizing enzyme (Le Roes-Hill and Prins 2016). Recently, omics-based approaches including metagenomics, metatranscriptomics, and metaproteomics have also revealed the potential of unculturable microorganisms, including fungi as sources of novel enzymes (Berini et al. 2017). These "omics"-based techniques could also be helpful in improving the yield and mass production of fungal enzymes.

9.7 Conclusion

Accumulation of azo dye stuff in soil and water is creating environmental pollution. Removal of azo dyes by conventional methods is found not to be very competent, especially in terms of complete removal of dye residue from the environment. However, high operating cost and generation of secondary pollutants are the constraints associated with azo dye removal. An eco-friendly alternative such as the use of microbes has been realized as a sustainable means of removing azo dye from different wastes. Among microbes, fungi have emerged as potential candidates to remove azo dyes from waste residues. They have an arsenal of enzymes with the capability to utilize various dye compounds. At the commercial level, the production of fungal azo dye-removing enzymes is facing hurdles. The fermentation processes and their optimization are the major areas of research that need to be further investigated whereas exploration of natural fungal strains with biotechnological values can also improve the situation.

References

Abadulla E, Tzanov T, Costa S, Robra KH, Cavaco-Paulo A, Gübitz GM (2000) Decolorization and detoxification of textile dyes with a laccase from *Trametes hirsuta*. Appl Environ Microbiol 66:3357–3362

Adnan LA, Yusoff ARM, Hadibarata T, Khudhair AB (2014) Biodegradation of bis-azo dye reactive Black 5 by white-rot fungus *Trametes gibbosa* sp. WRF 3 and its metabolite characterization. Water Air Soil Pollut 225(10)

Akar T, Demir TA, Kiran I, Ozcan A, Ozcan AS, Tunali S (2006) Biosorption potential of *Neurospora crassa* cells for decolorization of Acid Red 57 (AR57) dye. J Chem Technol Biotechnol 81(7):1100–1106

Akar T, Ozcan AS, Tunali S, Ozcan A (2008) Biosorption of a textile dye (Acid Blue 40) by cone biomass of *Thuja orientalis*: estimation of equilibrium, thermodynamic and kinetic parameters. Bioresour Technol 99(8):3057–3065

Akar ST, Akar T, Çabuk A (2009) Decolorization of a textile dye, reactive red 198 (rr198), by *Aspergillus parasiticus* fungal biosorbent. Braz J Chem Eng 26(2):399–405

Akdogan HA, Topuz MC, Urhan AA (2014) Studies on decolorization of reactive blue 19 textile dye by *Coprinus plicatilis*. J Environ Health Sci Eng 12(1)

Aksu Z, Cagatay SS (2006) Investigation of biosorption of gemazol turquise blue-G reactive dye by dried *Rhizopus arrhizus* in batch and continuous systems. Sep Purif Technol 48:24–35

Aksu Z, Kilic N, Ertugrul V, Donmez G (2007) Inhibitory effects of chromium (VI) and remazol black on chromium (VI) and dye stuff removals by *Trametes versicolor*. Enzyme Microb Technol 40:1167–1174

Ali H (2010) Biodegradation of synthetic dyes – a review. Water Air Soil Pollut 213:251–273

Ali M, Husain Q, Alam N, Ahmad M (2016) Enhanced catalytic activity and stability of ginger peroxidase immobilized on amino-functionalized silica-coated titanium dioxide nanocomposite: A cost-effective tool for bioremediation. Water Air Soil Pollut 228(1):22

Almeida EJR, Corso CR (2014) Comparative study of toxicity of azo dye procion red MX-5B following biosorption and biodegradation treatments with the fungi *Aspergillus niger* and *Aspergillus terreus*. Chemosphere 112:317–322

Aretxage A, Romero S, Sarr M, Vincent T (2001) Adsorption step in the biological degradation of a textile dye. Biotechnol Prog 17:664–668

Asad S, Amoozegar MA, Pourbabaee AA, Sarbolouki MN, Dastgheib SMM (2007) Decolorization of textile azo dyes by newly isolated halophilic and halotolerant bacteria. Bioresour Technol 98(11):2082–2088

Asgher M, Yasmeen Q, Iqbal HMN (2013) Enhanced decolorization of Solar brilliant red 80 textile dye by an indigenous white rot fungus *Schizophyllum commune* IBL-06. Saudi J Biol Sci 20(4):347–352

Asma D, Kahraman S, Cing S, Yesilada O (2006) Adsorptive removal of textile dyes from aqueous solutions by dead fungal biomass. J Basic Microbiol 46(1):3–9

Asses N, Ayed L, Hkiri N, Hamdi M (2018) Congo red decolorization and detoxification by *Aspergillus niger*: removal mechanisms and dye degradation pathway. Biomed Res Int 2018:1–9

Aust SD, Swaner PR, Stalh JD (2003) Detoxification and metabolism of chemicals by white rot fungi. In: Pesticide decontamination and detoxification. Oxford University Press, Washington, DC, pp 3–14

Baccar R, Blánquez P, Bouzid J, Feki J, Attiya H, Sarrà M (2011) Decolorization of a tannery dye: From fungal screening to bioreactor application. Biochem Eng J 56(3):184–189

Bankole PO, Adekunle AA, Obidi OF, Chandanshive VV, Govindwar SP (2018) Biodegradation and detoxification of scarlet RR dye by a newly isolated filamentous fungus, *Peyronellaea prosopidis*. Sust Environ Res 28:214–222

Bayramoglu G, Arica MY (2007) Biosorption of benzidine based textile dyes "direct blue 1 and direct red 128" using native and heat-treated biomass of *Trametes versicolor*. J Hazard Mater 143:135–143

Berini F, Casciello C, Marcone GL, Marinelli F (2017) Metagenomics: novel enzyme from non-culturable microbes. FEMS Microbiol Lett 364:21

Bibi I, Bhatti HN, Asgher M (2011) Comparative study of natural and synthetic phenolic compounds as efficient laccase mediators for the transformation of cationic dye. Biochem Eng J 56:225–231

Bosco F, Mollea C, Ruggeri B (2016) Decolorization of Congo red by Phanerochaete chrysosporium: the role of biosorption and biodegradation. Environ Technol 17:45

Bourbonnais R, Paice MG (1992) Demethylation and delignification of Kraft pulp by Trametes versicolor laccase in the presence of 2,2-azinobis-(3-ethylbenzothiazoline-6-sulphonate). Appl Microbiol Biotechnol 36:823–827

Buitrón G, Quezada M, Moreno G (2004) Aerobic degradation of the azo dye acid red 151 in a sequencing batch biofilter. Bioresour Technol 92(2):143–149

Carmona-Ribeiro AM, Prieto T, Nantes IL (2015) Nanostructures for peroxidases. Front Mol Biosci 2(50)

Chakraborty S, Bask B, Dutta S, Bhunia B, Dey A (2013) Decolorization and biodegradation of Congo red dye by a novel white rot fungus *Alternaria alternate* CMERI F6. Bioresour Technol 147:662–666

Chang JS, Shan CC, Chen Y (2001a) Decolorization of azo dyes with immobilized *Pseudomonas luteola*. Process Biochem 36:757–763

Christiane A, Steeve M (2013) Biodegradation of reactive Blue 4 and Orange G by *Pycnoporus sanguineus* strain isolated in Gabon. J Bioremed Biodegr 4(7). https://doi.org/10.4172/2155-6199.1000206

Chang JS, Shan CC, Chen Y (2001b) Decolorization of azo dyes with immobilized *Pseudomonas luteola*. Process Biochem 36:757–763

Chen H (2006) Recent advances in azo dye degrading enzyme research. Curr Protein Pept Sci 7:101–111

Chen KC, Wu JY, Liou DJ, Hwang SCJ (2003) Decolorization of the textile dyes by newly isolated bacterial strains. J Biotechnol 101:57 68

Cho HY, Cho NS, Wilkolazka AJ, Rogalski J, Leonowicz A, Shin YS, Ohga S (2007) Effect of fungal laccase and new mediators, acetovanillone and acetosyringone on decolourization of dyes. J Fac Agric Kyushu Univ 52:275–280

Christie RM (2001) Condition and its potential in decolorization of azo dyes. Appl Microbiol Biotechnol 72:939–946

Coconi-Linares N, Oritz-Vazquez E, Fernandez F, Loske AM, Gomez-Lim M (2015) Recombinant expression of four oxidoreductases in Phanerochaete chrysosporium improves degradation of phenolic and non- phenolic substrates. J Biotechnol 209:76–84

Conesa A, Punt PJ, van den Hondel CA (2002) Fungal peroxidases: molecular aspects and applications. J Biotechnol 93(2):143–158

Couto SR, Sanromán MA (2006) Application of solid-state fermentation to food industry-a review. J Food Eng 76:291–302

Dabrowski A, Podkoscielny P, Hubicki ZP, Mariusz B (2005) Adsorption of phenolic compounds by activated carbon-a critical review. Chemosphere 58:1049–1070

Dhanjal NIK, Mittu B, Chauhan A, Gupta S (2013) Biodegradation of textile dyes using fungal isolates. J Environ Sci Technol 6(2):99–105

Dos Santos AB, Cervantes FJ, Van Lier JB (2007) Review paper on current technologies for decolourisation of textile wastewaters: perspective for anaerobic biotechnology. Bioresour Technol 98:2369–2385

Dwivedi U, Singh P, Pandey VP, Kumar A (2011) Structure-function relationship among bacterial, fungal and plant laccases. J Mol Catal B: Enzym 68:117–128

Enayatizamir N, Tabandeh F, Rodríguez-Couto S, Yakhchali B, Alikhani HA, Mohammadi L (2011) Biodegradation pathway and detoxification of the diazo dye Reactive Black 5 by *Phanerochaete chrysosporium*. Bioresour Technol 102(22):10359–10362

Erden E, Ucar CM, Gezer T, Pazarlioglu NK (2009) Screening for ligninolytic enzymes from autochthonous fungi and applications for decolorization of Remazol marine blue. Braz J Microbiol 40(2):346–353

Ezeronye OU, Okerentugba PO (1999) Performance and efficiency of a yeast biofilter for the treatment of a Nigerian fertilizer plant effluent. World J Microbiol Biotechnol 15:515–516

Fabbrini M, Galli C, Gentili P (2002) Comparing the catalytic efficiency of some mediators of laccase. J Mol Catal B Enzym 16:231–240

Fernández-Fueyo E, Ruiz-Dueñas FJ, Miki Y, Martínez MJ, Hammel KE, Martínez AT (2012) Lignin-degrading peroxidases from the genome of the selective ligninolytic fungus *Ceriporiopsis subvermispora*. J Biological Chem 287:16903–16916

Ferreira-Leitao VS, de Carvalho MEA, Bon EPS (2007) Lignin peroxidase efficiency for methylene blue decoloration: comparison to reported methods. Dyes Pigments 74:230–236

Forgacs E, Cserhati T, Oros G (2004) Removal of synthetic dyes from wastewaters: a review. Environ Int 30(7):953–971

Fu Y, Viraraghavan T (2001) Fungal decolorization of dye wastewater: a review. Bioresour Technol 79:251–262

Furukawa T, Bello FO, Horsfall L (2014) Microbial enzyme system for lignin degradation and their transcriptional regulation. Front Biol 9:448–471

Galhaup C, Goller S, Peterbauer CK, Strauss J, Haltrich D (2002) Characterization of the major laccase isoenzyme from *Trametes pubescens* and regulation of its synthesis by metal ions. J Microbiol 148:2159–2169

Ghobadi Z, Nejad S, Borghei M, Yaghmaei S (2019) Biodegradation of synthetic dye using partially purified and characterized laccase and its proposed mechanism. Int J Environ Sci Technol. https://doi.org/10.1007/s13762-019-02226-5

Gold MH, Alic M (1993) Molecular biology of the lignin-degrading *basidiomycete Phanerochaete chrysosporium*. Microbiol Rev 57:605–622

Gomes E, Aguiar AP, Carvalho CC, Bonfá MRB, Silva R, Boscolo M (2009) Ligninases production by Basidiomycetes strains on lignocellulosic agricultural residues and their application in the decolorization of synthetic dyes. Braz J Microbiol 40(1):31–39

Gomez-Toribio V, Martinez AT, Martinez MJ, Guillen F (2001) Oxidation of hydroquinones by the versatile ligninolytic peroxidase from *Pleurotus eryngii*. H2O2 generation and the influence of Mn2+. Eur J Biochem 268:4787–4793

Gowri RS, Vijayaraghavan R, Meenambigai P (2014) Microbial degradation of reactive dyes-a review. Int j of current microbio and. Appl Sci 3:421–436

Grassi E, Scodeller P, Filiel N, Carballo R, Levin L (2011) Potential of *Trametes trogii* culture fluids and its purified laccase for the decolorization of different types of recalcitrant dyes without the addition of redox mediators. Int J Biodeteriorat Biodegradat 65(4):635–643

Grierson JR, Link JM, Mathis CA (1989) A radiosynthesis of [F-18] fluoromisonidazole. J Nucl Med 30(3):343–350

Grinhut T, Salame TM, Chen Y, Hadar Y (2011) Involvement of ligninolytic enzymes and Fenton-like reaction in humic acid degradation by *Trametes* sp. Appl Microbiol Biotechnol 91:1131–1140

Guo JB, Zhou JT, Wang D, Tian CP, Wang P, Uddin MS, Yu H (2007) Biocatalyst effects of immobilized anthraquinone on the anaerobic reduction of azo dyes by the salt-tolerant bacteria. Water Resour 41:426 432

Hadibarata T, Yusoff ARM, Aris A, Salmiati HT, Kristanti RA (2012) Decolorization of Azo, Triphenylmethane and anthraquinone dyes by laccase of a newly isolated *Armillaria* sp. F022. Water Air Soil Pollut 223(3):1045–1054

Hadibarata T, Amalina L, Yusoff ARM, Yuniarto A, Fikri MM, Zubir A, Khudhair AB, Naser A (2013) Microbial decolorization of an azo dye reactive black 5 using white-rot fungus *Pleurotus eryngii* F032. Water Air Soil Pollut 224:1595

Hai F, Yamamoto K, Fukushi K (2007) Hybrid treatment system for dye wastewater. Crit Rev Environ Sci Technol 37:315–377

Hammel KE, Cullen D (2008) Role of fungal peroxidases in biological ligninolysis. Curr Opin Plant Biol 11(3):349–355

Hanapi SZ, Abdelgalil SA, Hatti-kaul R, Aziz R, Enshasy HAE (2018) Isolation of a new efficient dye decolorizing white rot fungus *Cerrena* Sp. WICC F39. J Sci Ind Res 77:399–405

He XL, Song C, Li Y, Wang N, Xu L, Han X, Wei D (2018) Efficient degradation of azo dyes by a newly isolated fungus *Trichoderma tomentosum* under non-sterile conditions. Ecotoxicol Environ Saf 150:232–239

Hofrichter M (2002) Review: lignin conversion by manganese peroxidase (MnP). Enzyme Microbiol Technol 30(4):454–466

Hsueh CC, Chen BY (2008) Exploring effects of chemical structure on azo dye decolorization characteristics by *Pseudomonas luteola*. J Hazard Mater 154:703–710

Husain M, Husain Q (2011) Applications of redox mediators in the treatment of organic pollutants by using oxido-reductive enzymes: a review. Crit Rev Environ Sci Technol 38:1–42

Husain Q, Karim Z, Banday ZZ (2010) Decolorization of textile effluent by soluble fenugreek (*Trigonella foenum graecum L*) seed peroxidase. Water Air Soil Pollut 212:319–328

Jadhav JP, Kalyani DC, Telke AA, Phugare SS, Govindwar SP (2010) Evaluation of the efficacy of a bacterial consortium for the removal of color, reduction of heavy metals, and toxicity from textile dye effluent. Bioresour Technol 101(1):165–173

Jadhav I, Vasniwal R, Shrivastava D, Jadhav K (2016) Microorganism-based treatment of azo dyes. J Environ Sci Technol 9:188–197

Janusz G, Czurylo A, Frac M, Rola B, Sulei J, Pawlik A, Siwulski M, Rogalski J (2015) Laccase production and metabolic diversity among *Flammulina velutipes* strains. J Microbiol Biotechnol 31:121–133

Kant R (2012) Textile dyeing industry an environmental hazard. Nat Sci 04:22–26

Kaushik P, Malik A (2009) Fungal dye decolourization: recent advances and future potential. Environ Int 35:127–141

Kasinath A, Novotný C, Svobodová K, Patel KC, Šašek V (2003) Decolorization of synthetic dyes by *Irpex lacteus* in liquid cultures and packed-bed bioreactor. Enzyme Microb Technol 32(1):167–173

Khalaf MA (2008) Biosorption of reactive dye from textile wastewater by non-viable biomass of *Aspergillus niger* and *Spirogyra* sp. Bioresour Technol 99(14):6631–6634

Khan R, Bhawana P, Fulekar MH (2013) Microbial decolorization and degradation of synthetic dyes: a review. Rev Environ Sci Biotechnol 12:75–97

Khlifi R, Belbahri L, Woodward S, Ellouz M, Dhouib A, Sayadi S, Mechichi T (2010) Decolourization and detoxification of textile industry wastewater by the laccase-mediator system. J Hazard Mater 175:802–808

Kilic NK, Nielsen JL, Yuce M, Donmez G (2007) Characterization of a simple bacterial consortium for effective treatment of wastewaters with reactive dyes and Cr(VI). Chemosphere 67:826–831

Kiran I, Akar T, Ozcan AS, Ozcan A, Tunali S (2006) Biosorption kinetics and isotherm studies of Acid Red 57 by dried *Cephalosporium aphidicola* cells from aqueous solutions. Biochem Eng J 31(3):197–203

Knapp JS, Vantoch-Wood EJ, Zhang F (2001) Use of Wood-rotting fungi for the decolorization of dyes and industrial effluents, Chapter. Cambridge University Press, Cambridge, pp 253–261. ISBN 0 521 78119 1

Kocyigit H, Ugurlu A (2015) Biological decolorization of reactive azo dye by anaerobic/aerobic sequencing batch reactor system. Global NEST 17:210–219

Krishnamoorthy R, Arul Jose P, Ranjith M, Anandham R, Suganya K, Prabhakaran J, Thiyageshwari S, Johnson J, Gopal NO, Kumutha K (2018) Decolourisation and degradation of azo dyes by mixed fungal culture consisted of *Dichotomomyces cejpii* MRCH 1-2 and *Phoma tropica* MRCH 1-3. J Environ Chem Eng 6(1):588–595

Kumari K, Abraham T (2007) Biosorption of anionic textile dyes by nonviable biomass of fungi and yeast. Bioresour Technol 98(9):1704–1710

Kunamneni A, Permaul K, Singh S (2005) Amylase production in solid state fermentation by the thermophilic fungus *Thermomyces lanuginosus*. J Biosci Bioeng 100:168–171

Kurade MB, Waghmode TR, Kagalkar AN, Govindwar SP (2012) Decolorization of textile industry effluent containing disperse dye scarlet RR by a newly developed bacterial-yeast consortium BL-GG. Chem Eng J 184:33–41

Kurade MB, Waghmode TR, Khandare RV, Jeon BH, Govindwar SP (2016) Biodegradation and detoxification of textile dye disperse red 54 by *Brevibacillus laterosporus* and determination of its metabolic fate. J Biosci Bioeng 121:442–449

Lasram S, Oueslati S, Valero A, Marin S, Ghorbel A, Sanchis V (2010) Water activity and temperature effects on fungal growth and ochratoxin A production by ochratoxigenic *Aspergillus carbonarius* isolated from Tunisian grapes. J Food Sci 75(2):M89–M97

Lavanya C, Dhankar R, Chhikara S, Sheoran S (2014) Degradation of toxic dyes: a review. Int J Curr Microbiol Appl Sci 6:189–199

Le Roes-Hill M, Prins A (2016) Biotechnological potential of oxidative enzymes from Actinobacteria. Actinobacteria-Basics and Biotechnological Applications. https://doi.org/10.5772/61321

Levin L, Papinutti L, Forchiassin F (2004) Evaluation of Argentinean white-rot fungi for their ability to produce lignin modifying enzymes and decolorize industrial dyes. Bioresour Technol 94:169–176

Longo MA, Deive FJ, Domínguez A, Sanromán MA (2008) Solid-state fermentation for food and feed application. In: Current Developments in Solid-State Fermentation. Asiatech Publishers, New Delhi, pp 379–411

Mahmoud MS, Mostafa M, Mohamed SA, Mahmoud N (2017) Bioremediation of red azo dye from aqueous solutions by *Aspergillus niger* strain isolated from textile wastewater. J Environ Chem Eng 5:547–554

Malakootian M, Heidar RM (2018) Reactive orange 16 dye adsorption from aqueous solutions by *Psyllium* seed powder as a low-cost biosorbent: kinetic and equilibrium studies. J Appl Water Sci 8:212

Martinez AT (2002) Molecular biology and structure–function of lignin degrading heme peroxidases. Enzyme Microb Technol 30:425–444

Martínez MJ, Böckle B, Camarero S, Ruiz-Dueñas FJ, Guillén FA, Martínez AT (1996) MnP isoenzymes produced by two *Pleurotus* species in liquid culture and during wheat-straw solid-state fermentation. ACS Symp Ser 655:183–196

Martorell MM, Pajot HF, de Figueroa LIC (2012) Dye decolorizing yeasts isolated from Las Yungas rainforest dye assimilation and removal used as selection criteria. Int J Biodeter Biodegr 66:25–32

Mendoza L, Jonstrup M, Kaul RH, Mattiasson B (2011) Azo dye decolorization by a laccase mediator system in a membrane reactor: enzyme and mediator reusability. Enzyme Microb Technol 49:478–484

Mishra A, Kumar S, Pandey KA (2011) Laccase production and simultaneous decolorization of synthetic dyes in unique inexpensive medium by new isolates of white rot fungus. Int J Biodeterior Biodegrad 65:487–493

Mondal P, Majumder CB, Mohanty B (2006) Laboratory based approaches for arsenic remediation from contaminated water: recent developments. J Hazard Mater 137:464–479

Mugdha A, Usha M (2012) Enzymatic treatment of waste containing dyestuffs using different delivery systems. Sci Rev Chem Commun 2:31–40

Mukherjee A, Kumar DA (2018) A review on applications of lignolytic enzymes of fungi. J Pharm Pharm Sci 7:484–493

Murugesan K, Arulmani M, Nam IH, Chang SY, Kalaichelvan PT (2006) Purification and characterization of laccase produced by a white rot fungus *Pleurotus sajor-caju* under submerged culture. Appl Microb Biotechnol 72:939–946

Nilsson I, Moller A, Mattiasson B, Rubindamayugi MST, Welander U (2006) Decolorization of synthetic and real textile waste water by the use of white -rot fungi. Enzyme Microb Technol 38:94–100

Oliveira SF, Rodriguesda JM, Kasuya LMC, Ladeira LO, Junior AC (2018) Enzymatic extract containing lignin peroxidase immobilized on carbon nanotubes: potential biocatalyst in dye decolorization. Saudi J Biol Sci 25:651–659

Oturkar CC, Nemade HN, Mulik PM, Patole MS, Hawaldar RR, Gawai KR (2011) Mechanistic investigation of decolorization and degradation of reactive red 120 by *Bacillus lentus* BI377. Bioresour Technol 102:758–764

Pandey A (2003) Solid state fermentation. Biochem Eng J 13:81–84

Pandey A, Singh P, Iyengar L (2007) Bacterial decolorization and degradation of azo dyes. Int J Biodeter Biodegr 59:73–76

Pandey V, Awasthi M, Singh S, Tiwari S, Dwivedi UN (2017) A comprehensive review on function and application of plant peroxidases. Biochem Anal Biochem 6:2–16

Pandi A, Kuppuswami GM, Ramudu KN, Palanivel S (2018) A sustainable approach for degradation of leather dyes by a new fungal laccase. J Clean Prod 211:590–597

Pazarlioglu NK, Urek RO, Ergun F (2005) Biodecolorization of direct Blue-15 by immobilized *Phanerochaete chrysosporium*. Process Biochem 40:1923–1929

Pereira L, Coelho AV, Viegas CA, Correia Dos Santos MM, Robalo MP, Martins LO (2009) Enzymatic biotransformation of the azo dye Sudan Orange G with bacterial CotA-laccase. J Biotechnol 139:68–77

Pietikainen J, Pettersson M, Baath E (2005) Comparison of temperature effects on soil respiration and bacterial and fungal growth rates. FEMS Microbiol Ecol 52(1):49–58

Podgornik H, Grgic I, Perdih A (1999) Decolorization rate of dyes using lignin peroxidases of *Phanerochaete chrysosporium*. Chemosphere 38(6):1353–1359

Pointing SB (2001) Feasibility of bioremediation by white-rot fungi. Appl Microbiol Biotechnol 57:20–33

Pramanik S, Chaudhuri S (2018) Laccase activity and azo dye decolorization potential of *Podoscypha elegans*. Mycobiology 46(1):79–83

Prasad ASA, Rao KVB (2013) Aerobic biodegradation of azo dye by *Bacillus cohnii* MTCC 3616; an obligately alkaliphilic bacterium and toxicity evaluation of metabolites by different bioassay systems. Appl Microbiol Biotechnol 97:7469–7481

Prigione V, Tigini V, Pezzella C, Anastasi A, Sannia G, Varese GC (2008) Decolourisation and detoxification of textile effluents by fungal biosorption. Water Res 42(12):2911–2920

Przystas W, Zablocka-Godlewska E, Grabinska-Sota E (2015) Efficacy of fungal decolorization of a mixture of dyes belonging to different classes. Braz J Microbiol 46(2):415–424

Qin X, Luo H, Zhang X, Yao B, Ma F, Su X (2018) Dye-decolorizing peroxidases in Irpex lacteus combining the catalytic properties of heme peroxidases and laccase play important roles in ligninolytic system. Biotechnol Biofuels 11:302

Qu Y, Shi S, Ma F, Yan B (2010) Decolorization of reactive dark blue K-R by the synergism of fungus and bacterium using response surface methodology. Bioresour Technol 101:8016–8023

Rafatullah M, Sulaiman O, Hashim R, Ahmad A (2010) Adsorption of methylene blue on low cost adsorbents: a review. J Hazard Mater 177:70–80

Rani B, Kumar V, Singh J, Bisht S, Teotia P, Sharma S, Kela R (2014) Bioremediation of dyes by fungi isolated from contaminated dye effluent sites for bio-usability. Braz J Microbiol 45:1055–1063

Rawat D, Sharma RS, Karmakar S, Arora LS, Mishra V (2018) Ecotoxic potential of a presumably non-toxic azo dye. Ecotoxicol Environ Saf 148:528–537

Rekik H, Zarai Jaouadi N, Bouacem K, Zenati B, Kourdali S, Badis A, Annane R, Bouanane-Darenfed A, Bejar S, Jaouadi B (2019) Physical and enzymatic properties of a new manganese peroxidase from the white-rot fungus *Trametes pubescens* strain i8 for lignin biodegradation and textile-dyes biodecolorization. Int J Biol Macromol 15:514–525

Robinson T, Chandran B, Nigam P (2001) Studies on the production of enzymes by white rot fungi for the decolorisation of textile dyes. Enzyme Microb Technol 29:575–579

Romano S (2018) Extending the "one strain many compounds" (OSMAC) principle to marine microorganisms. Mar Drugs 16:1–29

Ruiz HA, Rodrı´guez-Jasso RM, Rodrı´guez R, Contreras-Esquivel JC, Aguilar CN (2012) Pectinase production from lemon peel pomace as support and carbon source in solid-state fermentation column tray bioreactor. Biochem Eng J 65:90–95

Ruiz-Duenas F, Fernandez E, Jesus M, Angel M, Martinez T (2011) *Pleurotus ostreatus* heme peroxidases: an in silico analysis from the genome sequence to the enzyme molecular structure. C R Biol 334:795–805

Sabu C, Augur C, Swati C, Pandey A (2006) Tannase production by *Lactobacillus* sp. ASR-S1 under solid-state fermentation. Process Biochem 41:575–580

Sani RK, Banerjee UC (1999) Screening for organisms applicable to the decolorization of triphenyl aniline dyes and optimization of biotransformation conditions in stirred tank reactor. Indian J Environ Ecoplan 2:1–9

Saratale RG, Saratale GD, Kalyani DC, Chang JS, Govindwar SP (2009) Enhance decolorization and biodegradation of textile azo dye scarlet R by using developed microbial consortium-GR. Bioresour Technol 100:2493–2500

Selvakumar S, Manivasagan R, Chinnappan K (2013) Biodegradation and decolourization of textile dye wastewater using *Ganoderma lucidum*. 3 Biotech 3:71–79

Sen K, Pakshirajan K, Santra SB (2012) Modeling the biomass growth and enzyme secretion by the white rot fungus *Phanerochaete chrysosporium*: a stochastic-based approach. Appl Biochem Biotechnol 167(4):705–713

Sen SK, Raut S, Bandyopadhyay P, Raut S (2016) Fungal decolouration and degradation of azo dyes: a review. Fungal Biol Rev 30:112–133

Sharma P, Singh L, Dilbagh N (2009) Biodegradation of Orange II dye by Phanerochaete chrysosporium in simulated wastewater. J Sci Ind Res 68:157–161

Shedbalkar U, Jadhav JP (2011) Detoxification of malachite green and textile industrial effluent by *Penicillium ochrochloron*. Biotechnol Bioprocess Eng 16:196–204

Singh H (2006) Mycoremediation-fungal bioremediation. Wiley Interscience, Hoboken, NJ, pp 421–471. ISBN-13: 978-0-471-75501-2

Singh D, Chen S (2008) The white-rot fungus *Phanerochaete chrysosporium*: conditions for the production of lignin-degrading enzymes. Appl Microbiol Biotechnol 81:399–417

Singh PK, Shukla P (2015) Systems biology as an approach for deciphering microbial interactions. Briefings Funct Genom 14:166–168

Singh S, Pakshirajan K, Daverey A (2010) Enhanced decolourization of Direct Red-80 dye by the white rot fungus *Phanerochaete chrysosporium* employing sequential design of experiments. Biodegradation 21(4):501–511

Singh R, Kumar M, Mittal A, Mehta PK (2016) Microbial enzymes: industrial progress in 21st century. 3 Biotech 6(2)

Songulashvili G, Jimenex-Tobon G, Jaspers C, Penninel MJ (2011) High production of laccase by *Ganoderma Lucidum* 447 in submerged cultivation on ethanol production supplemented with Cu^{2+}. Mycosphere 2:507–513

Songulashvili G, Spindler D, Jimenex-Tobon GA, Jaspers C, Kerns G, Penninchk MJ (2015) Production of high level of laccase by submerged fermentation at 120-L scale of *Cerrena unicolor* C-139 on wheat bran. Curr Rev Biol:121–125

Spadaro JT, Gold MH, Renganathan V (1992) Degradation of azo dyes by the lignin-degrading fungus *Phanerochaete chrysosporium*. Appl Environ Microbiol 58:2397–2401

Tan L, Li H, Ning S, Hao J (2014) Aerobic decolorization and degradation of acid orange G (AOG) by suspended growing cells and immobilized cells of a yeast strain *Candida tropicalis* TL-F1. Appl Biochem Biotechnol 174:1651–1667

Telke AA, Ghodake GS, Kayani DC, Dhanve RS, Govindwar SP (2011) Biochemical characteristics of a textile dye degrading extracellular laccase from a *Bacillus* sp. ADR. Bioresour Technol 102:1752–1756

Theerachat M, Emond S, Cambon E, Bordes F, Marty A, Nicaud J (2012) Engineering and production of laccase from *Trametes versicolor* in the yeast *Yarrowia lipolytica*. Bioresour Technol 125:267–274
Tien M, Kirk TK (1983) Lignin-degrading enzyme from the hymenomycete *Phanerochaete chrysosporium*. Burds Sci 221:661–663
Torres E, Bustos-Jaimes I, Le Borgne S (2003) Potential use of oxidative enzymes for the detoxification of organic pollutants. Appl Catal Environ 46:1–15
Van der Zee FP, Villaverde S (2005) Combined anaerobic–aerobic treatment of azo dyes—a short review of bioreactor studies. Water Resour 39:1425–1440
Vanhulle S, Trovaslet M, Enaud E, Lucas M, Taghavi S, Van der Lelie D, Van Aken B, Foret M, Onderwater RC, Wesenberg D, Agathos SN, Schneider YJ, Corbisier AM (2008) Decolorization, cytotoxicity, and genotoxicity reduction during a combined ozonation /fungal treatment of dye-contaminated wastewater. Environ Sci Technol 4:584–589
Vantamuri AB, Kaliwal BB (2016) Production of laccase by newly isolated *Marasmius* Sp. BBKAV79 in solid state fermentation and its antiproliferative activity. Int J Pharma Sci Res 12:4978–4987
Vaithanomsat P, Apiwatanapiwat W, Petchoy O, Chedchant J (2010) Production of ligninolytic enzymes by white-rot Fungus sp. KAPI0039 and their application for reactive dye removal. Int J Chem Eng 2010:1–6
Vendruscolo F, Bu¨hler MMR, Carvalho JC, Oliveira D, Moritz DE, Schmidell W, Ninow JL (2016) A reality on the production and application of microbial pigments. Appl Biochem Biotechnol 178:211–223
Wang J, Liu GF, Hong L, Jin RF, Zhou JT, Lei TM (2012) Biodegradation of Acid Orange 7 and its auto-oxidative decolorization product in membrane-aerated biofilm reactor. Int J Biodeter Biodegrad 67:73–77
Wells A, Teria M, Eve T (2006) Green oxidations with laccase mediator systems. Biochem Soc Trans 34:304–308
Wesenberg D, Kyriakides I, Agathos SN (2003) White-rot fungi and their enzymes for the treatment of industrial dye effluents. Int J Chemtechnol Appl 2(2):126–136
Wong DWS (2009) Structure and action mechanism of ligninolytic enzymes. Appl Biochem Biotechnol 157:174–209
Wong Y, Yu J (1999) Laccase-catalyzed decolorization of synthetic dyes. Water Res 33:3512–3520
Xin B, Xia Y, Zhang Y, Aslam H, Liu C, Chen S (2012) A feasible method for growing fungal pellets in column reactor inoculated with mycelium fragments and their application for bleach plant effluent (BPE) by marine fungi. Biotechnol Lett 18:105–106
Xiong XJ, Meng XJ, Zheng TL (2010) Biosorption of C.I. Direct Blue 199 from aqueous solution by nonviable *Aspergillus niger*. J Hazard Mater 175(1-3):241–246
Xu M, Guo J, Cen Y, Zhong X, Cao W, Sun G (2005) Shewanella decolorization is sp. Nov., a dye decolorizing bacterium isolated from activated sludge of a waste-water treatment plant. Int J Syst Evol Microbiol 55:363–368
Yang J, Yang X, Lin Y, Ng TB, Lin J, Ye X (2015) Laccase-catalyzed decolorization of malachite green: performance optimization and degradation mechanism. PLoS One 28:1–14
Yasmeen Q, Asgher M, Munir AS, Nawaz H (2013) Optimization of ligninolytic enzymes production through response surface methodology. Bioresources 8(1):944–968
Yeddou-Mezenner N (2010) Kinetics and mechanism of dye biosorption onto an untreated antibiotic waste. Desalination 262:251–259
Yemendzhiev H, Alexieva Z, Krastanov A (2014) Decolorization of synthetic dye reactive Blue 4 by mycelial culture of White-Rot Fungi 1. Biotechnol Biotechnol Equip 23(3):1337–1339
You SJ, Teng JY (2009) Anaerobic decolorization bacteria for the treatment of azo dye in a sequential anaerobic and aerobic membrane bioreactor. J Taiwan Inst Chem Eng 40:500–505
Zavarzina AG, Lisov AV, Leontievsky AA (2018) JGR biogeoscience: the role of ligninolytic enzymes Laccase and a versatile peroxidase of the white-rot fungus *Lentinus tigrinus* in biotransformation of soil humic matter: comparative in vivo study. Biogeochem N Org Matter 123:2727–2742

Zeinab H, Kheiralla Said M, Badr El-Din Saad M, Malek A, Douaa H, Aziz A (2013) Optimization of cultural conditions for lignin peroxidase production by *Phanerochaete chrysosporium* and *Pleurotus ostreatus*. Acad J Biotechnol 1:087–095

Zeroual Y, Kim BS, Kim CS, Blaghen M, Lee KM (2006) Biosorption of bromophenol blue from aqueous solutions by *Rhizopus stolonifer* biomass. Water Air Soil Pollut 177(1-4):135–146

Zhao X, Hardin IR, Hwang HM (2005) Biodegradation of a model azo disperse dye by the white rot fungus *Pleurotus ostreatus*. Int Biodeter Biodegr 57:1–6

Zheng F, Cui BK, Wu XJ, Meng G, Liu HX, Si J (2016) Immobilization of laccase onto chitosan beads to enhance its capacity to degrade synthetic dyes. Int J Biodeterior Biodegrad 110:69–78

Zhuo R, Ma L, Fan F, Gong Y, Wan X, Jiang M, Zhang X, Yang Y (2011) Decolorization of different dyes by a newly isolated white-rot fungi strain *Ganoderma* sp.En3 and cloning and functional analysis of its laccase gene. J Hazard Mater 192(2):855–873

Microbial Enzymes in Biocontrol of Phytopathogens

10

Priya Mishra, Jitendra Mishra, S. K. Dwivedi, and Naveen Kumar Arora

Abstract

Microbial enzymes not only work as a biocatalysts in essential metabolic reactions but also help in the survival of microbes in a particular niche. Rhizosphere microbes are well known for their ability to promote plant growth and control phytopathogens. Rhizosphere microbes may enhance plant tolerance towards phytopathogens by several means. Amongst all known mechanisms of biocontrol, secretion of the lytic enzymes is recognized as an efficient way to deter phytopathogens residing in the vicinity of the rhizosphere. Rhizosphere microbes produce chitinases, cellulases, proteases, and β-glucanases in response to phytopathogen attack. These biocontrol enzymes utilize different mechanisms involved in the elimination of phytopathogens and indirectly support plant's growth and survival. The biocontrol ability of these enzymes makes them a good choice as biocontrol tools. However, large-scale industrial production of these enzymes could also be useful in making good quality of biocontrol products in the form of biopesticides. Although microbial enzymes have a great potential in the field of biocontrol, some constraints such as production cost, formulation design, quality, shelf life, and stability in field conditions are the major issues that need to be researched thoroughly.

P. Mishra · S. K. Dwivedi · N. K. Arora (✉)
Department of Environmental Science, Babasaheb Bhimrao Ambedkar (Central) University, Lucknow, India

J. Mishra
DST-CPR, Babasaheb Bhimrao Ambedkar (Central) University, Lucknow, India

N. K. Arora et al. (eds.), *Microbial Enzymes: Roles and Applications in Industries*, Microorganisms for Sustainability 11,
https://doi.org/10.1007/978-981-15-1710-5_10

10.1 Introduction

Phytopathogens including bacteria, fungi, viruses, and nematodes have always hampered crop productivity. There have been several incidences where a variety of crops were devastated by phytopathogens. During 1840s, the Irish potato famine (late blight of potato) in Europe was recorded as an epidemic (Bourke 1964). Later, in 1845, M. J. Berkeley had proven that the causative agent of this deadly disease was a water mold *Phytophthora infestans* (Berkeley 1846). In 1895, F. C. Stewart identified bacterial wilt of sweet corn also known as Stewart's wilt disease caused by *Phytomonas stewartii*. The disease was declared to be an epidemic in the year 1932 and severely destructed the entire field of Golden Bantam sweet corn in many parts of the United States (Ivanoff and Keitt 1937). In another incidence, *Helminthosporium maydis* causing southern corn leaf blight disease affected about 80% of US corn crop with an estimated loss of one billion dollar and was declared as an epidemic in the year 1970–1971 in the province of Florida, Georgia, Alabama, and Mississippi (Tatum 1971). According to Miller et al. (2017), crop losses from evolving phytopathogens and pests have emerged as a threat to global food security. However, Strange and Scott (2005) earlier reported that about 10–20% loss of global agricultural production is caused by phytopathogens, which deprived 800 million people from getting adequate food. The United Nations Food and Agriculture Organization (FAO) estimated that pests and diseases are responsible for about 25% of crop loss (Martinelli et al. 2015).

Preventing crops from phytopathogens has always been challenging. For more than a century, different approaches to control phytopathogens were evaluated. However, amidst the different methods used, strategies based on chemical or synthetic products have been the mainstay. The primary reason for the high usage of synthetic chemicals is their immediate effect against several types of pests. The unprecedented use of chemical pesticides resulted in a doubling of its consumption in some of the countries (Zhang et al. 2011). However, it was the efforts of some environmentalists and ecologists who envisaged the dark side of their use in the environment (Gupta and Ali 2008). Exposure of pesticides on human health and its other consequences can be realized by a study of World Health Organization (WHO) which stated that every year three million agricultural workers in the developing world experienced acute poisoning from pesticides and about 18,000 died (Miller 2004). Apart from this, the other listed negative effects of pesticides on the environment are groundwater pollution, loss in soil fertility and biodiversity (Mahmood et al. 2016).

The use of biological control agents (BCA) in the eradication of phytopathogens has proven to be more useful than chemical control. In 1919, the first evidence of "classical biocontrol" was reported, which involved control of pests by the use of predators or parasitoids (Smith 1919). Later research on plant-microbe interaction turned the classical biological control into a more effective robust approach also utilizing microorganisms (Tsujibo et al. 2003). From that beginning to the current time, enormous research has proven the role of rhizosphere microbes in the suppression of plant diseases. Rhizosphere microbes use a vast array of mechanisms to eliminate phytopathogen from plants. They produce antibiotics, enzymes, and volatile organic compounds (VOCs) such as hydrogen cyanide (HCN) allelochemicals for effective management of phytopathogens. Enzymes such as lipase, protease, laccase/ligninase, cellulose, glucanase, and chitinase have also been evaluated for their biocontrol activity against several pathogenic bacteria and fungi (Chernin and Chet 2002). Microbial enzymes have a great biocontrol potential. These enzymes may confer plant protection from a wide range of phytopathogens. Preparation of biocontrol products by using biocontrol enzymes producing microbial strain(s) or adding extracellular crude enzyme could enhance the use and efficacy of biopesticide product. This approach also has a great biotechnological potential in preventing crop loss due to phytopathogens. However, extended application and performance of microbial enzyme-based biocontrol products at the field level are still not much studied and greatly depend on application strategies, formulation method, and types of strains used. In this chapter the role of some potential microbial enzymes in the biocontrol of phytopathogens is discussed.

10.2 Microbial Enzymes

Microbial enzymes offer great biocatalytic potential in different commercial sectors. From ancient time, microbial enzymes are being used in the production of various food products including bread, wine, vinegar, pickle, and curd. Microbial enzymes gained interest for their extensive use in industries because of their stability, ease of production, and high biocatalytic activity. Development in fermentation methods also assured abundant production of microbial enzymes to the industries. Nowadays, microbial enzymes have several applications in different industries including pharmaceutical, baking, dairy, beverage, feed, biopolymer, paper and pulp, leather, textile, cosmetics, detergents, organic synthesis, and waste management. According to International Union of Biochemistry (IUB) enzymes are categorized into six types, (1) oxidoreductase, (2) transferase, (3) hydrolase, (4) lyase, (5) isomerase, and (6) ligase, and microbes can produce enzymes of all six categories. In 2018, the global market of enzyme reached $5.5 billion and is estimated further to reach $7 billion by 2023 (BCC Research 2018). Some examples of industrially produced microbial enzymes are amylase, arylsulfatase, β-glucosidase, cellulase, chitinase, dehydrogenase, phosphatase, protease, lipase, laccase, pectinase, xylanase, phytase, ureases, etc. Amongst them, many play an essential role in ecosystem functioning where they serve the purpose of organic matter decomposition, biotransformation of complex organic molecules, and control of soilborne phytopathogens.

10.3 Major Microbial Enzymes in Biocontrol

There are certain enzymes which can degrade or lyse cell wall of phytopathogens. This phenomenon is widespread in the rhizosphere region where plant growth-promoting microbes (PGPM) repel or destroy phytopathogens by secretion of lytic enzymes and indirectly help in plant growth and development. The detailed study on microbial hydrolases and other lytic enzymes confirms their biocontrol activity against several phytopathogens (Table 10.1). Investigation of some fungal and bacterial enzymes revealed that they can inhibit or modify cell-wall synthesis, perforate cell membrane, or degrade cell wall of host or plant pathogens (Roberti et al. 2002; Mota et al. 2017) and are collectively known as biocontrol enzymes. In this section, the role and mechanisms of microbial biocontrol enzymes are being described.

10.3.1 Chitinase

Chitinase (EC 3.2.1.14) hydrolyzes β-l,4-glycosidic linkages of chitin which is a poly-β-1,4-N-acetylglucosamine (GlcNAc). GlcNAc is the second most abundant organic compound next to cellulose and is the major constituent of arthropod exoskeletons, tendons, and linings of their respiratory, excretory, and digestive systems and cell walls of a variety of fungi (Clark and Smith 1936). Chitinases have been divided into two types on the basis of their functioning: endochitinases and exochitinases. Endochitinases (EC 3.2.1.14) randomly cleave internal points over the entire length and produce dimer diacetyl-chitobiose and N-acetyl glucosamine multimer such as chitotriose and chitotetraose (Fig. 10.1). Exochitinases are of two types: (1) chitobiosidases (EC 3.2.1.29) which cleave non-reducing ends of chitin and produce diacetylchitobiose in a stepwise fashion and (2) β-l,4-glucosaminidases (EC 3.2.1.30) which cleave oligomers obtained by endochitinases into monomers of N-acetyl glucosamine (Sahai and Manocha 1993). The chitinases are extensively distributed in living organisms and perform various biological processes; for example, they provide self-defense against chitin-containing pathogens of higher plants as well as animals, and help in degradation of the old cuticle, chitin assimilation, and preparation of protoplast from fungi (Flach et al. 1992; Bhattacharya et al. 2007). Chitinase activity has been explored in many organisms including bacteria, fungi, yeast, insects, and vertebrates (Jeuniaux 1961; Elango et al. 1982; Suslow and Jones 1990; Kramer and Muthukrishnan 1997; Hartl et al. 2012).

Chitinases have been documented as notable enzymes showing extraordinary role in the biocontrol of phytopathogens. Chitinases from bacteria and fungi are reported with fungicidal and insecticidal activities. A number of bacterial genera including *Streptomyces, Pseudomonas, Bacillus, Escherichia, Alteromonas,* and *Aeromonas* have been identified with chitinolytic activity (Sitrit et al. 1995; Watanabe et al. 1999; Tsujibo et al. 2003). Actinomycetes are broadly studied group of microorganisms known to produce various secondary metabolites and enzymes which are commercially used in medical and agricultural fields. Several species of *Streptomyces* such as *S. lydicus, S. aureofaciens, S. griseus,* and *S. halstedii* have been reported with chitinolytic activity (Mahadevan and Crawford 1997; Tanabe et al. 2000; Tsujibo et al. 2003; Joo 2005). Beyer and Diekmann (1985) showed the chitinolytic activity of *Streptomyces* sp. ATCC 11238 in the degradation of the cell wall of *Penicillium chrysogenum.* Hoster et al. (2005) found that several strains of *Streptomyces* can not only promote plant growth but also show biocontrol activity

Table 10.1 Some microbial enzymes reported with biocontrol activity against phytopathogens

S. no.	Name of enzyme	Producing microbes	Phytopathogen	References
1.	Chitinase	*Enterobacter agglomerans*	*Rhizoctonia solani*	Sitrit et al. (1995)
2.	Chitinase	*Bacillus or Streptomyces*	*Fusarium culmorum, Sclerotia sclerotiorum, Guignardia bidwellii*, and *Botrytis cinerea*	Hoster et al. (2005)
3.	Chitinase	*Aeromonas caviae*	*Sclerotium rolfsii, R. solani* and *Fusarium oxysporum f.sp. Vasinfectum*	Inbar and Chet (1991)
4.	Chitinase	*Bacillus thuringiensis* var. *israelensis*	*S. rolfsii*	Reyes-ramírez et al. (2004)
5.	Chitinase	*Bacillus sp. BPR7*	*Macrophomina phaseolina, F. oxysporum, F. solani, Sclerotinia sclerotiorum, R. solani* and *Colletotrichum* sp.	Kumar et al. (2012)
6.	Chitinase	*Pseudomonas* spp.	*F. oxysporum f.* sp. *redolens* and *Gaeumannomyces graminis* var. *tritici*	Sundheim et al. (1988)
7.	Chitinase	*Pseudomonas*	*Macrophomina* sp., *Aspergillus* sp. and *Phytophthora* sp.	Saraf et al. (2008)
8.	Chitinase	*Pseudomonas putida* and *Bacillus subtilis*	*M. phaseolina*	Sharma et al. (2018)
9.	Chitinase	*Serratia plymuthica R51*	*F. oxysporum, S. rolfsii, P. infestans, R. solani, Pythium myriotylum, Colletotrichum acutatum and Corynespora cassiicola*	John and Radhakrishnan (2018)
10.	Chitinase	*Trichoderma asperellum*	*Rigidoporus microporus*	Sakpetch et al. (2018)
11.	Chitinase	*Geobacillus thermodenitrificans* and *Bacillus aerius*	*Phytophthora capsici*	San Fulgencio et al. (2018)
12.	Chitinase	*Bacillus* sp. B25	*Fusarium verticillioides*	Douriet-Gámez et al. (2018)
13.	Chitinase	*Trichoderma harzianum Rifai T24*	*S. rolfsii*	El-Katatny et al. (2001)
14.	Chitinase	*Pseudomonas fluorescens*	*R. solani*	Nagarajkumar et al. (2004)
15.	Chitinase	*Bacillus* sp.	*F. oxysporum* f. sp. *Lycopersici*	Jangir et al. (2018)
16.	Chitinase	*Paenibacillus ehimensis*	*F. culmorum* and *Drechslera sorokiniana*	Aktuganov et al. (2008)

(continued)

Table 10.1 (continued)

S. no.	Name of enzyme	Producing microbes	Phytopathogen	References
17.	Chitinase	*Serratia marcescens, Micromonospora carbonacea* and *Streptomyces viridodiasticus*	*Sclerotinia minor*	El-Tarabily et al. (2000)
18.	Chitinase	*Bacillus* sp.	*R. solani, M. phaseolina*	El-Bendary et al. (2016)
19.	Chitinase	Fluorescent *Pseudomonas*	*P. capsici* and *R. solani*	Arora et al. (2007)
20.	Cellulase	*Trichoderma longibrachiatum*	*Pythium ultimum*	Migheli et al. (1998)
21.	Cellulase	*Trichoderma* sp.	*S. rolfsii* and *Fusarium cicero*	Anand and Reddy (2009)
22.	Cellulase	*Trichoderma harzianum*	*P. ultimum*	Thrane et al. (1997)
23.	Glucanase	*Bacillus* sp.	*R. solani, M. phaseolina*	El-Bendary et al. (2016)
24.	β-1,3-Glucanase	Fluorescent *Pseudomonas*	*P. capsici* and *R. solani*	Arora et al. (2007)
25.	β-1,3-Glucanase	*Pseudomonas stutzeri*	*Fusarium solani*	Lim and Kim et al. (1995)
26.	β-1,3-Glucanase	*Pseudomonas cepacia*	*S. rolfsii, R. solani,* and *P. ultimum*	Fridlender et al. (1993)
27.	β-1,3-Glucanase	*P. fluorescens*	*R. solani*	Nagarajkumar et al. (2004)
28.	β-1,3-Glucanase	*P. ehimensis*	*F. culmorum* and *D. sorokiniana*	Aktuganov et al. (2008)
29.	α-1,3-Glucanases	*T. harzianum*	*Aspergillus niger, Botrytis cinerea, C. acutatum, F. oxysporum, Penicillium aurantiogriseum,* and *R. solani*	Ait-Lahsen et al. (2001)
30.	β-1,3-Glucanase	*T. harzianum*	*P. ultimum*	Thrane et al. (1997)
31.	β-1,3-Glucanase	*Paenibacillus terrae*	*Magnaporthe oryzae, Exserohilum turcicum, Xanthomonas campestris pv. glycines,* and *R. solani*	Yu et al. (2019)
32.	β-1,3-1,4-Glucanase	*Paenibacillus polymyxa*	*B. cinerea*	Li et al. (2015)
33.	β-1,3-Glucanase	*Bacillus* sp.	*F. oxysporum f. sp. lycopersici*	Jangir et al. (2018)
34.	β-1,3-Glucanase	*S. marcescens, M. carbonacea,* and *S. viridodiasticus*	*S. minor*	El-Tarabily et al. (2000)

(continued)

Table 10.1 (continued)

S. no.	Name of enzyme	Producing microbes	Phytopathogen	References
35.	β-1,3-Glucanase	*Streptomyces* sp.	*M. oryzae*	Shao et al. (2018)
36.	Protease	*Bacillus* sp.	*F. oxysporum* f. sp. *lycopersici*	Jangir et al. (2018)
37.	Serine protease	*Stenotrophomonas maltophilia*	*P. ultimum*	Dunne et al. (2000)
38.	Aspartic protease	*T. harzianum*	*B. cinerea, Mucor circinelloides, Aspergillus fumigatus, Aspergillus flavus, R. solani,* and *Candida albicans*	Deng et al. (2018)
40.	Protease	*T. harzianum*	*F. oxysporum, Colletotrichum capsici, Gloeocercospora sorghi,* and *Colletotrichum truncatum*	Sharma et al. (2016)
41.	Protease	*Bacillus* sp.	*R. solani, M. phaseolina*	El-Bendary et al. (2016)

via production of various enzymes, i.e., cellulases, xylanases, chitinases, lipases, and catalases (Hoster et al. 2005). In a study, Yandigeri et al. (2015) reported chitinolytic activity of *Streptomyces vinaceusdrappus* S5 MW2 isolated from water sample of Chilka Lake in India. The strain showed antifungal activity against the sclerotia-producing *Rhizoctonia solani. S. lydicus* has been reported to produce extracellular chitinase with an essential role in in vivo antifungal biocontrol activity (Mahadevan and Crawford 1997).

Bacterial chitinases with biocontrol activity are also well defined. Several species of *Pseudomonas* have been reported to produce extracellular chitinases. Folders et al. (2001) reported synthesis of an extracellular chitinase by a novel pathway through *chiC* gene in *Pseudomonas aeruginosa*. Mishra and Arora (2012) reported the role of extracellular chitinase isolated from *P. aeruginosa* in the destruction of fungus *Xanthomonas campestris* causing black-rot disease. *Pseudomonas* strains isolated from the rhizosphere of chickpea and green gram were found to produce hydrolytic chitinases and cellulases with antagonistic activity against *R. solani* and *Pythium aphanidermatum* (Sindhu and Dadarwal 2001). Several soil bacilli are also known to act as fungal antagonists and degrade and lyse their cell wall. *Bacillus licheniformis* and *Bacillus thuringiensis* can secrete extracellular chitinase, which works against several phytopathogenic fungi (Gomaa 2012). Tasharrofi et al. (2011) reported chitinase production in *Bacillus pumilus*. However in a study Saber et al. (2015) found that *Bacillus subtilis*-produced chitinase showed biocontrol against *R. solani* causing disease in potato. In the dual-culture in vitro antagonistic study, chitinase

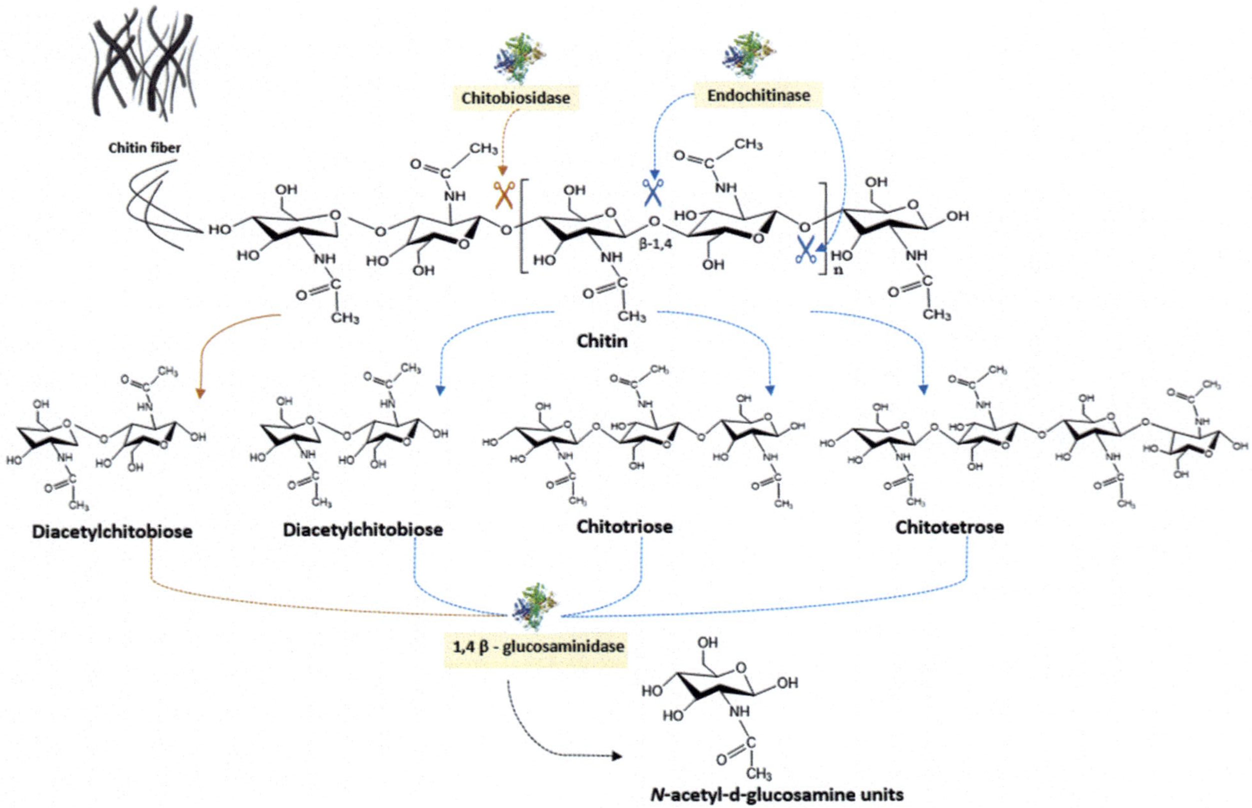

Fig. 10.1 Mechanism of exo- and endochitinases

from *Paenibacillus* sp. inhibited *Fusarium oxysporum*, *Alternaria burnsii*, and *R. solani* fungi by 52.5, 75.0, and 85.71%, respectively (El-Sayed et al. 2019). In another study, *Corallococcus* sp. was found to produce chitin hydrolase CcCti1 which hydrolyzed chitin into *N*-acetylated chitohexaose and inhibited the growth of phytopathogen *Magnaporthe oryzae* in a dose-dependent manner (Li et al. 2019).

Genomic studies have revealed the localization of the gene for chitinase enzyme in many microorganisms (Daimon et al. 2005). For instance, through transposon mutagenesis, which is a potent way to identify genetic elements regulating specific phenotypes, genes for chitinolytic activity were spotted in *A. caviae* CB101 bacterium. It was noticed that bacterium contains (1) *nagA* and *nagB*, (2) *ftsX* and *exeL*, and (3) *varA* and *rpoH* gene sequences and some other unknown genes responsible for chitinolytic activity (Li et al. 2007). Role of *chiA* gene in chitinolytic activity was also identified in a bacterial strain *Stenotrophomonas maltophilia* 34S1 and it was found that mutation in *chiA* gene reduced the ability of *S. maltophilia* to control summer patch disease on Kentucky bluegrass (Kobayashi et al. 2002).

Entomopathogenic fungi have gained considerable attention in the biological control of plant insects (Fig. 10.2). Some of the well-known entomopathogenic fungi which are being used from a long time to control insects and pests are *Metarhizium anisopliae*, *Beauveria* spp., and *Verticillium lecanii* (Strasser et al. 2000; Tsujibo et al. 2003). Several studies confirmed that chitinase and chitobiase are predominately produced by the entomopathogenic fungus (Leger et al. 1991; Valadares-Inglis and Peberdy 1997). Amongst other fungi, *Trichoderma harzianum* is the most commonly used biocontrol agent known for its ability to inhibit different bacterial and fungal plant pathogens (Leger et al. 1991; Kang et al. 1999). *T. harzianum* utilizes endo- and extracellular chitinases as the principal mechanism of biocontrol. The endochitinases and chitobiosidases from *T. harzianum* have been reported to show antifungal activity against fungal pathogens including *Botrytis cinerea, Fusarium solani, Ustilago avenae, Uncinula necator, Saccharomyces cerevisiae, Fusarium graminearum,* and *Pythium ultimum* (Lorito et al. 1993). Apart from *T. harzianum*, *Aspergillus niger* is reported with chitinolytic activity against plant pathogenic strains of *Fusarium solani, Fusarium culmorum, and R. solani* (Brzezinska and Jankiewicz 2012).

There are few reports on the presence of genes for chitinases in baculoviruses which are commonly used as viral biopesticides; however, in this case, the exact significance of chitinase in its role in insect infection is not fully elucidated (Gopalakrishnan et al. 1995; Sitrit et al. 1995; Szewczyk et al. 2006; Karabörklü et al. 2018).

10.3.2 Cellulases

Cellulases belong to the glycoside hydrolase (GH) family and cleave β-1,4-D glucan linkages of cellulose which is a major polysaccharide component of the cell wall of plants and also present in certain bacteria, fungi, and few protozoans including *Naegleria gruberi, Dictyostelium discoideum,* and *Acanthamoeba castellanii* (Medie et al. 2012). Hydrolysis of cellulose produces cello-oligosaccharide, cellobiose, and

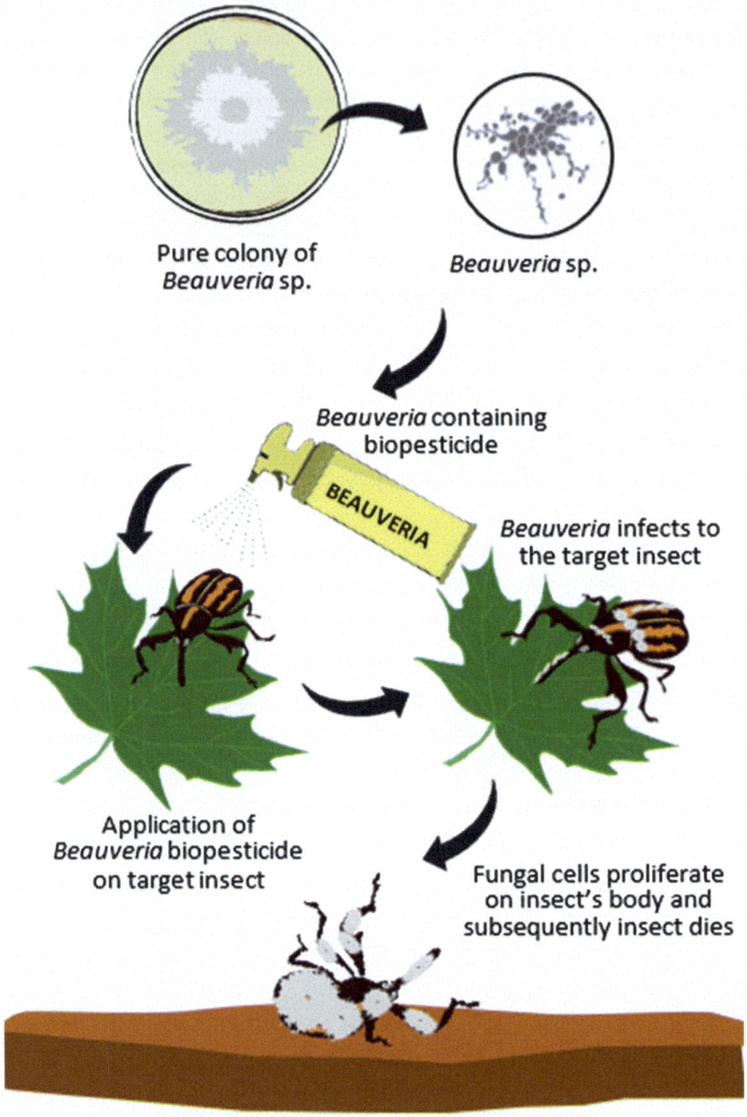

Fig. 10.2 Use of entomopathogenic fungi in the biocontrol of insects

glucose (Fig. 10.3). Cellulase enzyme has three major types; exoglucanase, also known as cellobiohydrolases (EC 3.2.1.91), endo-β-1,4 glucanases (EC 3.2.14), and β-glucosidase (EC 3.2.1.21) also known as cellobiase. All three cellulases take part in the synergistic conversion of cellulose to glucose (Schülein 1988; Sukumaran et al. 2005; Zhang and Zhang 2013). An oxidative kind of cellulase has also been reported, which possesses the ability to depolymerize cellulose with the help of free radical reactions (Medie et al. 2012). According to carbohydrate-active enzyme

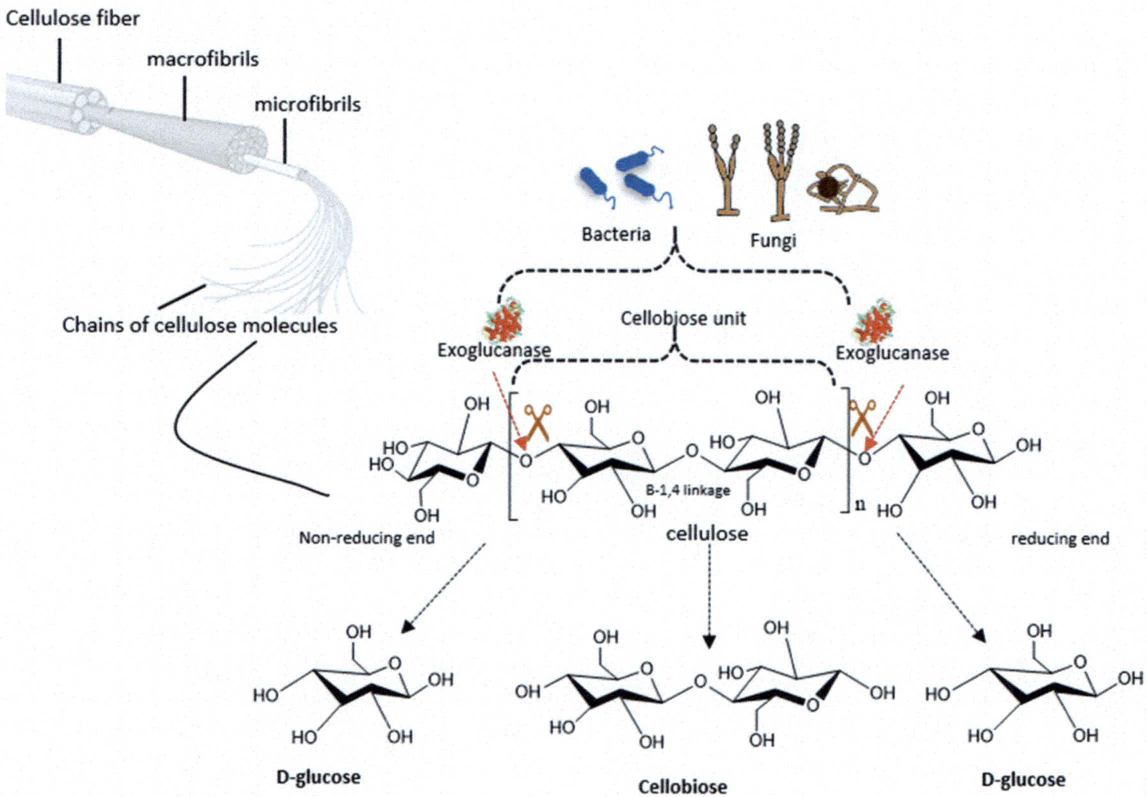

Fig. 10.3 Mechanism of cellulases

database (CAZy), endoglucanases are found in the GH families 5–8, 12, 16, 44, 45, 48, 51, 64, 71,74, 81, 87, 124, and 128. Exoglucanases or cellobiohydrolases are found in GH families 5–7 and 48 and β-glucosidases in GH families 1, 3, 4, 17, 30, and 116 (Juturu and Wu 2014). Biocontrol activity of cellulases is widely reported in some bacterial and fungal genera. *B. subtilis* has been reported with the production of cellulase and glucanase along with chitinase for biocontrol of fungal pathogen *Colletotrichum gloeosporioides* causing anthracnose disease of chili (Ashwini and Srividya 2014). Endophytic bacteria are also identified to control plant pathogens by releasing different types of cellulases. Endophytic *Bacillus pumilus JK*-SX001 is reported to produce extracellular cellulase and protease, which inhibited three fungal pathogens *Cytospora chrysosperma*, *Phomopsis macrospora*, and *Fusicoccum aesculi*, all causing poplar canker disease (Ren et al. 2013). Cho et al. (2007) found that endophytic *P. polymyxa* GS01, *Bacillus* sp. GS07, and *Pseudomonas poae* JA01 showed potential biocontrol activity against phytopathogenic fungi including *R. solani, F. oxysporum, P. ultimum*, and *Phytophthora capsici*. Some species of *Micromonospora* have also been recognized to play an essential role in biocontrol. For example, cellulase from *Micromonospora carbonacea* bacterium showed biocontrol of *Phytophthora cinnamomi*, the causal organism of root rot in Banksia grandis (El-Tarabily et al. 1996).

Fungal cellulases are more preferred than bacterial cellulases due to stronger penetration ability into the cellulosic material and are widely used in biotechnological applications (Wei et al. 2009). Similar to chitinase, different types of cellulases are also reported from *Trichoderma* spp. showing biocontrol activity. The cellulase and chitinase from *Trichoderma* sp. have been reported in in vivo biocontrol of *Sclerotium rolfsii* and *Fusarium cicero* (Anand and Reddy 2009). Transformation study in *T. longibrachiatum* (CECT2606) showed overexpression of β-1,4-endoglucanase gene responsible for biocontrol of *P. ultimum* in cucumber (Migheli et al. 1998). The cellulase genes *Thph1* and *Thph2* of *T. harzianum* are reported in in vitro antagonistic activity against *F. graminearum*, causing *Fusarium* stalk rot in maize (Quesada-Ocampo et al. 2016). These genes also showed a relation in the activation of 2,4-dihydroxy-7-methoxy-2H-1,4-benzoxazin-3(4H)-one (DIMBOA) which is known for its potential to inhibit pathogenicity and mycotoxin-related proteins in pathogen *F. graminearum* (Quesada-Ocampo et al. 2016). The research also confirmed the role of *Thph1* and *Thph2* in triggering the formation of reactive oxygen species (ROS), increasing the level of cytosolic calcium in leaves and modulation of induced systemic resistance (ISR) against foliar disease in maize (Saravanakumar et al. 2016).

Fungus *Pythium oligandrum,* is known for its extraordinary trait of mycoparasitism against more than 50 fungal and oomycete species (Gabrielová et al. 2018). The fungus utilizes its chitinases, cellulases, proteases, and glucanases as major components of mycoparasitism (Benhamou et al. 2012). The fungus has been successfully used in the biocontrol of soilborne phytopathogen *Phytophthora parasitica* via secretion of the extracellular cellulolytic enzyme (Picard et al. 2000). In the last few years, the ability of some strains of yeast in the biocontrol of phytopathogens by producing cellulases has been reported. For instance, a yeast *Wickerhamomyces anomalus* showed in vitro and in vivo biocontrol activity against phytopathogens *B. cinerea* and *Penicillium digitatum* and the genes *WaEXG1* and *WaEXG2* were found to be associated with the production of exoglucanases in the inhibition of phytopathogenic fungi (Parafati et al. 2017). In a study, Chen et al. (2018a) reported biocontrol potential of 24 strains of yeast against gray mold *B. cinerea* (teleomorph: *Botryotinia fuckeliana*) through the production of cellulase, chitinase, and protease. Further, they found that amongst all *Galactomyces candidum* JYC1146, *Aureobasidium pullulans* JYC1278, and *A. pullulans* JYC1291 showed maximum inhibition of *B. cinerea* (69.6%, 33.7%, and 23.6%, respectively).

Apart from bacteria and fungi, cellulases from actinomycetes have also been reported in the biocontrol of plant pathogens. For instance, cellulase from *Streptomyces rubrolavendulae S4* were found to show antagonistic activity against fungal pathogen *P. aphanidermatum* causing damping off disease in plants (Loliam et al. 2013). For more details on biocontrol enzyme of actinomycetes, one can see literature by Behie et al. (2017) and Olanrewaju and Babalola (2019).

10.3.3 Proteases

Proteases, also termed as peptidases or proteinases, cleave the peptide bond present between the amino acid residues in a polypeptide chain. Based on the reaction, chemical nature of the catalytic site, and evolutionary relationships, proteases can be classified into two major types: (1) endopeptidases, which cleave internally the amino acids and (2) exopeptidases, which remove amino acids either from amino-terminal or carboxy-terminal end of the protein. Exopeptidases are further subdivided into (1) aminopeptidases that remove amino acid from amino-terminal or (2) carboxypeptidases that remove at carboxy-terminal, respectively. Some other types of proteases are also reported such as serine proteases (EC 3.4.21), cysteine proteases (EC 3.4.22), aspartic proteases (EC3.4.23), metalloproteases (EC 3.4.24), and threonine proteases (EC 3.4.25). These proteases are categorized based on the architecture of their catalytic site (Garcia-Carreon 1997; Clark and Pazdernik 2016).

Apart from different applications in industries, proteases have also been recognized for their wide applicability in the field of agriculture. Earlier work on biocontrol demonstrated the ability of lytic enzymes including extracellular protease in antagonizing several plant pathogens (Sacherer et al. 1994; Dunne et al. 1997; Whipps 2001). There are several reports which show the role of proteases obtained from soil bacilli in biocontrol of phytopathogenic fungi and bacteria. An antagonistic strain of *Bacillus amyloliquefaciens* isolated from the rhizosphere of jute showed protease-mediated biocontrol activity against *Macrophomina phaseolina, F. oxysporum, Fusarium semitectum*, and *Alternaria alternata* (Majumdar and Chakraborty 2017). In a study, Essghaier et al. (2009) reported production of salt-tolerant protease from *B. pumilus* M3-16 strain responsible for biocontrol activity against phytopathogenic fungus *B. cinerea.* Wei et al. (2010) reported the role of protease from *Bacillus* sp. AR156 and GJ24 strains in the inhibition of a nematode *Meloidogyne incognita* and they suggested that in vitro protease activity could be used as an important parameter in the selection of BCAs against root-knot nematodes. The use of microbial proteases with nematicidal activity in biocontrol of nematodes in the rhizosphere was also realized by other workers (Lian et al. 2007; Illakkiam et al. 2013). For example, Siddiqui and Shaukat (2005) also reported the potential of extracellular protease producing biocontrol agent (BCA) *Pseudomonas fluorescens* CHA0 in the management of root-knot disease causing nematode *M. incognita* in tomato and soybean. In another study, a nematode-associated bacterium *P. fluorescens* (pf36) isolated from the rhizosphere of the banana plant showed expression of three nematicidal protease genes, i.e., *pase1*, *pase4*, and *pase6*, involved in the protection of plant from phytopathogenic nematode *Radopholus similis* (Chen et al. 2018b).

Amongst fungi, vigorous research has been done on biocontrol traits of *T. harzianum* (Schuster and Schmoll 2010). It has been found that proteases and chitinases are the two major enzymes of *Trichoderma* attributed for the biocontrol of plant pathogens (Benítez et al. 2004). Different strains of *Trichoderma* are reported which use their extracellular proteases against *Fusarium* sp., *Colletotrichum* sp., *Gloeocercospora* sp. and *Botrytis* sp. (Elad and Kapat 1999; Schirmböck et al. 1994; Jayalakshmi et al. 2009; Sharma et al. 2016). The alkaline and serine proteases from *T. harzianum* and *Trichoderma virens* have been characterized and were found to be effective against phytopathogen *R. solani* (Benítez et al. 1998; Pozo et al. 2004). Proteases with nematicidal activity are also reported in some strains of *Trichoderma*, *Monacrosporium microscaphoides*, and *Arthrobotrys oligospora* (Tunlid et al. 1992; Suárez et al. 2005; Wang et al. 2006). Proteases of entomopathogenic fungi are also explored for insect control (Leger et al. 1992). Extracellular proteases of entomopathogenic fungi can easily hydrolyze proteinaceous insect cuticle; hence, they are widely used as potential bioagents for preventing crop loss due to insect attack. In last few years recombinant proteases with enhanced antifungal activity have also been investigated against *Penicillium expansum*, *B. cinerea*, *Monilinia fructicola*, and *A. alternata* (Banani et al. 2014; Fan et al. 2014).

10.3.4 β-1,3-Glucanase

β-1,3-Glucanases are glycoside hydrolases which cleave long chains of β-1,3-glucan and are widely dispersed in plants, fungi, and bacteria. There are two types of β-1,3-glucanases: (1) exo β-1,3-glucanases (EC 3.2.1.58) which act randomly, inside a glucan chain, and (2) endo β-1,3-glucanases (EC 3.2.1.39) which release glucose residues from the non-reducing end (Gueguen et al. 1997). β-1,3-Glucanases are more extensively known for their role in modifying glucan and maintaining the rigidity of the cell-wall structure of fungi (Mouyna et al. 2013). However, the role of β-1,3-glucanase in biological control of soilborne plant pathogens is also very much explored, and nowadays β-1,3-glucanase or other glucanase-producing microbes are being opted as efficient BCA. According to Aimanianda et al. (2017), β-1,3-glucanases' ability to modify fungal cell-wall β-(1,3)-glucan polymer can be successfully employed in the development of BCA.

β-1,3-Glucanases from *Pseudomonas cepacia* have been reported for the biocontrol of the plant pathogenic *S. rolfsii, R. solani*, and *P. ultimum* (Fridlender et al. 1993). Recently, Yu et al. (2019) reported that β-1,3-glucanases from *Paenibacillus terrae* have the potential to biocontrol *M. oryzae, Exserohilum turcicum, X. campestris* pv. glycines, and *R. solani* fungi causing rice blast, corn spot disease, soybean bacterial spot disease, and rice sheath blight disease, respectively. There have also been reports where β-1,3-glucanases and secondary metabolites such as lipopeptide antibiotic, viscosinamide, and,2,4-diacetylphloroglucinol acted synergistically to inhibit phytopathogens (Nielsen et al. 1998). For instance, in a study β-1,3-glucanase along with an antibiotic extract from *B. subtilis* conferred protection against rice blast and sheath blight diseases caused by *Pyricularia grisea* and *R. solani*, respectively (Leelasuphakul et al. 2006). Arora et al. (2007) have revealed that β-1,3-glucanases and chitinase from fluorescent pseudomonads inhibit *R. solani* and *P. capsici*. In a study, *Bglu1* and *Bglu2* genes from *Bacillus velezensis* for β-1,3 and 1,4-glucanase production gene were cloned and expressed in *Escherichia coli* and subsequently purified enzyme β-1,3 and 1,4-glucanases were found to exert antifungal activity against three phytopathogenic fungi, i.e., *Helicobasidium purpureum, Cryphonectria parasitica,* and *Cylindrocladium quinqueseptatum* (Xu et al. 2016). β-1,3-Glucanase of endophytic bacteria has also been evaluated in the biocontrol of fungal pathogens. In a study, Jha (2019) found that endophytic *P. aeruginosa* and *Pseudomonas pseudoalcaligenes* along with activation of isolates phenolics and flavonoids also secreted β-1,3-glucanase and catalase in paddy and helped in the development of preformed defense against fungal pathogen *Pyricularia grisea* responsible for the fungal blast in paddy. Shao et al. (2018) showed *the* potential of *Streptomyces* sp. in combating rice blast disease in seedling caused by *M. oryzae*.

Mycoparasitic species of *Trichoderma* are best recognized for the production of complex proteins that are directly involved in parasitizing host fungus (Ramada et al. 2016). However, proteomic studies provided ample evidence of the growth of *Trichoderma* on media supplemented with cell-wall extract of plant pathogenic fungi (Grinyer et al. 2005). Further, it has also been investigated that β-1,3-glucanases of *Trichoderma* can degrade glucan chain in the cell wall of several fungal pathogens such as *R. solani, B. cinerea,* and *Fusarium* sp. (Monteiro and Ulhoa 2006). There are several reports where species of *Trichoderma* were found to secrete β-1,3-glucanase for protecting the plant from fungal pathogens. β-1,3-Glucanase from *T. harzianum* has been used in the biocontrol of *S. rolfsii* (El-Katatny et al. 2001). However, *Trichoderma asperellum* along with chitinase and protease produced β-glucanases against *F. graminearum* causing stalk rot of maize. (Li et al. 2016) Similarly, Khare et al. (2018) reported that lytic enzymes β-1,3-glucanases and chitinase along with peptaibols from *T. cerinum* can be effectively used in the management of *F. oxysporum*. In another study, Baiyee et al. (2019) reported involvement of β-1,3-glucanases and chitinase of *Trichoderma spirale* in the biocontrol of *Corynespora cassiicola* or *Curvularia aeria* causative agent of leaf spot disease in lettuce. Wonglom et al. (2019) reported biocontrol of *Sclerotium* sp. by β-1,3-glucanase produced by *Trichoderma*. Apart from *Trichoderma*, *Chaetomium globosum* and *Chaetomium cupreum* have also been reported for the secretion of β-1,3-glucanase with biocontrol activity (Jiang et al. 2017a, b). A significant success related with biocontrol trait of fungal β-1,3-glucanase is the expression of its gene in crop plants which provides increased tolerance to fungal diseases. The technique was successful in the strawberry plant for providing tolerance against *Rosellinia necatrix*, a soilborne pathogen causing root and crown rot in strawberry (Mercado et al. 2015).

10.4 Mass Production and Optimization

Mass production of microbial biocontrol enzymes is a significant area of research. Enhancing the commercial production of microbial enzymes needs selective approaches where maximum yield could be obtained by low capital cost. Use of low-cost raw materials as substrates, improved bioprocess technologies, and bioengineering of microbes are some of the multifaceted approaches that can be harnessed for maximum production of microbial enzymes (Lynd et al. 2002; Sukumaran et al. 2005). Fermentation and bioprocessing technologies are an essential component of mass production of microbial enzymes (Sarkar et al. 2010). Submerged fermentation (SmF) is preferred for bacterial enzymes because these microbes require higher water potential in comparison to fungi (Chahal 1983). Solid-state fermentation (SSF) is mostly implemented for fungal enzyme production and requires less water potential (Troller and Christian 1978; Babu and Satyanarayana 1996). However, there are few reports available where bacteria such as *Bacillus* spp. can use SSF for enzyme production (Chaari et al. 2012; Kapilan and Arasaratnam 2011). The significant advantage of SSF is that it provides higher yield, which is linked with increased biomass production and low breakdown of the product (Hansen et al. 2015). SSF also provides low cost of operation with less skilled manpower and use of cheap agricultural and animal residues and biomass as substrate (Pandey et al. 2000). However, at the industrial level, SmF is more widely used at a routine basis because process parameters (such as temperature, agitation, aeration, foam, and pH) can be handled easily during the entire process (Vaidyanathan et al. 1999). According to Subramaniyam and Vimala (2012), yield can be optimized by the careful selection of substrates. SSF utilizes different solid substrates, including agricultural and animal wastes, which are nutrient rich. Agricultural waste materials such as wheat bran, rice straw, bagasse, fruit pulp, coconut coir, hay, and vegetable waste can decrease the cost of production (Pandey et al. 1999). Fermentative production requires specific concentrations of media components, for instance, nutrients, ions, and growth hormones, as they are important bioprocess tools for medium design (Çalık et al. 2001).

For industrial production of chitinase, bacteria are mostly preferred in comparison to fungi because chitinase activity is higher in the mycelium of fungi rather than the culture medium (Lopes et al. 2008). Contrary to this, in the case of cellulases, fungi are better performers than bacteria because they (bacteria) have lower penetration potential and inefficacy to use a wide range of substrates for cellulase production. Dewi et al. (2016) reported medium optimization of β-glucanase production by *B. subtilis* SAHA 32.6 strain used in biological control of oil palm pathogen and it was found that production is influenced by inoculum size, oat β-glucan, and yeast extract used as substrate. Sharma et al. (2017) described the optimization of production parameters for microbial alkaline proteases, and they inferred that pH of production medium, ionic strength, temperature, and mechanical handling are the major factors which maximize production. In another study, Singh and Bajaj (2016) evaluated the production of a thermostable and wide-range pH-stable protease from *B. subtilis* K-1 strain using cost-effective agricultural residues. They found that for enhanced production of protease, the most significant variables were incubation time and type of agricultural wastes used as substrate. Further optimization of these variables by the central composite design of response surface methodology (RSM) showed a substantial protease yield enhancement (112%). Likewise, RSM and other statistical and modeling techniques are also available, which work as empirical models or tools and help in the optimization of production conditions of enzymes (Hao et al. 2012). For example, Meriem and Mahmoud (2017) reported chitinase production by *Streptomyces griseorubens* by using a Plackett–Burman design (PBD) model. The results of PBD experiments revealed that that syrup of date, colloidal chitin, PO4 (K_2HPO_4, KH_2PO_4), and yeast extract had important effects on the production of chitinase. In another study, the production of chitinase by *Chitinolyticbacter meiyuanensis* was assessed via Plackett–Burman (P-B) and response surface methodology (RSM) and it was found that both models can be useful in the prediction of nutrient demand for maximum production of chitinase (Hao et al. 2012).

10.5 Future Directions and Conclusion

Biocontrol enzymes are the significant products which can be used for the protection of crop from devastating phytopathogens. Although in comparison to other industrial enzymes, production and commercial applicability of biocontrol enzymes are not well established, in the near future, their use may expand, especially in the development of biocontrol products. Some of the issues being faced in the production of biocontrol enzymes are lack of efficient strains, high production costs, inadequate formulation design, and instability at various conditions. Recently, researchers tried to overcome the shortcomings related with their production, and it has been found that utilization of agro-wastes and animal material can cut down the cost of the hydrolytic enzymes (Viayaraghavan et al. 2019; Sindhu et al. 2019). Recently, Sarker et al. (2019) reported that by using an aqueous extract of cow dung manure, vermicompost, de-oiled neem cake, sugar, sugarcane molasses, baker's yeast powder, and crab shell powder in their different combinations production of chitinase can be enhanced. Meruvu and Meruvu (2019) found that by using fermented wheat bran and shrimp shells, waste yield of chitinase can be enhanced by 4.24-fold or 31% in *Citrobacter freundii* harit D11 strain. Use of statistical surface methodology in the optimization of protease production in *Bacillus* species was described by Suberu et al. (2019). For improving the ability of enzyme production genetic modification techniques are more promising than physical and chemical approaches. In a study, Hafez et al. (2019) also noticed that by the introduction of a recombinant gene P2 in *S. griseorubens* E44G strain, its chitinolytic activity may be increased by 1.39-fold. With the genetic improvement of empathogenic fungal enzymes, the efficiency of fungal biopesticides can also be improved (Lovett and St. Leger 2018). It has also been realized that rhizosphere microbes are potential candidates for secreting hydrolytic enzymes; however, in the last few years studies also affirmed the role of marine microbes in the production of hydrolyses withstanding extreme temperatures and salinity (de Veras et al. 2018). The potential of viral chitinases belonging to the GH18 family has also been evaluated for fungicidal and insecticidal properties (Berini et al. 2018). Biotechnology of recombinant enzymes has also expanded the multifarious use of microbial enzymes in the direction of crop protection (de Sousa et al. 2019).

Although considerable progress has been made in optimization, screening, and strain improvement of microbes used for the production of an enzyme with biocontrol trait, negligible work is done for the development of biocontrol enzyme-containing product. In the real sense, formulation design is the biggest challenge in the development of any biocontrol product (Arora and Mishra 2016; Mishra and Arora 2018). As in conventional approach of bioformulations or biopesticide preparation, the target remains to deliver biocontrol agent itself, and no other ingredients are utilized to act as a biocontrol agent; hence incorporating microbial enzyme in the formulation product may provide additional advantage and will also increase its activity against target pest or phytopathogens. However, presently there are no such biocontrol products available for commercial use, but in the near future, more research on microbial enzymes with biocontrol activity may enable their utilization in the commercial development of biopesticides. Interestingly, with the advent of nanotechnology, we have succeeded in the designing of nanoparticle-based formulations loaded with a microbial enzyme which can be used in the biocontrol of host-specific pathogens (Chinnaperumal et al. 2018).

Lastly, it must be noted that microbial enzymes with biocontrol traits are of immense value to plant defense from phytopathogens. Furthermore, considering some regulatory constraints such as registration and safety assessment, the use of microbial enzyme may be extended in the manufacturing of the new generation biocontrol products showing host-specific and broad-spectrum activity.

References

Aimanianda V, Simenel C, Garnaud C, Clavaud C, Tada R, Barbin L, Mouyna I, Heddergott C, Popolo L, Ohya Y, Delepierre M (2017) The dual activity responsible for the elongation and branching of β-(1, 3)-glucan in the fungal cell wall. MBio 8(3):e00619–e00617

Ait-Lahsen H, Soler A, Rey M, de la Cruz J, Monte E, Llobell A (2001) An antifungal exo-α-1, 3-glucanase (AGN13. 1) from the biocontrol fungus *Trichoderma harzianum*. Appl Environ Microbiol 67:5833–5839

Aktuganov G, Melentjev A, Galimzianova N, Khalikova E, Korpela T, Susi P (2008) Wide-range antifungal antagonism of *Paenibacillus ehimensis* IB-Xb and its dependence on chitinase and β-1, 3-glucanase production. Can J Microbiol 54:577–587

Anand S, Reddy J (2009) Biocontrol potential of Trichoderma sp. against plant pathogens. Int J Agric Sci 1:30

Arora NK, Mishra J (2016) Prospecting the roles of metabolites and additives in future bioformulations for sustainable agriculture. Appl Soil Ecol 107:405–407

Arora NK, Kim MJ, Kang SC, Maheshwari DK (2007) Role of chitinase and β-1, 3-glucanase activities produced by a fluorescent pseudomonad and in vitro inhibition of *Phytophthora capsici* and *Rhizoctonia solani*. Can J Microbiol 53:207–212

Ashwini N, Srividya S (2014) Potentiality of *Bacillus subtilis* as biocontrol agent for management of anthracnose disease of chilli caused by *Colletotrichum gloeosporioides* OGC1. 3 Biotech 4:127–136

Babu K, Satyanarayana T (1996) Production of bacterial enzymes by solid state fermentation. J Sci Ind Res 55:464–467

Baiyee B, Pornsuriya C, Ito S-i, Sunpapao A (2019) *Trichoderma spirale* T76-1 displays biocontrol activity against leaf spot on lettuce (*Lactuca sativa* L.) caused by *Corynespora cassiicola* or *Curvularia aeria*. Biol Control 129:195–200

Banani H, Spadaro D, Zhang D, Matic S, Garibaldi A, Gullino ML (2014) Biocontrol activity of an alkaline serine protease from *Aureobasidium pullulans* expressed in *Pichia pastoris* against four postharvest pathogens on apple. Int J Food Microbiol 182:1–8

BCC Research (2018) Global markets for enzymes in industrial applications. Business Communications Company (BCC) Inc., Norwalk, CT

Behie SW, Bonet B, Zacharia VM, McClung DJ, Traxler MF (2017) Molecules to ecosystems: Actinomycete natural products in situ. Front Microbiol 7:2149

Benhamou N, le Floch G, Vallance J, Gerbore J, Grizard D, Rey P (2012) *Pythium oligandrum*: an example of opportunistic success. Microbiology 158(Pt 11):2679–2694

Benítez T, Delgado-Jarana J, Rincon A, Rey M, Limón C (1998) Biofungicides: *Trichoderma* as a biocontrol agent against phytopathogenic fungi. Recent Res DevMicrobiol 2:129–150

Benítez T, Rincón AM, Limón MC, Codon AC (2004) Biocontrol mechanisms of *Trichoderma* strains. Int Microbiol 7(4):249–260

Berini F, Katz C, Gruzdev N, Casartelli M, Tettamanti G, Marinelli F (2018) Microbial and viral chitinases: attractive biopesticides for integrated pest management. Biotechnol Adv 36:818–838

Berkeley MJ (1846) Observations, botanical and physiological, on the Potato Murrain. J Hortic Soc London 1:9–34

Beyer M, Diekmann H (1985) The chitinase system of *Streptomyces* sp. ATCC 11238 and its significance for fungal cell wall degradation. Appl Microbiol Biotechnol 23:140–146

Bhattacharya D, Nagpure A, Gupta RK (2007) Bacterial chitinases: properties and potential. Crit Rev Biotechnol 27:21–28

Bourke PA (1964) Emergence of potato blight, 1843–46. Nature 203:805

Brzezinska MS, Jankiewicz U (2012) Production of antifungal chitinase by Aspergillus niger LOCK 62 and its potential role in the biological control. Curr Microbiol 65:666–672

Çalık P, Çalık G, Özdamar TH (2001) Bioprocess development for serine alkaline protease production: a review. Rev Chem Eng 17:1–62

Chaari F, Kamoun A, Bhiri F, Blibech M, Ellouze-Ghorbel R, Ellouz-Chaabouni S (2012) Statistical optimization for the production of lichenase by a newly isolated *Bacillus licheniformis* UEB CF in solid state fermentation using pea pomace as a novel solid support. Ind Crop Prod 40:192–198

Chahal D (1983) Growth characteristics of microorganisms in solid state fermentation for upgrading of protein values of lignocelluloses and cellulase production. Foundations of biochemical engineering kinetics and thermodynamics in biological systems American Chemical Society, Washington DC ACS symposium series. ACS Publications, pp 421–442

Chen PH, Chen RY, Chou JY (2018a) Screening and evaluation of yeast antagonists for biological control of *Botrytis cinerea* on strawberry fruits. Mycobiology 46:33–46

Chen D, Wang D, Xu C, Chen C, Li J, Wu W, Huang X, Xie H (2018b) Nematicidal protease genes screened from a soil metagenomic library to control *Radopholus similis* mediated by *Pseudomonas fluorescens* pf36. Appl Microbiol Biotechnol 102:3301–3314

Chernin L, Chet I (2002) Microbial enzymes in the biocontrol of plant pathogens and pests. Enzymes in the environment. CRC Press, Boca Raton, FL, pp 179–234

Chinnaperumal K, Govindasamy B, Paramasivam D, Dilipkumar A, Dhayalan A, Vadivel A, Sengodan K, Pachiappan P (2018) Bio-pesticidal effects of *Trichoderma viride* formulated titanium dioxide nanoparticle and their physiological and biochemical changes on *Helicoverpa armigera* (Hub.). Pestic Biochem Physiol 149:26–36

Cho KM, Hong SY, Lee SM, Kim YH, Kahng GG, Lim YP, Kim H, Yun HD (2007) Endophytic bacterial communities in ginseng and their antifungal activity against pathogens. Microb Ecol 54:341–351

Clark DP, Pazdernik NJ (2016) Chapter 9: Proteomics. In: Clark DP, Pazdernik NJ (eds) Biotechnology, 2nd edn. Academic Cell, Boston, MA, pp 295–333

Clark GL, Smith AF (1936) X-ray diffraction studies of chitin, chitosan, and derivatives. J Phys Chem 40:863–879

Daimon T, Katsuma S, Iwanaga M, Kang W, Shimada T (2005) The BmChi-h gene, a bacterial-type chitinase gene of *Bombyx mori,* encodes a functional exochitinase that plays a role in the chitin degradation during the molting process. Insect Biochem Mol Biol 35:1112–1123

de Sousa M, Melo VMM, Hissa DC, Manzo RM, Mammarella EJ, Antunes A, García JL, Pessela BC, Goncalves LRB (2019) One-step immobilization and stabilization of a recombinant *Enterococcus faecium* DBFIQ E36 L-arabinose isomerase for D-tagatose synthesis. Appl Biochem Biotechnol 188(2):310–325

de Veras BO, dos Santos YQ, Diniz KM, Carelli GSC, dos Santos EA (2018) Screening of protease, cellulase, amylase and xylanase from the salt-tolerant and thermostable marine *Bacillus subtilis* strain SR60. F1000Research:7

Deng J-J, Huang W-Q, Li Z-W, Lu D-L, Zhang Y, Luo X-c (2018) Biocontrol activity of recombinant aspartic protease from *Trichoderma harzianum* against pathogenic fungi. Enzyme Microb Technol 112:35–42

Dewi RTK, Mubarik NR, Suhartono MT (2016) Medium optimization of β-glucanase production by *Bacillus subtilis* SAHA 32.6 used as biological control of oil palm pathogen. Emir J Food Agric 28:116–125

Douriet-Gámez NR, Maldonado-Mendoza IE, Ibarra-Laclette E, Blom J, Calderón-Vázquez CL (2018) Genomic analysis of *Bacillus* sp. Strain B25, a biocontrol agent of maize pathogen *Fusarium verticillioides*. Curr Microbiol 75:247–255

Dunne C, Crowley JJ, Moënne-Loccoz Y, Dowling DN, O'Gara F (1997) Biological control of *Pythium ultimum* by *Stenotrophomonas maltophilia* W81 is mediated by an extracellular proteolytic activity. Microbiology 143:3921–3931

Dunne C, Moënne-Loccoz Y, de Bruijn FJ, O'Gara F (2000) Overproduction of an inducible extracellular serine protease improves biological control of *Pythium ultimum* by *Stenotrophomonas maltophilia* strain W81. Microbiology 146:2069–2078

Elad Y, Kapat A (1999) The role of *Trichoderma harzianum* protease in the biocontrol of *Botrytis cinerea.* Eur J Plant Pathol 105(2):177–189

Elango N, Correa JU, Cabib E (1982) Secretory character of yeast chitinase. J Biol Chem 257:1398–1400

El-Bendary MA, Hamed HA, Moharam ME (2016) Potential of *Bacillus* isolates as bio-control agents against some fungal phytopathogens. Biocatal Agric Biotechnol 5:173–178

El-Katatny M, Gudelj M, Robra K-H, Elnaghy M, Gübitz G (2001) Characterization of a chitinase and an endo-β-1, 3-glucanase from *Trichoderma harzianum* Rifai T24 involved in control of the phytopathogen *Sclerotium rolfsii*. Appl Microbiol Biotechnol 56:137–143

El-Sayed MH, Nassar OM, Nasr HA (2019) Efficacy of thermophilic soil-isolated *Paenibacillus* sp. NBR10 as a chitinolytic and biocontrol bacterium-in vitro study. Egypt J Bot 59(1:195–208

El-Tarabily KA, Sykes ML, Kurtböke ID, Hardy GESJ, Barbosa AM, Dekker RF (1996) Synergistic effects of a cellulase-producing *Micromonospora carbonacea* and an antibiotic-producing *Streptomyces violascens* on the suppression of Phytophthora cinnamomi root rot of *Banksia grandis*. Can J Bot 74:618–624

El-Tarabily K, Soliman M, Nassar A, Al-Hassani H, Sivasithamparam K, McKenna F, Hardy GSJ (2000) Biological control of *Sclerotinia minor* using a chitinolytic bacterium and actinomycetes. Plant Pathol 49:573–583

Essghaier B, Bejji M, Jijakli H, Boudabous A, Sadfi-Zouaoui N (2009) High salt-tolerant protease from a potential biocontrol agent *Bacillus pumilus* M3-16. Ann Microbiol 59:553

Fan H, Liu Z, Zhang R, Wang N, Dou K, Mijiti G, Diao G, Wang Z (2014) Functional analysis of a subtilisin-like serine protease gene from biocontrol fungus *Trichoderma harzianum.* J Microbiol 52(2):129–138

Flach J, Pilet P-E, Jolles P (1992) What's new in chitinase research? Experientia 48:701–716

Folders J, Algra J, Roelofs MS, Van Loon LC, Tommassen J, Bitter W (2001) Characterization of *Pseudomonas aeruginosa* chitinase, a gradually secreted protein. J Bacteriol 183:7044–7052

Fridlender M, Inbar J, Chet I (1993) Biological control of soilborne plant pathogens by a β-1, 3 glucanase-producing *Pseudomonas cepacia*. Soil Biol Biochem 25:1211–1221

Gabrielová A, Mencl K, Suchánek M, Klimeš R, Hubka V, Kolařík M (2018) The oomycete *Pythium oligandrum* can suppress and kill the causative agents of dermatophytoses. Mycopathologia 183:751–764

Garcia-Carreon FL (1997) Classification of proteases without tears. Biochem Educ 25:161–167

Gomaa EZ (2012) Chitinase production by *Bacillus thuringiensis* and *Bacillus licheniformis*: their potential in antifungal biocontrol. J Microbiol 50:103–111

Gopalakrishnan B, Muthukrishnan S, Kramer KJ (1995) Baculovirus-mediated expression of a Manduca sexta chitinase gene: properties of the recombinant protein. Insect Biochem Mol Biol 25:255–265

Grinyer J, Hunt S, McKay M, Herbert BR, Nevalainen H (2005) Proteomic response of the biological control fungus *Trichoderma atroviride* to growth on the cell walls of *Rhizoctonia solani.* Curr Genet 47(6):381–388

Gueguen Y, Voorhorst WG, van der Oost J, de Vos WM (1997) Molecular and biochemical characterization of an Endo-β-1, 3-glucanase of the hyperthermophilic archaeon *Pyrococcus furiosus.* J Biol Chem 272:31258–31264

Gupta VK, Ali I (2008) Removal of endosulfan and methoxychlor from water on carbon slurry. Environ Sci Technol 42:766–770

Hafez EE, Rashad YM, Abdulkhair WM, Al-Askar AA, Ghoneem KM, Baka ZA, Shabana YM (2019) Improving the chitinolytic activity of *Streptomyces griseorubens* e44g by mutagenesis. J Microbiol Biotechnol Food Sci 8

Hansen GH, Lübeck M, Frisvad JC, Lübeck PS, Andersen B (2015) Production of cellulolytic enzymes from Ascomycetes: comparison of solid state and submerged fermentation. Process Biochem 50:1327–1341

Hao Z, Cai Y, Liao X, Zhang X, Fang Z, Zhang D (2012) Optimization of nutrition factors on chitinase production from a newly isolated *Chitiolyticbacter meiyuanensis* SYBC-H1. Braz J Microbiol 43:177–186

Hartl L, Zach S, Seidl-Seiboth V (2012) Fungal chitinases: diversity, mechanistic properties and biotechnological potential. Appl Microbiol Biotechnol 93:533–543

Hoster F, Schmitz JE, Daniel R (2005) Enrichment of chitinolytic microorganisms: isolation and characterization of a chitinase exhibiting antifungal activity against phytopathogenic fungi from a novel Streptomyces strain. Appl Microbiol Biotechnol 66:434–442

Illakkiam D, Anuj NL, Ponraj P, Shankar M, Rajendhran J, Gunasekaran P (2013) Proteolytic enzyme mediated antagonistic potential of *Pseudomonas aeruginosa* against *Macrophomina phaseolina*. IJEB 51(11):1024–1031

Inbar J, Chet I (1991) Evidence that chitinase produced by *Aeromonas caviae* is involved in the biological control of soil-borne plant pathogens by this bacterium. Soil Biol Biochem 23:973–978

Ivanoff S, Keitt G (1937) The occurrence of aerial bacterial strands on blossoms, fruits, and shoots blighted by *Erwinia amylovora.* Phytopathology 27:702–709

Jangir M, Pathak R, Sharma S, Sharma S (2018) Biocontrol mechanisms of *Bacillus* sp., isolated from tomato rhizosphere, against *Fusarium oxysporum* f. sp. lycopersici. Biol Control 123:60–70

Jayalakshmi SK, Raju S, Rani SU, Benagi VI, Sreeramulu K (2009) *Trichoderma harzianum* L1 as a potential source for lytic enzymes and elicitor of defense responses in chickpea (*Cicer arietinum* L.) against wilt disease caused by *Fusarium oxysporum* f. sp. ciceri. Aust J Crop Sci 3(1):44

Jeuniaux C (1961) Chitinase: an addition to the list of hydrolases in the digestive tract of vertebrates. Nature 192:135–136

Jha Y (2019) Endophytic bacteria mediated anti-autophagy and induced catalase, β-1, 3-glucanases gene in paddy after infection with pathogen *Pyricularia grisea*. Indian Phytopathol 72(1):99–106

Jiang C, Song J, Cong H, Zhang J, Yang Q (2017a) Expression and characterization of a novel antifungal Exo-β-1, 3-glucanase from *Chaetomium cupreum*. Appl Biochem Biotechnol 182(1):261–275

Jiang C, Song J, Zhang J, Yang Q (2017b) Identification and characterization of the major antifungal substance against *Fusarium sporotrichioides* from *Chaetomium globosum*. World J Microbiol Biotechnol 33(6):108

John CJ, Radhakrishnan E (2018) Chemicobiological insight into anti-phytopathogenic properties of Rhizospheric *Serratia plymuthica* R51. Proc Natl Acad Sci India Sect B: Biol Sci 88:1629–1635

Joo G-J (2005) Purification and characterization of an extracellular chitinase from the antifungal biocontrol agent *Streptomyces halstedii*. Biotechnol Lett 27(19):1483–1486

Juturu V, Wu JC (2014) Microbial cellulases: engineering, production and applications. Renew Sustain Energy Rev 33:188–203

Kang SC, Park S, Lee DG (1999) Purification and characterization of a novel chitinase from the entomopathogenic fungus, *Metarhizium anisopliae*. J Invertebr Pathol 73:276–281

Kapilan R, Arasaratnam V (2011) Paddy husk as support for solid state fermentation to produce xylanase from *Bacillus pumilus*. Ric Sci 18:36–45

Karabörklü S, Azizoglu U, Azizoglu ZB (2018) Recombinant entomopathogenic agents: a review of biotechnological approaches to pest insect control. World J Microbiol Biotechnol 34:14

Khare E, Kumar S, Kim K (2018) Role of peptaibols and lytic enzymes of *Trichoderma cerinum* Gur1 in biocontrol of *Fusarium oxysporum* and chickpea wilt. Environ Sust 1:39–47

Kobayashi DY, Reedy RM, Bick J, Oudemans PV (2002) Characterization of a chitinase gene from *Stenotrophomonas maltophilia* strain 34S1 and its involvement in biological control. Appl Environ Microbiol 68:1047–1054

Kramer KJ, Muthukrishnan S (1997) Insect chitinases: molecular biology and potential use as biopesticides. Insect Biochem Mol Biol 27(11):887–900

Kumar P, Dubey RC, Maheshwari DK (2012) *Bacillus* strains isolated from rhizosphere showed plant growth promoting and antagonistic activity against phytopathogens. Microbiol Res 167:493–499

Leelasuphakul W, Sivanunsakul P, Phongpaichit S (2006) Purification, characterization and synergistic activity of β-1, 3-glucanase and antibiotic extract from an antagonistic *Bacillus subtilis* NSRS 89-24 against rice blast and sheath blight. Enzyme Microb Technol 38:990–997

Leger RS, Cooper R, Charnley A (1991) Characterization of chitinase and chitobiase produced by the entomopathogenic fungus *Metarhizium anisopliae*. J Invertebr Pathol 58:415–426

Leger RJ, Staples RC, Roberts DW (1992) Cloning and regulatory analysis of starvation-stress gene, ssgA, encoding a hydrophobin-like protein from the entomopathogenic fungus, *Metarhizium anisopliae*. Gene 120(1):119–124

Li Q, Xiao X, Wang F (2007) Screening of genes involved in chitinase production in *Aeromonas caviae* CB101 via transposon mutagenesis. J Appl Microbiol 102:640–649

Li J, Liu W, Luo L, Dong D, Liu T, Zhang T, Lu C, Liu D, Zhang D, Wu H (2015) Expression of *Paenibacillus polymyxa* β-1, 3-1, 4-glucanase in *Streptomyces lydicus* A01 improves its biocontrol effect against *Botrytis cinerea*. Biol Control 90:141–147

Li Y, Sun R, Yu J, Saravanakumar K, Chen J (2016) Antagonistic and biocontrol potential of *Trichoderma asperellum* ZJSX5003 against the maize stalk rot pathogen *Fusarium graminearum*. Indian J Microbiol 56:318–327

Li Z, Xia C, Wang Y, Li X, Qiao Y, Li C, Zhou J, Zhang L, Ye X, Huang Y, Cui Z (2019) Identification of an endo-chitinase from *Corallococcus* sp. EGB and evaluation of its antifungal properties. Int J Biol Macromol 132:1235–1243

Lian LH, Tian BY, Xiong R, Zhu MZ, Xu J, Zhang KQ (2007) Proteases from *Bacillus*: a new insight into the mechanism of action for rhizobacterial suppression of nematode populations. Lett Appl Microbiol 45(3):262–269

Lim H-S, Kim S-D (1995) The role and characterization of β-1, 3-glucanase in biocontrol of *Fusarium solani* by *Pseudomonas stutzeri* YPL-1. J Microbiol 33:295–301

Loliam B, Morinaga T, Chaiyanan S (2013) Biocontrol of *Pythium aphanidermatum* by the cellulolytic actinomycetes *Streptomyces rubrolavendulae* S4. Sci Asia 39:584–590

Lopes MA, Gomes DS, Koblitz MGB, Pirovani CP, de Mattos Cascardo JC, Góes-Neto A, Micheli F (2008) Use of response surface methodology to examine chitinase regulation in the basidiomycete *Moniliophthora perniciosa*. Mycol Res 112:399–406

Lorito M, Harman G, Hayes C, Broadway R, Tronsmo A, Woo SL, Di Pietro A (1993) Chitinolytic enzymes produced by *Trichoderma harzianum*: antifungal activity of purified endochitinase and chitobiosidase. Phytopathology 83:302–307

Lovett B, St. Leger RJ (2018) Genetically engineering better fungal biopesticides. Pest Manag Sci 74:781–789

Lynd LR, Weimer PJ, Van Zyl WH, Pretorius IS (2002) Microbial cellulose utilization: fundamentals and biotechnology. Microbiol Mol Biol Rev 66:506–577

Mahadevan B, Crawford DL (1997) Properties of the chitinase of the antifungal biocontrol agent *Streptomyces lydicus* WYEC108. Enzyme Microb Technol 20:489–493

Mahmood I, Imadi SR, Shazadi K, Gul A, Hakeem KR (2016) Effects of pesticides on environment. In: Hakeem KR, Akhtar MS, Abdullah SNA (eds) Plant, soil and microbes: volume 1: implications in crop science. Springer International Publishing, Cham, pp 253–269

Majumdar S, Chakraborty U (2017) Optimization of protease production from plant growth promoting *Bacillus amyloliquefaciens* showing antagonistic activity against phytopathogens. Int J Pharm Bio Sci 8(2):635–642

Martinelli F, Scalenghe R, Davino S, Panno S, Scuderi G, Ruisi P, Villa P, Stroppiana D, Boschetti M, Goulart LR (2015) Advanced methods of plant disease detection. Rev Agron Sust Dev 35:1–25

Medie FM, Davies GJ, Drancourt M, Henrissat B (2012) Genome analyses highlight the different biological roles of cellulases. Nat Rev Microbiol 10:227

Mercado JA, Barceló M, Pliego C, Rey M, Caballcro JL, Muñoz-Blanco J, Ruano-Rosa D, López-Herrera C, de los Santos B, Romero-Muñoz F, Pliego-Alfaro F (2015) Expression of the β-1, 3-glucanase gene bgn13. 1 from *Trichoderma harzianum* in strawberry increases tolerance to crown rot diseases but interferes with plant growth. Transgenic Res 24(6):979–989

Meriem G, Mahmoud K (2017) Optimization of chitinase production by a new *Streptomyces griseorubens* C9 isolate using response surface methodology. Ann Microbiol 67:175–183

Meruvu H, Meruvu HBT (2019) Biosynthesis and characterization of an anticoagulant chitinase from Fermented Wheat Bran & Shrimp Shells' substratum. Trends Peptide Protein Sci 3:5-1-9

Migheli Q, González-Candelas L, Dealessi L, Camponogara A, Ramón-Vidal D (1998) Transformants of *Trichoderma longibrachiatum* overexpressing the β-1, 4-endoglucanase gene egl1 show enhanced biocontrol of *Pythium ultimum* on cucumber. Phytopathology 88:673–677

Miller GT (2004) Sustaining the Earth. Chapter 9, 6th edn. Thompson Learning, Inc., Pacific Grove, CA, pp 211–216

Miller RNG, Costa Alves GS, Van Sluys M-A (2017) Plant immunity: unravelling the complexity of plant responses to biotic stresses. Ann Bot 119:681–687

Mishra S, Arora NK (2012) Management of black rot in cabbage by rhizospheric *Pseudomonas* species and analysis of 2,4-diacetylphloroglucinol by qRT-PCR. Biol Control 61:32–39

Mishra J, Arora NK (2018) Secondary metabolites of fluorescent pseudomonads in biocontrol of phytopathogens for sustainable agriculture. Appl Soil Ecol 125:35–45

Monteiro VN, Ulhoa CJ (2006) Biochemical characterization of a β-1, 3-glucanase from *Trichoderma koningii* induced by cell wall of *Rhizoctonia solani*. Curr Microbiol 52(2):92–96

Mota MS, Gomes CB, Souza Júnior IT, Moura AB (2017) Bacterial selection for biological control of plant disease: criterion determination and validation. Braz J Microbiol 48(1):62–70

Mouyna I, Hartl L, Latgé JP (2013) β-1, 3-glucan modifying enzymes in *Aspergillus fumigatus*. Front Microbiol 4:81

Nagarajkumar M, Bhaskaran R, Velazhahan R (2004) Involvement of secondary metabolites and extracellular lytic enzymes produced by *Pseudomonas fluorescens* in inhibition of *Rhizoctonia solani*, the rice sheath blight pathogen. Microbiol Res 159:73–81

Nielsen MN, Sørensen J, Fels J, Pedersen HC (1998) Secondary metabolite- and Endochitinase-dependent antagonism toward plant-pathogenic microfungi of *Pseudomonas fluorescens* isolates from sugar beet rhizosphere. Appl Environ Microbiol 64:3563–3569

Olanrewaju OS, Babalola OO (2019) *Streptomyces*: implications and interactions in plant growth promotion. Appl Microbiol Biotechnol 103:1179–1188

Pandey A, Selvakumar P, Soccol CR, Nigam P (1999) Solid state fermentation for the production of industrial enzymes. Curr Sci:149–162

Pandey A, Soccol CR, Mitchell D (2000) New developments in solid state fermentation: I-bioprocesses and products. Process Biochem 35:1153–1169

Parafati L, Cirvilleri G, Restuccia C, Wisniewski M (2017) Potential role of exoglucanase genes (WaEXG1 and WaEXG2) in the biocontrol activity of *Wickerhamomyces anomalus*. Microb Ecol 73:876–884

Picard K, Ponchet M, Blein JP, Rey P, Tirilly Y, Benhamou N (2000) Oligandrin. A proteinaceous molecule produced by the Mycoparasite *Pythium oligandrum* induces resistance to *Phytophthora parasitica* infection in tomato plants. Plant Physiol 124:379–396

Pozo MJ, Baek JM, Garcıa JM, Kenerley CM (2004) Functional analysis of tvsp1, a serine protease-encoding gene in the biocontrol agent *Trichoderma virens*. Fungal Genet Biol 41(3):336–348

Quesada-Ocampo L, Al-Haddad J, Scruggs A, Buell C, Trail F (2016) Susceptibility of maize to stalk rot caused by *Fusarium graminearum* deoxynivalenol and zearalenone mutants. Phytopathology 106:920–927

Ramada MH, Steindorff AS, Bloch C Jr, Ulhoa CJ (2016) Secretome analysis of the mycoparasitic fungus *Trichoderma harzianum* ALL 42 cultivated in different media supplemented with *Fusarium solani* cell wall or glucose. Proteomics 16(3):477–490

Ren JH, Li H, Wang YF, Ye JR, Yan AQ, Wu XQ (2013) Biocontrol potential of an endophytic *Bacillus pumilus* JK-SX001 against poplar canker. Biol Control 67:421–430

Reyes-Ramírez A, Escudero-Abarca B, Aguilar-Uscanga G, Hayward-Jones P, Barboza-Corona JE (2004) Antifungal activity of *Bacillus thuringiensis* chitinase and its potential for the biocontrol of phytopathogenic fungi in soybean seeds. J Food Sci 69:131–134

Roberti R, Zakrisson E, Flamigni F, De Vero L, Cesari A (2002) Antagonistic fungi producing hydrolytic enzymes, active in degrading the cell wall of some foot rot pathogens (Fusarium spp.) of wheat/Antagonistische Pilze, die hydrolytische Enzyme erzeugen, welche zellwandabbauende Aktivität an einigen Erregern der Halmbasiskrankheiten (*Fusarium* spp.) des Weizens zeigen. Zeitschrift Für Pflanzenkrankheiten Und Pflanzenschutz/J Plant Dis Protect 109(1):101–108

Saber WI, Ghoneem KM, Al-Askar AA, Rashad YM, Ali AA, Rashad EM (2015) Chitinase production by *Bacillus subtilis* ATCC 11774 and its effect on biocontrol of Rhizoctonia diseases of potato. Acta Biol Hung 66:436–448

Sacherer P, Défago G, Haas D (1994) Extracellular protease and phospholipase C are controlled by the global regulatory gene gacA in the biocontrol strain Pseudomonas fluorescens CHA0. FEMS Microbiol Lett 116(2):155–160

Sahai A, Manocha M (1993) Chitinases of fungi and plants: their involvement in morphogenesis and host—parasite interaction. FEMS Microbiol Rev 11:317–338

Sakpetch P, Aran H, Kuwahara Y, Komeda H, Asano Y (2018) Isolation of indigenous antagonistic microorganism to inhibit *Rigidoporus microporus* and other plant pathogens and analysis of the bioactive compounds. Biol Control 124:53–60

San Fulgencio NS, Suárez-Estrella F, López M, Jurado M, López-González J, Moreno J (2018) Biotic aspects involved in the control of damping-off producing agents: the role of the thermotolerant microbiota isolated from composting of plant waste. Biol Control 124:82–91

Saraf M, Thakker A, Patel BV (2008) Biocontrol activity of different species of *Pseudomonas* against phytopathogenic fungi in vivo and in vitro conditions. Int J Biotechnol Biochem 4(3):223–233

Saravanakumar K, Yu C, Dou K, Wang M, Li Y, Chen J (2016) Synergistic effect of *Trichoderma*-derived antifungal metabolites and cell wall degrading enzymes on enhanced biocontrol of *Fusarium oxysporum* f. sp. cucumerinum. Biol Control 94:37–46

Sarkar S, Pramanik A, Mitra A, Mukherjee J (2010) Bioprocessing data for the production of marine enzymes. Mar Drugs 8:1323–1372

Sarker KN, Mohapatra PKD, Dutta S (2019) Use of low cost natural resources for enhanced chitinase production and optimization using CCD and RSM: a new initiative for bio-control of plant pathogen. Indian Phytopathol:1–20

Schirmböck M, Lorito M, Wang YL, Hayes CK, Arisan-Atac I, Scala F, Harman GE, Kubicek CP (1994) Parallel formation and synergism of hydrolytic enzymes and peptaibol antibiotics, molecular mechanisms involved in the antagonistic action of *Trichoderma harzianum* against phytopathogenic fungi. Appl Environ Microbiol 60(12):4364–4370

Schülein M (1988) Cellulases of *Trichoderma reesei*. Methods Enzymol Elsevier, pp 234–242

Schuster A, Schmoll M (2010) Biology and biotechnology of *Trichoderma*. Appl Microbiol Biotechnol 87(3):787–799

Shao Z, Li Z, Fu Y, Wen Y, Wei S (2018) Induction of defense responses against *Magnaporthe oryzae* in rice seedling by a new potential biocontrol agent *Streptomyces* JD211. J Basic Microbiol 58:686–697

Sharma V, Salwan R, Sharma PN (2016) Differential response of extracellular proteases of *Trichoderma harzianum* against fungal phytopathogens. Curr Microbiol 73:419–425

Sharma KM, Kumar R, Panwar S, Kumar A (2017) Microbial alkaline proteases: optimization of production parameters and their properties. J Genet Eng Biotechnol 15:115–126

Sharma CK, Vishnoi VK, Dubey RC, Maheshwari DK (2018) A twin rhizospheric bacterial consortium induces systemic resistance to a phytopathogen *Macrophomina phaseolina* in mung bean. Rhizosphere 5:71–75

Siddiqui IA, Shaukat SS (2005) Phenylacetic acid-producing *Rhizoctonia solani* represses the biosynthesis of nematicidal compounds in vitro and influences biocontrol of *Meloidogyne incognita* in tomato by *Pseudomonas fluorescens* strain CHA0 and its GM derivatives. J Appl Microbiol 98(1):43–55

Sindhu S, Dadarwal K (2001) Chitinolytic and cellulolytic *Pseudomonas* sp. antagonistic to fungal pathogens enhances nodulation by Mesorhizobium sp. Cicer in chickpea. Microbiol Res 156:353–358

Sindhu R, Gnansounou E, Rebello S, Binod P, Varjani S, Thakur IS, Nair RB, Pandey A (2019) Conversion of food and kitchen waste to value-added products. J Environ Manage 241:619–630

Singh S, Bajaj BK (2016) Bioprocess optimization for production of thermoalkali-stable protease from *Bacillus subtilis* K-1 under solid-state fermentation. Prep Biochem Biotechnol 46:717–724

Sitrit Y et al (1995) Cloning and primary structure of the chiA gene from Aeromonas caviae. J Bacteriol 177(14):4187–4189

Smith HS (1919) On some phases of insect control by the biological method. J Econ Entomol 12:288–292

Strange RN, Scott PR (2005) Plant disease: a threat to global food security. Annu Rev Phytopathol 43:83–116

Strasser H, Vey A, Butt TM (2000) Are there any risks in using entomopathogenic fungi for pest control, with particular reference to the bioactive metabolites of *Metarhizium, Tolypocladium* and *Beauveria* species? Biocontrol Sci Technol 10:717–735

Suárez MB, Sanz L, Chamorro MI, Rey M, González FJ, Llobell A, Monte E (2005) Proteomic analysis of secreted proteins from *Trichoderma harzianum*: identification of a fungal cell wall-induced aspartic protease. Fungal Genet Biol 42:924–934

Suberu Y, Akande I, Samuel T, Lawal A, Olaniran A (2019) Optimization of protease production in indigenous *Bacillus* species isolated from soil samples in Lagos, Nigeria using response surface methodology. Biocatal Agric Biotechnol 18:101011

Subramaniyam R, Vimala R (2012) Solid state and submerged fermentation for the production of bioactive substances: a comparative study. Int J Sci Nat 3:480–486

Sukumaran RK, Singhania RR, Pandey A (2005) Microbial cellulases-production, applications and challenges. J Scient Ind Res 64:832–844

Sundheim L, Poplawsky AR, Ellingboe AH (1988) Molecular cloning of two chitinase genes from *Serratia marcescens* and their expression in *Pseudomonas* species. Physiol Mol Plant Pathol 33:483–491

Suslow TV, Jones JD. Novel chitinase-producing bacteria and plants. U.S. Patent 4,940,840, issued July 10, 1990

Szewczyk B, Hoyos-Carvajal L, Paluszek M, Skrzecz I, De Souza ML (2006) Baculoviruses—re-emerging biopesticides. Biotechnol Adv 24:143–160

Tanabe T, Kawase T, Watanabe T, Uchida Y, Mitsutomi M (2000) Purification and characterization of a 49-kDa chitinase from *Streptomyces griseus* HUT 6037. J Biosci Bioeng 89:27–32

Tasharrofi N, Adrangi S, Fazeli M, Rastegar H, Khoshayand MR, Faramarzi MA (2011) Optimization of chitinase production by *Bacillus pumilus* using Plackett-Burman design and response surface methodology. Iran J Pharm Res: IJPR 10:759

Tatum L (1971) The southern corn leaf blight epidemic. Science 171:1113–1116

Thrane C, Tronsmo A, Jensen DF (1997) Endo-1, 3-β-glucanase and cellulase from *Trichoderma harzianum*: purification and partial characterization, induction of and biological activity against plant pathogenic Pythium spp. Eur J Plant Pathol 103:331–344

Troller J, Christian J (1978) Microbial growth. Water Activity Food:86–102

Tsujibo H et al (2003) Characterization of chitinase genes from an alkaliphilic actinomycete, *Nocardiopsis prasina* OPC-131. Appl Environ Microbiol 69(2):894–900

Tunlid A, Jansson H-B, Nordbring-Hertz B (1992) Fungal attachment to nematodes. Mycol Res 96(6):401–412

Vaidyanathan S, Macaloney G, Vaughan J, McNeil B, Harvey LM (1999) Monitoring of submerged bioprocesses. Crit Rev Biotechnol 19:277–316

Valadares-Inglis MC, Peberdy JF (1997) Location of chitinolytic enzymes in protoplasts and whole cells of the entomopathogenic fungus Metarhizium anisopliae. Mycol Res 101:1393–1396

Viayaraghavan P, Jeba Kumar S, Valan Arasu M, Al-Dhabi NA (2019) Simultaneous production of commercial enzymes using agro industrial residues by statistical approach. J Sci Food Agric 99:2685–2696

Wang M, Yang J, Zhang KQ (2006) Characterization of an extracellular protease and its cDNA from the nematode-trapping fungus *Monacrosporium microscaphoides*. Can J Microbiol 52(2):130–139

Watanabe T, Kanai R, Kawase T, Tanabe T, Mitsutomi M, Sakuda S, Miyashita K (1999) Family 19 chitinases of Streptomyces species: characterization and distribution. Microbiology 145:3353–3363

Wei H, Xu Q, Taylor LE, Baker JO, Tucker MP, Ding S-Y (2009) Natural paradigms of plant cell wall degradation. Curr Opin Biotechnol 20(3):330–338

Wei LH, Xue QY, Wei BQ, Wang YM, Li SM, Chen LF, Guo JH (2010) Screening of antagonistic bacterial strains against *Meloidogyne incognita* using protease activity. Biocontrol Sci Technol 20(7):739–750

Whipps JM (2001) Microbial interactions and biocontrol in the rhizosphere. J Exp Bot 52(suppl_1):487–511

Wonglom P, Daengsuwan W, Ito SI, Sunpapao A (2019) Biological control of Sclerotium fruit rot of snake fruit and stem rot of lettuce by *Trichoderma* sp. T76-12/2 and the mechanisms involved. Physiol Mol Plant Pathol 107:1–7

Xu T, Zhu T, Li S (2016) β-1, 3-1, 4-glucanase gene from *Bacillus velezensis* ZJ20 exerts antifungal effect on plant pathogenic fungi. World J Microbiol Biotechnol 32:26

Yandigeri MS, Malviya N, Solanki MK, Shrivastava P, Sivakumar G (2015) Chitinolytic *Streptomyces vinaceusdrappus* S5MW2 isolated from Chilika lake, India enhances plant growth and biocontrol efficacy through chitin supplementation against *Rhizoctonia solani*. World J Microbiol Biotechnol 31:1217–1225

Yu WQ, Zheng GP, Yan FC, Liu WZ, Liu WX (2019) *Paenibacillus terrae* NK3-4: a potential biocontrol agent that produces β-1, 3-glucanase. Biol Control 129:92–101

Zhang XZ, Zhang YHP (2013) Cellulases: characteristics, sources, production, and applications. Bioproc Technol Biorefinery Sust Prod Fuels Chem Polymers 1:131–146

Zhang W, Jiang F, Ou J (2011) Global pesticide consumption and pollution: with China as a focus. Proc Int Acad Ecol Environ Sci 1:125

Commercial Application of Lignocellulose-Degrading Enzymes in a Biorefinery

11

Reinu E. Abraham and Munish Puri

Abstract

Enzymes play a significant role in several biotechnology-based industries for making the process cost effective. These enzymes are predominantly obtained from microbes (bacteria, fungi, and microalgae), plants and animals for serving intended applications. Recently, biofuels are advocated as clean and green, alternative source of energy to meet the growing demand of fossil fuels. However, biofuel production is not commercially scalable since lignocellulose biomass (LCB)-converting enzymes/processes are not cost effective. Several commercial enzymes including cellulase, xylanase, laminarinase and other ligninolytic enzymes are used to implement a synergistic action in the breakdown of LCB structure into pentose or hexose sugars, and other co-products. Enzyme dosage optimisation during the biomass processing and co-product production are considered amongst the major challenges in this biorefinery pathway. This chapter covers application of commercial enzyme preparations in LCB processing, and factors affecting co-product production in a biorefinery set-up to address processing challenges.

11.1 Introduction

Enzymes act as catalysts to enhance biochemical reactions and lead to product formation. The biochemical reaction process is more intense in comparison to any chemical reaction and the application of enzyme to any reaction is specific to the

R. E. Abraham · M. Puri (✉)
Centre for Marine Bioproducts Development, College of Medicine and Public Health, Flinders University, Bedford Park, South Australia, Australia
e-mail: munish.puri@flinders.edu.au

N. K. Arora et al. (eds.), *Microbial Enzymes: Roles and Applications in Industries*, Microorganisms for Sustainability 11,
https://doi.org/10.1007/978-981-15-1710-5_11

substrate (Choi et al. 2015; Arsalan and Younus 2018). They are primarily derived from microorganisms, plants and animals (Schimel et al. 2017; Park et al. 2018). Some of the properties that make enzyme more preferable over chemicals are their biodegradability, specificity, catalytic efficiency, high activity and selectivity (Li et al. 2018a; Margenot et al. 2018). Enzymes are finding a number of applications in various industries. Some of the recent advancements in enzyme technology include enzyme immobilisation, protein engineering and synthetic biology (Ehtesabi et al. 2018; Li et al. 2018b) for tailoring enzymes to a specific application.

Due to the biological catalytic activity of enzymes they are applied in industries such as food (baking, brewing, alcohol and juice), biotechnology (degradation of substrate and product development, cell rupture, modification of protein structure), textile and paper (treatment of fabrics and paper), nutraceutical and pharmaceutical (Spohner et al. 2015; Fatima and Khare 2018; Li et al. 2018a; Rosenthal and Lütz 2018). With their single biocatalytic mechanism, they help to enhance the property of product such as functionality, appearance, disruption or extraction of product (Huang et al. 2017). There are various challenges and drawbacks encountered while using an enzyme-based system. Most of the times it is either the cost or the efficiency of enzyme that brings the challenge to the product formation. The robustness of the substrate can demand addition of more enzymes for hydrolysis and the desired product can directly raise the production cost (Patel et al. 2016). Sometimes due to the inefficiency of the enzyme, the hydrolysis fails to occur effectively leading to lower conversion yield (Llorent-Martínez et al. 2017).

For various industrial applications, enzymes are currently isolated from microbes that naturally produce biocatalysts for the degradation of polysaccharides, proteins, starch or other substrates (Bernal et al. 2018). There are many leading commercial enzyme producers that are known at the global level. Amongst them, Novozymes, Chemworld, Genecor, Biocon and Enmex are major players. The majority of the enzymes used for biorefinery ("a sustainable process that integrates development of range of important value-added products from a renewable biomass") are cellulase, glucosidase, protease and amylase that are commercially available under different names depending upon the company or source. Cellulase is commercially obtained as Celluzyme 0.7Ta, Deterzyme CL-5, Endolase 5000L, Cellulast, Puradax EG 7000L and Carezyme 4500L. Protease is commercially available as Alcalase 2.5L, Bioproteasa L450 and L800, Enziprot 450L, Deterzyme L660, Savinase 16L, Type EX and Esperase 8.0L. Amylase enzyme is available as Purastar ST 15000L, Stainzyme 12L, Enziamilasa and Termamyl Ultra 300L (Burgos-Simón et al. 2017).

The activity of these enzymes depends on the environmental conditions from where the producing microbe has been isolated. For example the enzyme activity of microbes growing at thermophilic condition will be different from that growing in saline water, high tidal marine conditions, heavy metal areas, fuel tanks or moderate conditions (Asgher et al. 2014; Margenot et al. 2018; Meng et al. 2018). Thermostable enzymes have higher stability, reaction rate and substrate solubility compared to enzymes active at moderate temperatures. Certain approaches including genetic engineering, involving alteration of genome sequence of microbes, protein engineering or attaching the enzyme to a support (immobilisation), are used to improve the stability or alter the property of enzyme to suit the industrial requirements (Puri and Verma 2013; Sharma et al. 2018).

The enzymes used in a system can be reabsorbed rather than leaving within the liquid suspension. This can be accomplished by improving the stability of enzyme which will withstand the conditions such as temperature and pH during recycling, and provide easy separation with the same biochemical property (Ahmadi et al. 2018). This approach supports clean processing technology and extends economic conversion in industries such as food and fuel (Zhang et al. 2018). In biorefinery industries for fuel and bioproduct production the recoverable enzyme significantly reduced the enzyme loading in biomass conversion; however, the recovery technologies are not mature yet (Verma et al. 2013; Binod et al. 2018; Guo et al. 2018; Nadar et al. 2018). Some of the commonly used enzymes in a biorefinery are listed in Table 11.1.

11.2 Application in Lignocellulose Biorefinery

Enzymes play a significant role in the production of biofuel from biomass, essentially due to the presence of biological material such as lignocellulose and organic wastes (Liu et al. 2017). A schematic representation of biomass biorefinery is shown in Figs. 11.1 and 11.2. The lignocellulosic conversion to biofuel and biorefinery begins with collection and transportation of biomass. The collected biomass is subjected to drying and milling to reduce its size and thereafter the actual bioprocessing begins (Gu et al. 2018). The first step of bioprocessing includes the pretreatment of biomass that can be followed with chemical and enzymatic treatment under high pressure (Chen et al. 2017; Puri et al. 2012). This step is key in the processing as it results in the breakdown of the strong chemical bonds within the structure. The efficient conversion of lignocellulosic biomass into sugar largely depends on the pretreatment step (Ashraf and Schmidt 2018). Some of the commonly used pretreatment methods include chemical (acid, alkaline, organic ionic liquid pretreatment) and biological, high-pressure and microwave pretreatments. This step breaks the recalcitrant structure of lignocellulosic biomass and opens the structure.

The lignocellulosic biomass structure is composed of lignin, cellulose, hemicellulose, ash, proteins and volatile material (Karimi and Taherzadeh 2016). The composition and rigidity of the lignocellulosic biomass vary with the class of the feedstock such as softwood, hardwood, grasses and organic waste peel. The complexity within the structure also depends on the percentage of each component, largely the presence of lignin (Kumar et al. 2015). It is responsible for bringing hardness in the structures and mainly protecting the biomass from external factors and retaining the structural strength (Shen et al. 2018).

Lignin is a complex natural polymer which is composed of phenyl units including guaiacyl (G), syringyl (S) and p-hydroxyphenyl (H). They form the outer polymeric structure and the presence of S and G units is found in hardwood whereas softwood contains G units (Ralph et al. 2004). The percentage of lignin present in

Table 11.1 Various enzymes used commercially in the biorefinery of biomass and their sources

Enzymes	Source	Industrial application	References
Cellulase	*Aspergillus/ Trichoderma* sp.	Cellulose conversion to glucose	Chandel et al. (2012)
Hemicellulases	*Aspergillus* sp.	Enzymatic hydrolysis of rice straw to glucose	Zhao et al. (2018b)
Hemicellulases (endo-β-1,4-xylanases (EC 3.2.1.8))and cellulases ((exo-β-1,4-glucanase, or cellobiohydrolase (EC 3.2.1.91); endoglucanases (EC 3.2.1.4)), β-glucosidase, carboxymethylcellulase (CMCase), xylanase, and β-xylosidase	*Lichtheimia ramose*, *Penicillium oxalicum* GZ-2, *Penicillium citrinum* YS40–5	Sugarcane bagasse, rice bran, saccharification	Garcia et al. (2018), Liao et al. (2015), Ng et al. (2010)
Cellulolytic and hemicellulolytic enzymes	*Lichtheimia ramosa* (cultured from vegetable by-product)	Conversion of cellulose and hemicellulose	Garcia et al. (2015), Gonçalves et al. (2013)
Feruloyl esterase	*Alternaria alternata* PDA1 (plant pathogen)	Conversion of vegetable substrates for the biorefinery (suitable as industrial enzyme toolbox)	García-Calvo et al. (2018)
Sucrose isomerase (SI) [EC 5.499.11]	*Protaminobacter rubrum* (bacteria)	Produce isomaltulose from sucrose	Hellmers et al. (2018)
Proteases	Fish	Depolymerise chitin from shrimp waste	Sila et al. (2015)
Lipases (Lipozyme TL IM and Novozym 435)	*Candida rugosa* and *Rhizopus oryzae*	Production of biodiesel from soy bean oil, cooking oil into biodiesel	Lee et al. (2011)
Carbonic anhydrase	Enzyme mimics using ab initio modelling	Hydrate CO_2	Sahoo et al. (2018)
Xylanases, xylosidases, arabinofuranosidases and esterases	*Penicillium* (such as *P. purpurogenum*, *P. oxalicum* and *P. funiculosum*	Hemicellulose-degrading enzymes for biofuel and biorefinery	Yang et al. (2018)

the lignocellulosic biomass also affects the efficiency of pretreatment; the presence of higher lignin percentage will make the structure rigid and more difficult to break (Chandel et al. 2018).

These lignins are mainly removed from lignocellulosic biomass through chemical or high-pressure treatment but there are several lignolytic enzymes that are capable of digesting the outer structure (Wang and Jönsson 2018). Studies have reported that white-rot fungi have the ability to degrade the highly branched structure and

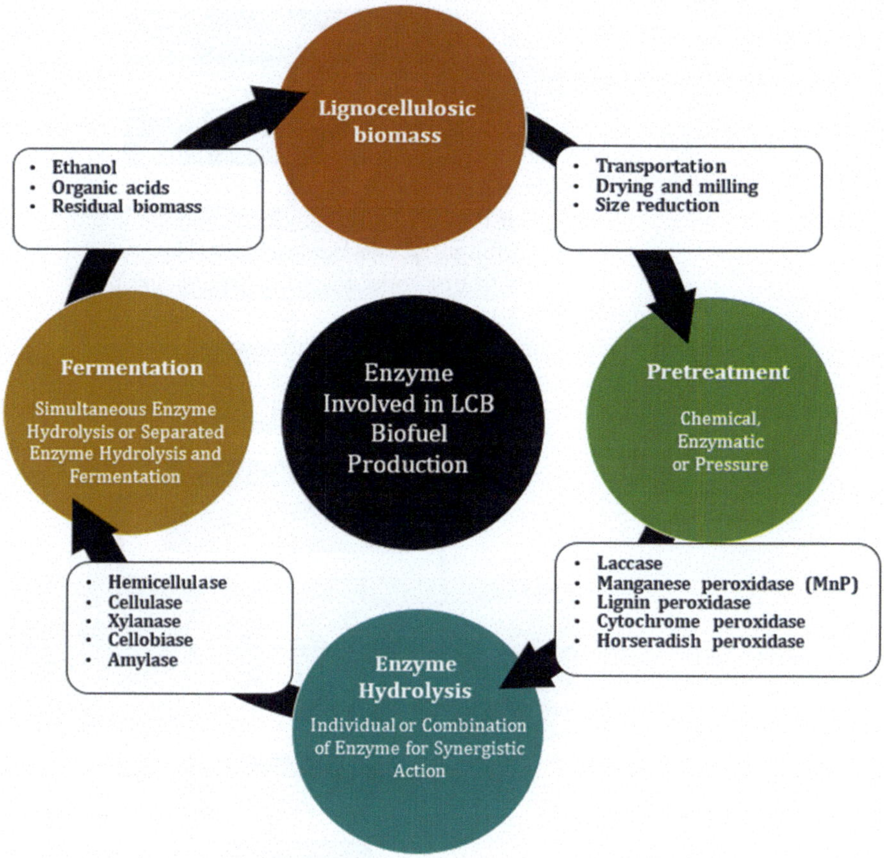

Fig. 11.1 Schematic conversion of lignocellulosic into bioproducts and role of enzymes

aromatic groups. These fungi produce extracellular oxidative enzymes such as peroxidases including manganese peroxidases (MnP), lignin peroxidases (LiP) and versatile peroxidases (VP) (de Gonzalo et al. 2016; Lambertz et al. 2016). These peroxidase enzymes are reported to efficiently degrade lignin into fragments by oxidising mediators and small oxidising agents that depolymerise lignin (Nousiainen et al. 2014). Reports have also suggested the occurrence of copper-based laccases but the composition of these enzymes varies with the fungal source and is restricted to fungi as the homologues were not observed in the genomic sequence of bacteria (Welinder 1992; Floudas et al. 2012). Fungal based oxidases also enhance the digestibility of lignin more efficiently. These include some hydrogen-based peroxidases such as glyoxal oxidases, carbohydrate oxidases and aryl alcohol oxidases. Dye-decolorizing peroxidases (DyPs, EC 1.11.1.19) were found recently that are haem-peroxidases sourced from bacteria to degrade the lignin structure (Colpa et al. 2014; Yoshida et al. 2012). Recently, studies were conducted using DyP-type peroxidases from bacteria and the ability to oxidise was lower in comparison to fungal based enzymes (Santos et al. 2014).

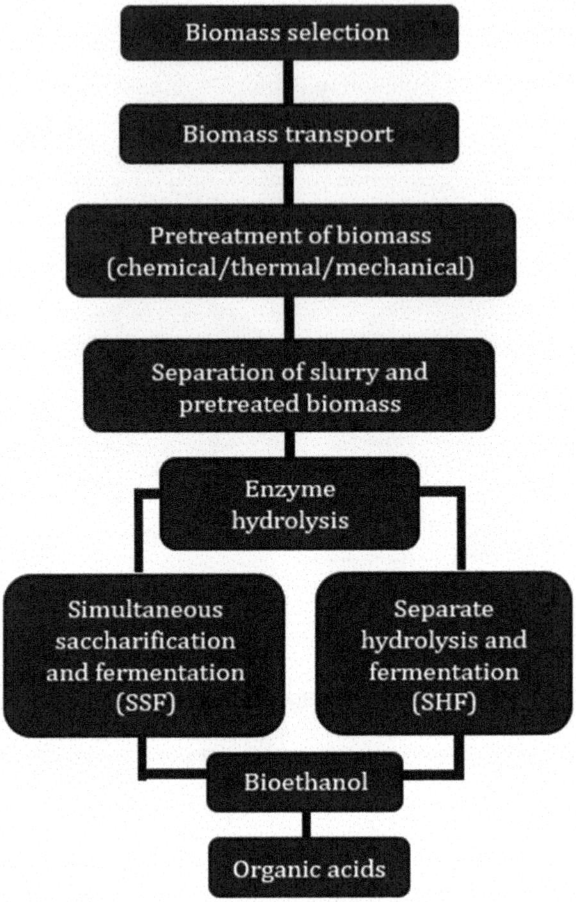

Fig. 11.2 Schematic process flow of biomass to biofuel

After pretreatment, breakdown of cellulose structure into simple reducing sugars through enzyme hydrolysis is achieved. Cellulase is an enzyme that is produced from microorganisms including bacteria and fungi (Xue et al. 2017). Cellulose, a polymer, made of a chain of glucose units that are linked with 1,4-β-D-glucosidic bond to each other, breaks down during enzyme hydrolysis. This process occurs with the synergistic action of cellulose-degrading enzymes including endo-β-1,4-glucanases, exo-β-1,4-glucanases (or cellobiohydrolase) and β-glucosidases (Selvam et al. 2017). The degradation of cellulose begins with the action of endo-β-1,4-glucanases that cuts the β-1,4-glucosidic linkage polymer into single-chain structure or oligosaccharides with new chain ends. The exo-β-1,4-glucanases (also known as cellobiohydrolase) further shorten the polymeric chain by cleaving reducing and non-reducing ends of oligosaccharides producing either single glucose unit or cellobiose (Yang et al. 2018). Cellobiase or β-glucosidase digests cellobiose into glucose but at times cellobiose acts as an inhibitor during the enzymatic hydrolysis.

Table 11.2 Microbial strains used in biorefinery to convert organic wastes into bioproducts

Source	Industrial application	References
Kluyveromyces marxianus	Ethanol production	Sousa et al. (2018)
Enterobacter, *Bacillus*, and *Clostridium*	Biohydrogen production	Mishra, et al. (2019)
Burkholderia sacchari DSM 17165	Convert glucose, xylose, arabinose and other reducing sugars into poly-3-hydroxybutyrate	Cesário et al. (2014)
Klebsiella oxytoca, *K. pneumoniae*, *Citrobacter freundii*, *Enterobacter agglomerans* and *C. butyricum*	Fermentation of glycerol in 1,3-propanediol and 2,3-butanediol	Lin et al. (2005), Syu (2001)

Some of the strains that are reported in literature and capable of fermenting the waste into bioproducts are described in Table 11.2.

There are other polysaccharides including pectin and hemicellulose and their presence varies with the composition of biomass. These polysaccharides are depolymerised using xylanase, pectinase and galactosidase; despite having cellulase into the system addition of these enzymes is required to enhance the synergistic action and conversion yield (Li et al. 2018c). Therefore, an enzyme complex having the efficiency to convert polysaccharide to reducing sugars is highly regarded in lignocellulosic conversion.

11.3 Factors Affecting Biofuel Production

Various technical and economic challenges are associated with the processing of second-generation biofuel. First barrier is encountered during the pretreatment of the lignocellulosic biomass that includes the efficient conversion and minimal loss of biomass during the pretreatment and washing. Second is the loading of the enzyme(s) which can alter the efficiency of pretreatment; additionally during the enzyme hydrolysis various inhibitory compounds such as furfurals and acetic acid are produced which compromise enzyme activity leading to incomplete hydrolysis (Sekoai et al. 2019). During the downstream processing these furfurals interfere with the conversion of hexose and pentose sugars into ethanol; additionally they can hinder the growth of yeast which converts sugars to bioethanol (Wang et al. 2018). These issues often lead to escalation of the production costs as troubleshooting such issues can at times be expensive.

The cost of biomass also affects industrial-scale biofuel production. The availability of cost-efficient lignocellulosic material containing a large amount of cellulose and hemicellulose with minimal quantity of lignin will enhance the scale-up process. With the increase of lignocellulosic rigidity, the degree of pretreatment will also increase (Dhyani and Bhaskar 2018). The composition of biomass varies in hardwood, softwood and grasses and henceforth its efficient conversion into useful products varies depending on the plant material (Kucharska et al. 2018). Moreover, the bioproducts that are produced during different stages of biofuel also count in the production and expense. Bioproducts such as organic acids, furfurals, cellulose fibres, alcohols (butanol and ethanol) and biomethane either are produced at different stages of lignocellulosic conversion or can be produced by modifying conversion pathway. These bioproducts at times hinder the efficient conversion to the main product and affect the final yield of the product. Therefore, a close monitoring of the conversion system is required to control the reaction to achieve higher conversion yield (Kumari and Singh 2018). Biocatalyis or co-immobilised enzyme or a cocktail of enzymes or co-culturing is also used in the conversion system to improve the product yield (Abraham et al. 2014; Hwangbo et al. 2019; Wang et al. 2019). The conversion of pretreated biomass into sugar hydrolysate occurs at a much higher rate in the presence of multiple biocatalyst/enzyme/culture system than in single-catalyst system.

11.4 Biorefinery and Enzyme Application

The growing demand of fossil fuel and depleting reserves has generated the concept of biorefinery from renewable sources (Mondou et al. 2018). With the availability of resources and development of technology, the biorefinery has been categorised into three generations: *first*, *second* and *third* (Binod et al. 2018; Leong et al. 2018). In first generation the source of raw material is food crops such as corn and sugarcane; but the increasing competition with food supply reduced its demand (Saladini et al. 2016). The utilisation of lignocellulosic wastes (lignocellulosic feedstock, forest residues, agricultural wastes and organic wastes) leads to second-generation technology and is considered more sustainable compared to first generation (Mitkidis et al. 2018; Pontes et al. 2018). The easy maintenance and less technological challenges in working with algal biomass have advanced to the next level and are considered as third generation of biofuel. The ability of algal biomass to produce high-value products is getting more demand in industry.

Algae are classified into two categories, macroalgae and micro-algae, and based on their structural composition and efficiency to produce a product they have different applications in industry (Saravanan et al. 2018; Veeranan et al. 2018). As a third-generation biomass their contribution to renewable energy generation includes the production of biohydrogen, biomethane, biodiesel and bioethanol (Leong et al. 2018). This is mainly achieved by micro-algae due to their efficiency in conversion and product development. Despite having less complex structure compared to terrestrial plants, algal biomass is often considered for pretreatment prior to its conversion to achieve higher fuel production. There are several pretreatment methods applied on the algal biomass but chemical or mechanical is the commonly applied method to improve the productivity (Leite et al. 2013). Enzyme-based pretreatment is also reported but the cost and specificity of enzyme limits its application as micro-algae have a biopolymeric structure that requires a synergistic action of enzyme cocktail to disrupt the structure (Fortier et al. 2017).

The micro-algae are composed of cellulose, hemicellulose and chitin. The absence of lignin makes the structure less complex compared to terrestrial plants and allows easier cell disruption (Abomohra et al. 2018; Tu et al. 2018). However, macroalgae are composed of polysaccharides such as alginate, cellulose, fucoidan, laminarin, polyphenol, steroids, pigment (depending on the species: green algae—chlorophyll, red algae—carotenoids, brown algae—phlorotannin), vitamins and minerals (Sudhakar et al. 2018; Vuong et al. 2018). Seaweeds are the common macroalgae that have unique chemical structure and most of its structural compounds (polysaccharides and pigments) are reported to have health benefits (Chia et al. 2018). Both micro- and macroalgae have high demand in industry for their polysaccharides and pigments as they have been proved for having high medicinal and health benefits (Zhao et al. 2018a). To disrupt the structure of biomass several commercial enzymes are used that include cellulase, xylanase, amylase, lipase, amyloglucosidase and protease. Some enzymes used for the digestion are extracellular while others are intracellular depending on the target and product demand.

Furthermore, the substrate specificity and target selectivity on the substrates make the enzymatic process more efficient and preferable than chemical treatment (Meller et al. 2017). Despite being cleaner it has few disadvantages such as cost which varies with its dosage in the processing and difficulty to recover from the reaction suspension. These challenges have brought the concept of reusing the enzymes (Terrasan et al. 2019). Enzyme immobilisation is a technique to attach them within a matrix or encapsulate within the support. This is achieved to improve the biocatalytic properties of enzyme such as molecule stability, resistance to higher temperature and shelf life and at times it can be reused (Voběrková et al. 2018). Immobilisation can help to reduce the cost of enzyme by reusing it in the system multiple times and some of the methods to immobilise enzymes are mentioned in Fig. 11.3.

The efficiency of immobilised enzyme preparations is highly dependent on the enzyme, attaching matrix and the nature of binding that should bind the enzyme without destroying its activity (Abraham et al. 2014; Singh et al. 2015). The immobilisation leads to improved chemical, mechanical, biochemical and kinetic properties and it depends on the reactive groups on the support and enzyme (Muley et al. 2018). The support used for the immobilisation can be a nanoparticle, metal organic frame, gels and nanotubes. Metals such as silica, carbon, gold, silver and zinc are commercially available and commonly used to prepare support (Liu et al. 2018).

In recent years metal-organic frameworks (MOFs) have gained interest in research due to the ability to hold high-value material within its 3-dimensional porous cage-like structure (Hu et al. 2018). The structure provides porosity and crystallinity due to its formation with metal cations and organic ligands. MOFs can be designed in different structures, pore size, surface area and functional groups. Due to the flexibility in the pore size (~10 nm), structure (cage or channel) and surface area they are successful in immobilising enzymes and bioactive drugs. This framework provides improved stability against chemicals, pH, temperature variation, concentration, aggregation and mechanical stress (Cui et al. 2018).

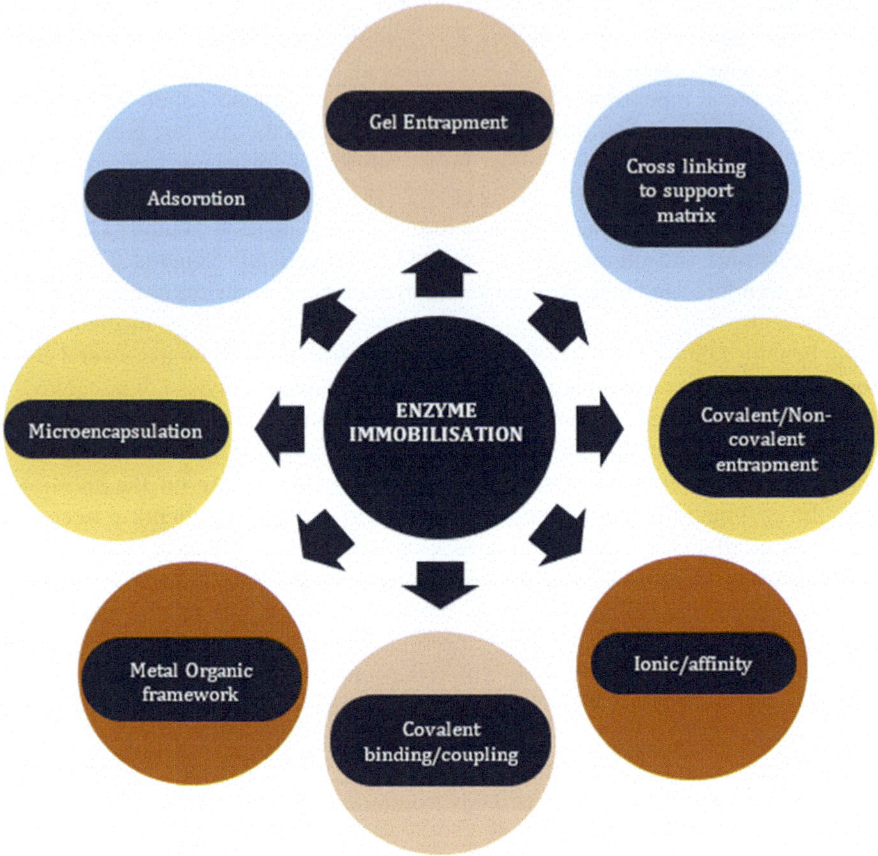

Fig. 11.3 Various methods used to immobilise an enzyme

11.5 Conclusion

The conversion of lignocellulosic biomass into high-value-added products requires an overall economic production process. Even though the biomass processing encounters several issues, the reduction of processing steps and further improvement in enzyme hydrolysis technology should be adopted. The commercial production of highly active enzymes with the ability to be reused can significantly bring a process shift in the current technology. Enzymes will replace chemicals in the biorefinery for a clean processing and for protecting the environment from toxic chemicals in the near future. This transformation will bring a lot of opportunities for the commercial production of highly active enzymes with multiple targets having easy separation and recycling properties.

References

Abomohra AE-F, El-Naggar AH, Baeshen AA (2018) Potential of macroalgae for biodiesel production: screening and evaluation studies. J Biosci Bioeng 125(2):231–237

Abraham RE, Verma M, Barrow CJ, Puri M (2014) Suitability of magnetic nanoparticle immobilised cellulases in enhancing enzymatic saccharification of pretreated hemp biomass. Biotechnol Biofuels 7:90

Ahmadi K, Razavi BS, Maharjan M, Kuzyakov Y, Kostka SJ, Carminati A, Zarebanadkouki M (2018) Effects of rhizosphere wettability on microbial biomass, enzyme activities and localization. Rhizosphere 7:35–42

Arsalan A, Younus H (2018) Enzymes and nanoparticles: modulation of enzymatic activity via nanoparticles. Int J Biol Macromol 118:1833–1847

Asgher M, Shahid M, Kamal S, Iqbal HMN (2014) Recent trends and valorization of immobilization strategies and ligninolytic enzymes by industrial biotechnology. J Mol Catal B: Enzym 101:56–66

Ashraf MT, Schmidt JE (2018) Process simulation and economic assessment of hydrothermal pretreatment and enzymatic hydrolysis of multi-feedstock lignocellulose – separate vs. combined processing. Bioresour Technol 249:835–843

Bernal C, Rodríguez K, Martínez R (2018) Integrating enzyme immobilization and protein engineering: an alternative path for the development of novel and improved industrial biocatalysts. Biotechnol Adv 36(5):1470–1480

Binod P, Gnansounou E, Sindhu R, Pandey A (2018) Enzymes for second generation biofuels: recent developments and future perspectives. Bioresour Technol Rep 5:317–325

Burgos-Simón C, Carrasco-Correa EJ, Beneito-Cambra M, Ramis-Ramos G, Simó-Alfonso EF (2017) Recognition and alignment of variables from UV–vis chromatograms and application to industrial enzyme digests classification. Chemom Intel Lab Syst 165:46–55

Cesário MT, Raposo RS, Almeida MCMD, van Keulen F, Ferreira BS, Telo JP, Fonseca MMRD (2014) Production of poly(3-hydroxybutyrate-co4-hydroxybutyrate) by *Burkholderia sacchari* using wheat straw hydrolysates and gamma-butyrolactone. Int J Biol Macromol 71:59–67

Chandel AK, Chandrasekhar G, Silva MB, Silva SSD (2012) The realm of cellulases in biorefinery development. Crit Rev Biotechnol 32(3):187–202

Chandel AK, Garlapati VK, Singh AK, Antunes FAF, da Silva SS (2018) The path forward for lignocellulose biorefineries: bottlenecks, solutions, and perspective on commercialization. Bioresour Technol 264:370–381

Chen H, Liu J, Chang X, Chen D, Xue Y, Liu P, Lin H, Han S (2017) A review on the pretreatment of lignocellulose for high-value chemicals. Fuel Process Technol 160:196–206

Chia SR, Show PL, Phang S-M, Ling TC, Ong HC (2018) Sustainable approach in phlorotannin recovery from macroalgae. J Biosci Bioeng 126(2):220–225

Choi J-M, Han S-S, Kim H-S (2015) Industrial applications of enzyme biocatalysis: current status and future aspects. Biotechnol Adv 33(7):1443–1454

Colpa DI, Fraaije MW, van Bloois E (2014) DyP-type peroxidases: a promising and versatile class of enzymes. J Ind Microbiol Biotechnol 41(1):1–7

Cui J, Ren S, Sun B, Jia S (2018) Optimization protocols and improved strategies for metal-organic frameworks for immobilizing enzymes: current development and future challenges. Coord Chem Rev 370:22–41

de Gonzalo G, Colpa DI, Habib MHM, Fraaije MW (2016) Bacterial enzymes involved in lignin degradation. J Biotechnol 236:110–119

Dhyani V, Bhaskar T (2018) A comprehensive review on the pyrolysis of lignocellulosic biomass. Renew Energy 129:695–716

Ehtesabi H, Bagheri Z, Eskandari F, Ahadian MM (2018) Molecular interaction between three-dimensional graphene aerogel and enzyme solution: effect on enzyme structure and function. J Mol Liq 265:565–571

Fatima SW, Khare SK (2018) Current insight and futuristic vistas of microbial transglutaminase in nutraceutical industry. Microbiol Res 215:7–14

Floudas D, Binder M, Riley R, Barry K, Blanchette RA, Henrissat B, Martínez AT, Otillar R, Spatafora JW, Yadav JS, Aerts A, Benoit I, Boyd A, Carlson A, Copeland A, Coutinho PM, de Vries RP, Ferreira P, Findley K, Foster B, Gaskell J, Glotzer D, Górecki P, Heitman J, Hesse C, Hori C, Igarashi K, Jurgens JA, Kallen N, Kersten P, Kohler A, Kües U, Kumar TKA, Kuo A, LaButti K, Larrondo LF, Lindquist E, Ling A, Lombard V, Lucas S, Lundell T, Martin R, McLaughlin DJ, Morgenstern I, Morin E, Murat C, Nagy LG, Nolan M, Ohm RA, Patyshakuliyeva A, Rokas A, Ruiz-Dueñas FJ, Sabat G, Salamov A, Samejima M, Schmutz J, Slot JC, St. John F, Stenlid J, Sun H, Sun S, Syed K, Tsang A, Wiebenga A, Young D, Pisabarro A, Eastwood DC, Martin F, Cullen D, Grigoriev IV, Hibbett DS (2012) The Paleozoic origin of enzymatic lignin decomposition reconstructed from 31 fungal genomes. Science 336(6089):1715–1719

Fortier M-OP, Roberts GW, Stagg-Williams SM, Sturm BSM (2017) Determination of the life cycle climate change impacts of land use and albedo change in algal biofuel production. Algal Res 28:270–281

Garcia NFL, da Silva Santos FR, Gonçalves FA, da Paz MF, Fonseca GG, Leite RSR (2015) Production of β-glucosidase on solid-state fermentation by *Lichtheimia ramosa* in agroindustrial residues: characterization and catalytic properties of the enzymatic extract. Electron J Biotechnol 18(4):314–319

Garcia NFL, Santos FRDS, Bocchini DA, Paz MFD, Fonseca GG, Leite RSR (2018) Catalytic properties of cellulases and hemicellulases produced by Lichtheimia ramosa: potential for sugarcane bagasse saccharification. Ind Crop Prod 122:49–56

García-Calvo L, Ullán RV, Fernández-Aguado M, García-Lino AM, Balaña-Fouce R, Barreiro C (2018) Secreted protein extract analyses present the plant pathogen *Alternaria alternata* as a suitable industrial enzyme toolbox. J Proteome 177:48–64

Gonçalves F, Leite R, Rodrigues A, Sanjinez-Argandoña E, Fonseca G (2013) Isolation, identification and characterization of a novel high level β-glucosidaseproducing *Lichtheimia ramosa* strain. Biocatal Agric Biotechnol 2:377–384

Gu H, An R, Bao J (2018) Pretreatment refining leads to constant particle size distribution of lignocellulose biomass in enzymatic hydrolysis. Chem Eng J 352:198–205

Guo H, Chang Y, Lee D-J (2018) Enzymatic saccharification of lignocellulosic biorefinery: research focuses. Bioresour Technol 252:198–215

Hellmers F, Takors R, Thum O (2018) Robust enzyme immobilizates for industrial isomalt production. Mol Cat 445:293–298

Hu Y, Dai L, Liu D, Du W, Wang Y (2018) Progress & prospect of metal-organic frameworks (MOFs) for enzyme immobilization (enzyme/MOFs). Renew Sustain Energy Rev 91:793–801

Huang G, Chen S, Dai C, Sun L, Sun W, Tang Y, Xiong F, He R, Ma H (2017) Effects of ultrasound on microbial growth and enzyme activity. Ultrason Sonochem 37:144–149

Hwangbo M, Tran JL, Chu K-H (2019) Effective one-step saccharification of lignocellulosic biomass using magnetite-biocatalysts containing saccharifying enzymes. Sci Total Environ 647:806–813

Karimi K, Taherzadeh MJ (2016) A critical review of analytical methods in pretreatment of lignocelluloses: composition, imaging, and crystallinity. Bioresour Technol 200:1008–1018

Kucharska K, Hołowacz I, Konopacka-Łyskawa D, Rybarczyk P, Kamiński M (2018) Key issues in modeling and optimization of lignocellulosic biomass fermentative conversion to gaseous biofuels. Renew Energy 129:384–408

Kumar G, Bakonyi P, Periyasamy S, Kim SH, Nemestóthy N, Bélafi-Bakó K (2015) Lignocellulose biohydrogen: practical challenges and recent progress. Renew Sustain Energy Rev 44:728–737

Kumari D, Singh R (2018) Pretreatment of lignocellulosic wastes for biofuel production: a critical review. Renew Sustain Energy Rev 90:877–891

Lambertz C, Ece S, Fischer R, Commandeur U (2016) Progress and obstacles in the production and application of recombinant lignin-degrading peroxidases. Bioengineered 7(3):145–154

Lee JH, Kim SB, Kang SW, Song YS, Park C, Han SO, Kim SW (2011) Biodiesel production by a mixture of *Candida rugosa* and *Rhizopus oryzae* lipases using a supercritical carbon dioxide process. Bioresour Technol 102(2):2105–2108

Leite GB, Abdelaziz AEM, Hallenbeck PC (2013) Algal biofuels: challenges and opportunities. Bioresour Technol 145:134–141

Leong W-H, Lim J-W, Lam M-K, Uemura Y, Ho Y-C (2018) Third generation biofuels: a nutritional perspective in enhancing microbial lipid production. Renew Sustain Energy Rev 91:950–961

Li G, Wang J-b, Reetz MT (2018a) Biocatalysts for the pharmaceutical industry created by structure-guided directed evolution of stereoselective enzymes. Bioorg Med Chem 26(7):1241–1251

Li K, Wang J, He Y, Abdulrazaq MA, Yan Y (2018b) Carbon nanotube-lipase hybrid nanoflowers with enhanced enzyme activity and enantioselectivity. J Biotechnol 281:87–98

Li Q, Loman AA, Callow NV, Islam SMM, Ju L-K (2018c) Leveraging pH profiles to direct enzyme production (cellulase, xylanase, polygalacturonase, pectinase, α-galactosidase, and invertase) by *Aspergillus foetidus*. Biochem Eng J 137:247–254

Liao H, Zheng H, Li S, Wei Z, Mei X, Ma H, Shen Q, Xu Y (2015) Functional diversity and properties of multiple xylanases from *Penicillium oxalicum* GZ-2. Sci Rep 5:12631–12631

Lin R, Liu H, Hao J, Cheng K, Liu D (2005) Enhancement of 1,3-propanediol production by *Klebsiella pneumoniae* with fumarate addition. Biotechnol Lett 27(22):1755–1759

Liu Y, Guo L, Wang L, Zhan W, Zhou H (2017) Irradiation pretreatment facilitates the achievement of high total sugars concentration from lignocellulose biomass. Bioresour Technol 232:270–277

Liu D-M, Chen J, Shi Y-P (2018) Advances on methods and easy separated support materials for enzymes immobilization. TrAC Trends Anal Chem 102:332–342

Llorent-Martínez EJ, Ortega-Barrales P, Zengin G, Mocan A, Simirgiotis MJ, Ceylan R, Uysal S, Aktumsek A (2017) Evaluation of antioxidant potential, enzyme inhibition activity and phenolic profile of *Lathyrus cicera* and *Lathyrus digitatus*: potential sources of bioactive compounds for the food industry. Food Chem Toxicol 107:609–619

Margenot AJ, Nakayama Y, Parikh SJ (2018) Methodological recommendations for optimizing assays of enzyme activities in soil samples. Soil Biol Biochem 125:350–360

Meller K, Szumski M, Buszewski B (2017) Microfluidic reactors with immobilized enzymes—characterization, dividing, perspectives. Sens Actuators B 244:84–106

Meng L, Li W, Bao M, Sun P (2018) Promoting the treatment of crude oil alkane pollution through the study of enzyme activity. Int J Biol Macromol 119:708–716

Mishra P, Krishnan S, Rana S, Singh L, Sakinah M, Ab Wahid Z (2019) Outlook of fermentative hydrogen production techniques: an overview of dark, photo and integrated dark-photo fermentative approach to biomass. Energ Strat Rev 24:27–37

Mitkidis G, Magoutas A, Kitsios F (2018) Market and economic feasibility analysis for the implementation of 2nd generation biofuels in Greece. Energ Strat Rev 19:85–98

Mondou M, Skogstad G, Bognar J (2018) What are the prospects for deploying advanced biofuels in Canada? Biomass Bioenergy 116:171–179

Muley AB, Thorat AS, Singhal RS, Harinath Babu K (2018) A tri-enzyme co-immobilized magnetic complex: process details, kinetics, thermodynamics and applications. Int J Biol Macromol 118:1781–1795

Nadar SS, Rao P, Rathod VK (2018) Enzyme assisted extraction of biomolecules as an approach to novel extraction technology: a review. Food Res Int 108:309–330

Ng IS, Li C-W, Chan S-P, Chir J-L, Chen PT, Tong C-G, Yu S-M, Ho T-HD (2010) High-level production of a thermoacidophilic β-glucosidase from *Penicillium citrinum* YS40-5 by solid-state fermentation with rice bran. Bioresour Technol 101(4):1310–1317

Nousiainen P, Kontro J, Manner H, Hatakka A, Sipilä J (2014) Phenolic mediators enhance the manganese peroxidase catalyzed oxidation of recalcitrant lignin model compounds and synthetic lignin. Fungal Genet Biol 72:137–149

Park MS, Oh S-Y, Lee S, Eimes JA, Lim YW (2018) Fungal diversity and enzyme activity associated with sailfin sandfish egg masses in Korea. Fungal Ecol 34:1–9

Patel AK, Singhania RR, Pandey A (2016) Novel enzymatic processes applied to the food industry. Curr Opin Food Sci 7:64–72
Pontes R, Romaní A, Michelin M, Domingues L, Teixeira J, Nunes J (2018) Comparative autohydrolysis study of two mixtures of forest and marginal land resources for co-production of biofuels and value-added compounds. Renew Energy 128:20–29
Puri M, Verma ML (2013) Enzyme immobilisation on nanomaterials for biofuel production. Trends Biotechnol 31:215–216
Puri M, Abraham RE, Barrow CJ (2012) Biofuel production: prospects, challenges and feedstocks. Renew Sustain Energy Rev 16:6022–6031
Ralph J, Lundquist K, Brunow G, Lu F, Kim H, Schatz PF, Marita JM, Hatfield RD, Ralph SA, Christensen JH, Boerjan W (2004) Lignins: natural polymers from oxidative coupling of 4-hydroxyphenyl-propanoids. Phytochem Rev 3(1):29–60
Rosenthal K, Lütz S (2018) Recent developments and challenges of biocatalytic processes in the pharmaceutical industry. Curr Opin Green Sustain Chem 11:58–64
Sahoo PC, Kumar M, Puri SK, Ramakumar SSV (2018) Enzyme inspired complexes for industrial CO2 capture: opportunities and challenges. J CO2 Util 24:419–429
Saladini F, Patrizi N, Pulselli FM, Marchettini N, Bastianoni S (2016) Guidelines for energy evaluation of first, second and third generation biofuels. Renew Sustain Energy Rev 66:221–227
Santos A, Mendes S, Brissos V, Martins LO (2014) New dye-decolorizing peroxidases from *Bacillus subtilis* and *Pseudomonas putida* MET94: towards biotechnological applications. Appl Microbiol Biotechnol 98(5):2053–2065
Saravanan AP, Mathimani T, Deviram G, Rajendran K, Pugazhendhi A (2018) Biofuel policy in India: a review of policy barriers in sustainable marketing of biofuel. J Clean Prod 193:734–747
Schimel J, Becerra CA, Blankinship J (2017) Estimating decay dynamics for enzyme activities in soils from different ecosystems. Soil Biol Biochem 114:5–11
Sekoai PT, Ouma CNM, du Preez SP, Modisha P, Engelbrecht N, Bessarabov DG, Ghimire A (2019) Application of nanoparticles in biofuels: an overview. Fuel 237:380–397
Selvam K, Senbagam D, Selvankumar T, Sudhakar C, Kamala-Kannan S, Senthilkumar B, Govarthanan M (2017) Cellulase enzyme: homology modeling, binding site identification and molecular docking. J Mol Struct 1150:61–67
Sharma B, Dangi AK, Shukla P (2018) Contemporary enzyme based technologies for bioremediation: a review. J Environ Manage 210:10–22
Shen Z, Zhang K, Si M, Liu M, Zhuo S, Liu D, Ren L, Yan X, Shi Y (2018) Synergy of lignocelluloses pretreatment by sodium carbonate and bacterium to enhance enzymatic hydrolysis of rice straw. Bioresour Technol 249:154–160
Sila A, Haddar A, Sayari N, Bougatef A (2015) Digestive alkaline proteases from the *Tunisian barbel* (*Barbus callensis*): characterization and application as a detergent additive, in chicken feather-degradation and as a dehairing agent. Iran J Fish Sci 14:672–686
Singh P, Morris H, Tivanski AV, Kohen A (2015) Determination of concentration and activity of immobilized enzymes. Anal Biochem 484:169–172
Sousa CCD, Gonçalves GTI, Falleiros LNSS (2018) Ethanol production using agroindustrial residues as fermentation substrates by *Kluyveromyces marxianus*. Ind Biotechnol 14(6):308–314
Spohner SC, Müller H, Quitmann H, Czermak P (2015) Expression of enzymes for the usage in food and feed industry with *Pichia pastoris*. J Biotechnol 202:118–134
Sudhakar K, Mamat R, Samykano M, Azmi WH, Ishak WFW, Yusaf T (2018) An overview of marine macroalgae as bioresource. Renew Sustain Energy Rev 91:165–179
Syu M-J (2001) Biological production of 2,3-butanediol. Appl Microbiol Biotechnol 55(1):10–18
Terrasan CRF, de Morais JWG, Contesini FJ (2019) Enzyme immobilization for oligosaccharide production. In: Melton L, Shahidi F, Varelis P (eds) Encyclopedia of food chemistry. Academic Press, Oxford, pp 415–423
Tu Q, Eckelman M, Zimmerman JB (2018) Harmonized algal biofuel life cycle assessment studies enable direct process train comparison. Appl Energy 224:494–509
Veeranan T, Kasirajaan R, Gurunathan B, Sahadevan R (2018) A novel approach for extraction of algal oil from marine macroalgae *Ulva fasciata*. Renew Energy 127:64–73

Verma ML, Barrow CJ, Puri M (2013) Nanobiotechnology as a novel paradigm for enzyme immobilisation and stabilisation with potential applications in biofuel production. Appl Microbiol Biotechnol 97:23–39

Voběrková S, Solčány V, Vršanská M, Adam V (2018) Immobilization of ligninolytic enzymes from white-rot fungi in cross-linked aggregates. Chemosphere 202:694–707

Vuong D, Kaplan M, Lacey HJ, Crombie A, Lacey E, Piggott AM (2018) A study of the chemical diversity of macroalgae from south eastern Australia. Fitoterapia 126:53–64

Wang Z, Jönsson LJ (2018) Comparison of catalytically non-productive adsorption of fungal proteins to lignins and pseudo-lignin using isobaric mass tagging. Bioresour Technol 268:393–401

Wang S, Sun X, Yuan Q (2018) Strategies for enhancing microbial tolerance to inhibitors for biofuel production: a review. Bioresour Technol 258:302–309

Wang J, Chio C, Chen X, Su E, Cao F, Jin Y, Qin W (2019) Efficient saccharification of agave biomass using Aspergillus niger produced low-cost enzyme cocktail with hyperactive pectinase activity. Bioresour Technol 272:26–33

Welinder KG (1992) Superfamily of plant, fungal and bacterial peroxidases. Curr Opin Struct Biol 2(3):388–393

Xue D-S, Liang L-y, Zheng G, Lin D-Q, Zhang Q-L, Yao S-J (2017) Expression of *Piromyces rhizinflata* cellulase in marine *Aspergillus niger* to enhance halostable cellulase activity by adjusting enzyme-composition. Biochem Eng J 117:156–161

Yang Y, Yang J, Liu J, Wang R, Liu L, Wang F, Yuan H (2018) The composition of accessory enzymes of *Penicillium chrysogenum* P33 revealed by secretome and synergistic effects with commercial cellulase on lignocellulose hydrolysis. Bioresour Technol 257:54–61

Yoshida T, Tsuge H, Hisabori T, Sugano Y (2012) Crystal structures of dye-decolorizing peroxidase with ascorbic acid and 2,6-dimethoxyphenol. FEBS Lett 586(24):4351–4356

Zhang Y, He S, Simpson BK (2018) Enzymes in food bioprocessing—novel food enzymes, applications, and related techniques. Curr Opin Food Sci 19:30–35

Zhao C, Yang C, Liu B, Lin L, Sarker SD, Nahar L, Yu H, Cao H, Xiao J (2018a) Bioactive compounds from marine macroalgae and their hypoglycemic benefits. Trends Food Sci Technol 72:1–12

Zhao J, Dong Z, Li J, Chen L, Bai Y, Jia Y, Shao T (2018b) Ensiling as pretreatment of rice straw: the effect of hemicellulase and *Lactobacillus plantar*um on hemicellulose degradation and cellulose conversion. Bioresour Technol 266:158–165

Microbial Cholesterol Oxidase: Industrial Applications

12

Akanksha, Vaibhav Mishra, and Kavindra Kumar Kesari

Abstract

Cholesterol oxidase (ChOx) belongs to the oxidoreductase family and catalyzes the oxidation of cholesterol compounds. ChOx produced by bacteria, streptomyces, fungi, higher plants, and insects is being used in pharmaceutical formulations, agriculture, and waste management. Nowadays, application of ChOx in the detection of clinical diseases has also been reported. The industrial use and demand for ChOx have increased due to its effective and accurate functionality of bioconversion of cholesterol compounds into enantiomerically pure compounds used in various industries. Furthermore, ChOx has also emerged as an insecticide and showed great potential in replacing conventional chemical agents used in the control of cotton pest. In this chapter, we accentuate on the structure, sources, production, and commercial applications of ChOx.

12.1 Introduction

Enzyme technology has emerged as the fastest growing branch in the field of industrial biotechnology. The countless efforts and vigorous research on microbial system physiology have enabled the wide use of microbial enzymes in different industries. According to an estimate more than 4000 enzymes have been discovered and amongst them roughly 200 are microbial in origin (Li et al. 2012). The extraordinary catalytic

Akanksha
Department of Microbiology, CSIR-Central Drug Research Institute, Lucknow, India

V. Mishra
Department of Biochemistry, King George Medical University, Lucknow, India

K. K. Kesari (✉)
Department of Applied Physics, Aalto University, Espoo, Finland

N. K. Arora et al. (eds.), *Microbial Enzymes: Roles and Applications in Industries*, Microorganisms for Sustainability 11,
https://doi.org/10.1007/978-981-15-1710-5_12

power, high degree of substrate specificity, and tolerance towards pH and temperature changes make microbial enzymes a good choice in industries. That is why enzymes like glucose oxidase, streptokinase, cholesterol oxidase, cellulase, lipase, and protease of microbial in origin are being used in the food, dairy, textile, biomedical, and agriculture industries in large quantities (www.specialtyenzymes.com). Presently, at a global level, a vast array of enzymes are floating in the market and actively contributing to the economy. DuPont and Novozymes accounted for 21% and 47% of market share of enzymes, respectively, in the year 2011 (Dewan 2011). In the same year, enzymes accounted for $1.2 billion of sale and this was further expected to rise to $2.2 billion by 2016 (Dewan 2012). Furthermore, in the same year, the food processing market made a turnover of $1.3 billion through enzymes (Prakash et al. 2013) and is expected to achieve about $4.1 billion by 2024 (www.marketwatch.com).

In the past 20 years, microbial cholesterol oxidases (ChOx) have gained attention, mainly due to its large use in biomedical field for determining free and bound cholesterol compounds (Salva et al. 1999). In clinical laboratories ChOx are being used in bile acid biosynthesis, detection of serum cholesterol, assessment of atherosclerosis, coronary heart disease, and other lipid disorders related with the risk of heart attack and thrombosis (Pollegioni et al. 2009; El-Naggar et al. 2017). For diagnostic purposes, ChOx is required for the detection or interruption of cell membrane cholesterol that is important for the maintenance of cell membrane structure, fluidity, permeability, and protein functions in eukaryotes (Espenshade and Hughes 2007).

The microbial production of ChOx is relatively simple and economically sustainable. North America, Europe, Japan, and China are the major countries which are manufacturing ChOx and making billions per year (https://www.pioneerreports.com/report/global-cholesterol-oxidase-market-research-report-2018/342260). This chapter is targeted to discuss the structure, sources, production, and application of ChOx enzymes.

12.2 Cholesterol Oxidase (ChOx)

ChOx catalyzes the oxidation of the most important steroid cholesterol by using electron acceptor (oxygen) in cholest-5-en-3-one and subsequent isomerization in cholest-4-en-3-one and form hydrogen peroxide (Srivastava et al. 2018a, b). ChOx enzyme belongs to the oxidoreductase family, which acts on the donor CHOH group in the presence of an electron acceptor. ChOx is also named as cholesterol: oxygen oxidoreductases, cholesterol-O_2 oxidoreductases, 3β-hydroxysteroid oxidoreductases, and 3β-hydroxysteroid: oxygen oxidoreductases. Additionally, ChOx is a monomeric flavoprotein that contains one molecule of flavin adenine dinucleotide (FAD) as a prosthetic group (Doukyu 2009; Srivastava et al. 2018a, b). The enzymes containing FAD have a consensus sequence of repeating glycine residues GXGXXG (Eventoff and Rossmann 1975; Ohlsson et al. 1974). Near-N-terminal FAD permits binding of the charged diphosphate moiety of the nucleotide (Hol et al. 1978). There are two distinct types of ChOx which belong to different protein families with no significant sequence homology and hence also differ in terms of structure, folding, and kinetic and thermodynamic properties (Chen et al. 2007).

12.2.1 Class I ChOx

It contains FAD redox cofactor which is non-covalently bound to the enzyme and belongs to the glucose-methanol-choline (GMC) oxidoreductase family. Mostly, Class I ChOx are found in actinomycetes such as *Streptomyces* sp. (Fig. 12.1). The sequence analysis in actinomycetes revealed that His447 and Glu361 residues are associated with the various chemical reactions like isomerization and oxidation (Yue et al. 1999). Navas et al. (2001) reported a comparison of amino acid sequences from class I enzymes from *Streptomyces* sp., *Rhodococcus* sp., and *Mycobacterium* sp. and found that Gly-X-Gly-X-X-Gly is the conserved sequence in the N-end of FAD region in the ChOx (Ohta et al. 1991). Besides cofactors diphosphate groups are present on the first α-helix near the N-terminus where the GXGXG glycine-conserved residues are located (Vrielink and Ghisla 2009).

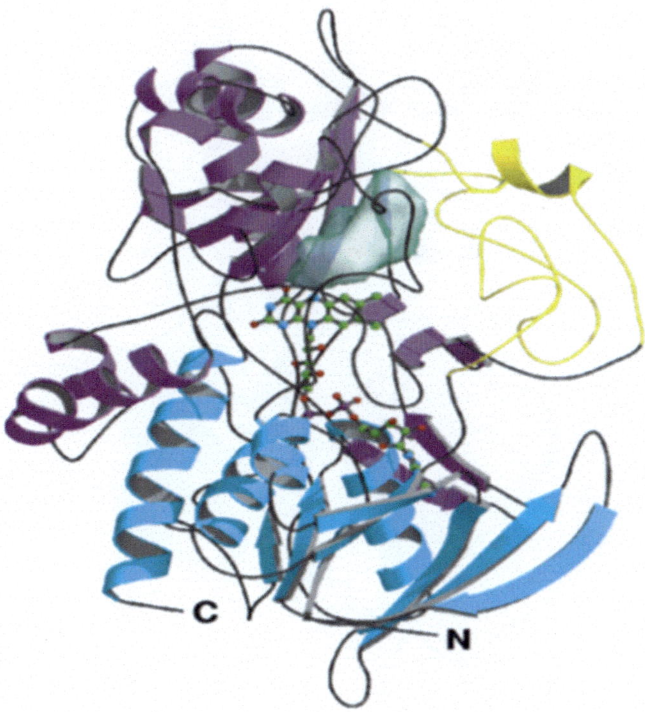

Fig. 12.1 Class I ChOx (source: Sampson and Vrielink 2003)

12.2.2 Class II ChOx

In the class II enzyme, the FAD cofactor is covalently linked to the enzyme (Croteau and Vrielink 1996; Sampson and Vrielink 2003). This enzyme has been found in *Rhodococcus erythropolis* and *Brevibacterium sterolicum* and pathogens such as *Burkholderia* sp., *Chromobacterium* sp., and *Pseudomonas aeruginosa*. By X-ray crystallography *B. sterolicum* ChOx structure has been found to be most indomitable due to its covalent bonds (Fig. 12.2) (Coulombe et al. 2001). Covalent bond of FAD contributes to the stability of the enzyme and is implicated in the redox potential (Caldinelli et al. 2005).

12.3 Sources of ChOx

ChOx was first isolated in 1944 by Turfitt (1944) from a bacterium *Nocardia erythropolis*, now known as *R. erythropolis*. After that Schatz et al. (1949) reported ChOx from *Mycobacterium* sp. and *Streptomyces* sp. However, ChOx activity of *R. erythropolis* isolated from various foods, e.g., butter, animal (pork and chicken)

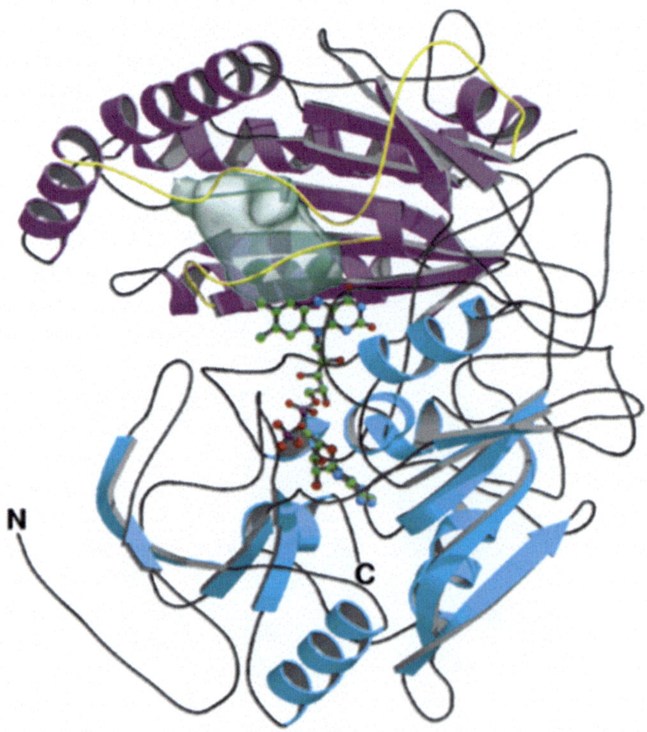

Fig. 12.2 Class II cholesterol oxidase from *Brevibacterium sterolicum* (source: Sampson and Vrielink 2003)

fats, and lipids, is also reported by Watanabe et al. (1986). ChOx have also been characterized and isolated from different bacterial genera belonging to non-food sources like soil, ocean, river water, ponds, and lakes (MacLachlan et al. 2000). Earlier, basidiomycetes fungi such as *Lentinus edodes, Oudemansiella radicate, Coprinus comatus,* and *Auricularia polytricha* were also reported with ChOx activity (Matsui et al. 1982). Apart from *R. erythropolis*, other bacteria known to produce ChOx are *Corynebacterium* sp. (Shirokane and Mizusawa 1977), *Arthrobacter* sp. (Liu et al. 1988; Wilmanska and Sedlaczek 1988), *Pseudomonas* sp. (Lee et al. 1989), *Mycobacterium* sp. (Smith et al. 1993), *Burkholderia* sp. (Doukyu and Aono 2001), *Enterobacter* sp. (Ye et al. 2008), *Bordetella* sp. (Lin et al. 2012), *Chryseobacterium* sp. (Reiss et al. 2014), and *Streptomyces* sp. (Srivastava et al. 2018a, b; Praveen et al. 2011). Both pathogenic and nonpathogenic bacteria are recognized for production of ChOx. Nonpathogenic microbes utilize ChOx in their cholesterol metabolism by using it (cholesterol) as a sole carbon source, while pathogenic ones need ChOx to degrade the structure of membrane of host macrophage (Doukyu et al. 2008; Pollegioni et al. 2009).

12.4 Mode of Action of ChOx

The ChOx enzyme performs a bifunctional reaction. First, it initiates the biological reaction in which D5-ene-3β-hydroxysteroids are oxidized with a trans A ± B ring to the D5-3-keto-steroid and then isomerization of D4-3-keto-steroid occurs (Smith and Brooks 1976). Subsequently, NAD-dependent dehydrogenases and oxygen regulate the mode of action (Edwards et al. 1976).

ChOx catalyzes its chemical reactions in three strides (Fig. 12.3). In the primary step, alcohol dehydrogenation of cholesterol ring occurs, which is involved in the redox process of FAD. Initially, cofactor FAD is oxidized and dehydrogenation converts it into reduced form. In the second catalytic step, the reduced form of FAD produces hydrogen peroxide and oxidized enzyme after interacting with di-oxygen. Finally, in the third step, the oxidized cholesterol goes through double-bond isomerization in the ring system and produces the end product cholest-4-en-3-one (Vrielink and Ghisla 2009).

12.5 Production of ChOx

Microorganisms produce both extra- and intracellular ChOx enzyme (Table 12.1). Various production methods and extraction techniques are available for obtaining a significant amount of ChOx enzyme. The techniques adopted for extraction depend on whether the enzyme is membrane bound/intracellular or extracellular. It is reported that isolation of membrane-bound or intracellular enzyme is tougher than extracellular form (Illanes 2008). The production of extracellular enzyme is easier, because enzyme excreted into the medium can be easily retrieved by simple filtration and centrifugation process, which removes all the solid cell debris and the left

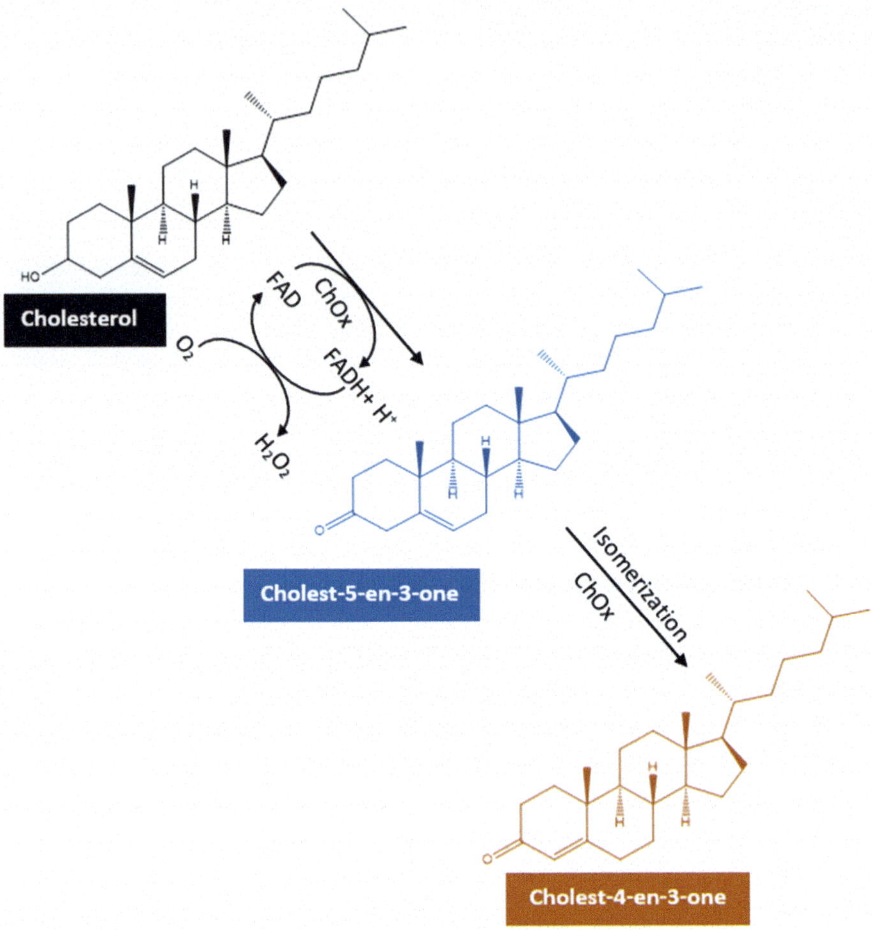

Fig. 12.3 Mechanism of cholesterol oxidase catalytic activity

is liquid portion containing free extracellular enzyme (MacLachlan et al. 2000). In a study, for the extraction of intracellular ChOx enzyme, the cells were mixed twice with 0.1 M potassium phosphate buffer (pH 7.0) containing 0.1% Triton X-100 for 1 h and it was found that by optimization of process parameter more than 90% of extracellular ChOx could be extracted (Yazdi et al. 2001). It has also been noticed that substrate optimization in medium has a greater effect on ChOx production. For example, Atrat et al. (1992) found that in *R. erythropolis* addition of 1 g/L of cholesterol increased the ChOx production by up to 3.3 U/g of cells. In another study, Kim et al. (2002) showed extracellular ChOx production of 3.14 U/mL after 24-h incubation by *Bacillus subtilis* SFF34 when the medium was supplemented with 0.2% cholesterol. Isobe et al. (2003a, b) reported high extracellular ChOx activity (up to 310 U/L) from a *γ-proteobacterium* cultured in cholesterol medium at 30 °C for 12 days. Lin et al. (2012) reported extracellular ChOx production (1700 U/L)

Table 12.1 Extracellular and intracellular ChOx produced by some bacteria

Type	Microorganism	References
Extracellular	*Arthrobacter simplex* *Bacillus sp.* *Pseudomonas spp.* *Schizophyllum commune*	Liu et al. (1988) Kim et al. (2002) Lee et al. (1989) Fukuyama and Miyake (1979)
Intracellular	*Corynebacterium cholesterolicum* *Mycobacterium spp.* *Nocardia rhodochrous*	Shirokane and Mizusawa (1977) Cheetham et al. (1982) Smith et al. (1993)
Extracellular/ intracellular	*Actinomyces lavendulae* *Nocardia erythropolis* *Streptomyces violascens*	Varma and Nene (2003) Sojo et al. (1997) Lartillot and Kedziora (1990)

from *Bordetella* sp. within 24-h culturing at pH 7.0, 260 rpm, and 37 °C. Furthermore, Varma and Nene (2003) reported both extra- and intracellular ChOx production from *Streptomyces lavendulae* by using 1.5% potato starch as a substrate. They showed that the use of potato starch provided extracellular enzyme ChOx activity of 2.0 U/mL in 72 h of fermentation while an intracellular ChOx was noticed for 104.3 U in 100 mL broth. Lartillot and Kedziora (1990) also showed high extracellular ChOx production in *Streptomyces* sp. According to Sojo et al. (1997) in *R. erythropolis* both extra- and intracellular activity of ChOx can be enhanced by using cholesterol as a substrate in spray-and-dry technique and a maximum ChOx up to 365.0 U/L and 1.70 U/g of cells can be obtained, respectively. Kreit et al. (1994) observed that ChOx production from *Rhodococcus* sp. was increased after addition of hexanotes or phytosterol compounds (provide carbon) in the media, and it affects both intracellular and extracellular ChOx production (400 U/L).

Extraction method and recovery system also influence the yield of ChOx. It has been found that the ultrafiltration and precipitation by inorganic salt may enhance recovery up to 90% and 83%, respectively (Varma and Nene 2003). Generally, low solubility of cholesterol in aqueous medium causes interruption in the recovery of ChOx. To overcome this limitation, an addition of surface-active agents into media is recommended. The most common surface-active agents are Tween-80 and Triton X-100 (Smith et al. 1993; Doukyu and Nihei 2015; Sahu et al. 2019). However, some other sterol-solubilizing agents like cyclodextrins or organic solvents like methanol, acetone, isopropanol, or 2-methoxyethanol can also enhance the recovery of ChOx (Giorgi et al. 2019).

12.6 Applications of ChOx

Microbial ChOx has multiple uses in the pharmaceutical, food, and agriculture sectors. In addition to this ChOx have a role in virulence and pathogenesis (Pei et al. 2006), steroid degradation (Brzostek et al. 2007), and biosensor application (Aparicio and Martin 2008; Mendes et al. 2007). Some defined uses of ChOx in different sectors are the following:

12.6.1 Clinical Diagnosis

Pharmaceutical industries and research laboratories utilize ChOx to determine the risk of cardiovascular diseases, thrombosis, and other lipid disorders. ChOx helps in the estimation of the serum cholesterol level (Richmond 1973; Allain et al. 1974), low-density lipid (LDL), and high-density lipid (HDL) in blood serum (Ernst and Cleeman 2002). In serum samples, cholesterol is present in ester form and after incubation of serum with cholesterol esterase free cholesterol is released. This free cholesterol is then oxidized by ChOx and further a peroxidase enzyme catalyzes to the hydrogen peroxide with an aromatic dye 4-aminoantipyrine (4-AP) and phenol which produce a red-color quinone-imine that is detected by spectrophotometer (Fig. 12.4).

12.6.2 ChOx Biosensor

Cholesterol determination from serum and food is very important for quality assurance. To detect the cholesterol from samples different types of biosensors have been in consideration. Nowadays, ChOx-containing biosensors are used to analyze the cholesterol level in food and in clinical samples (Vidal et al. 2004; Arya et al. 2008). Generally, in ChOx biosensors the level of oxygen and hydrogen peroxide determines the quality of biosensors. Currently, different types of ChOx-based biosensor are in use:

12.6.2.1 ChOx/Carbon Nanotube-Adorned Platinum (ChOx/CNT-Pt) Biosensor

The ChOx/CNT-Pt biosensor is developed by fixing ChOx in a sol-gel layer on electrodes of CNT-Pt. This electrode is formed by reduction technique and intercalation of platinum nanoparticles with graphite. This intercalating electrode increases the enzymatic activity and eases the conversion of H_2O_2. This immobilized electrode is preferable for cholesterol detection and generally provides very fast diagnosis of serum cholesterol (less than 20 s), and hence is most commonly used in clinical labs (Qiaocui et al. 2005).

12.6.2.2 Fourier Transformation Continuous Cycle Voltmeter [FFTCCV]

Combination of biosensor and FFTCCV is unique in using ChOx enzymes for cholesterol estimation. In this biosensor, the ChOx is fixed on several layers of carbon nanotubes and on nanoparticles of manganese dioxide which when kept on carbon rods acts as an electrode and is made up of Nafion synthetic polymers. The biosensor shows negligible inference of uric and ascorbic acid (Shi and Zhi 2005).

12.6.2.3 Poly-Pyrrole-Polyvinylsulfonate (PPy-PVS)

Novel PPy-PVS biosensors have been developed which use different types of polymers like polyindole, polyacetylene, polyaniline, and polythiophene. In PPy-PVS

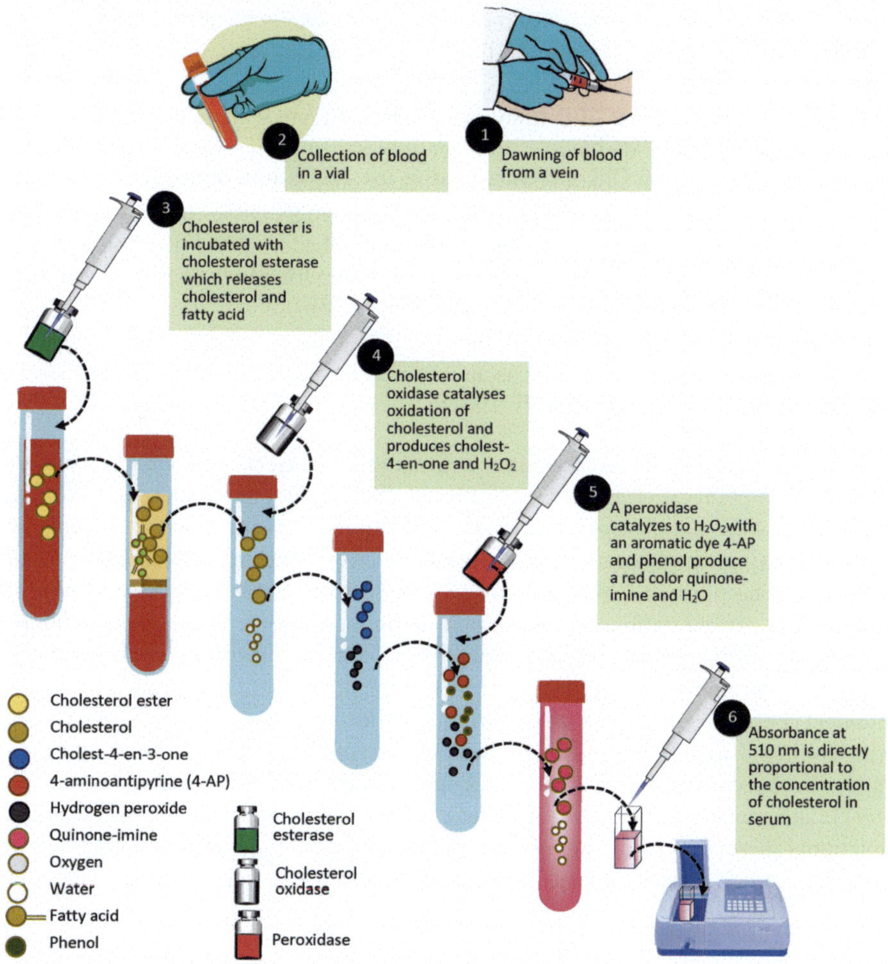

Fig. 12.4 Procedure of the detection of serum cholesterol by using ChOX

biosensors, platinum electrode bears a layer of poly-pyrrole-polyvinlysulfonate immobilized with ChOx, which determines the cholesterol level very accurately. The accuracy of this biosensor depends upon detection of H_2O_2 generated by cholesterol metabolism (Yıldırımoğlu et al. 2009).

12.6.2.4 N [3-(Trime-Thoxysilyl) Propyl]

To increase the sensitivity of biosensors, a novel method is projected, in which maximum use of horseshoe peroxidase (HRP) and ChOx helps in improving the quality of the sensor. In this biosensor single polymer N [3-(trime-thoxysilyl) propyl] aniline polymerizes and produces poly N [3-trime-thoxysilyl) propyl] aniline matrix. Furthermore, to stabilize the biosensor, ChOx and HRP enzymes are fixed on that matrix. This biosensor was developed to estimate the serum cholesterol level in lesser time (Bongiovanni et al. 2001).

12.6.2.5 Multiwall Carbon Nanotubes/Glassy Carbon Electrode (MWCNTs/GCE)

A new approach to form ChOx biosensor is to use ChOx immobilized on nanomaterials such as MnO_2 and MWCNTs on a GCE electrode surface to make a rapid detection and easier handling of biosensor. Carbon electrode biosensor is very effective in serum steroid analysis and under optimal conditions the detection limit of biosensor is 0.3 nanomolar (nM) (Norouzi et al. 2010).

12.6.3 Insecticidal Activity of ChOx

Insecticidal activity of ChOx against a Coleopteran insect *Anthonomus grandis* (boll weevil) (Purcell et al. 1993a, b, c) and few Lepidopteran such as *Heliothis virescens* (tobacco budworm), *Helicoverpa zea* (corn earworm), and *Pectinophora gossypiella* (pink bollworm) is being used for crop protection (Greenplate et al. 1995a, b). The purified ChOx protein was found to be active against boll weevil larvae (at a concentration of LC50 = 20.9 μg/mL), whereas the bioactivity of *Bacillus thuringiensis* proteins may vary from 1 μg/mL for tobacco budworm (*H. virescens* Fabricius) to 37 μg/mL against European corn borer (*Ostrinia nubilalis* Hubner) (Purcell et al. 1993a, b, c). After ingestion of ChOx enzyme, it associates with epithelial membrane in midgut which contains the sterol substrate and then catalyzes the conversion of cholesterol into cholest-4-en-3-one (Ghoshroy et al. 1997). The conversion of cholesterol into cholest-4-en-3-one results in physical and functional disruption of the membrane and ultimately causes developmental arrest and death of boll weevil larvae (Purcell et al. 1993a, b, c).

12.6.4 Transformation of Steroids

Steroid metabolism is a complex process. Several enzymatic steps are needed to convert steroid into desired intermediate products and CO_2 and water as final end products. As microbes use their ChOx for metabolizing steroid compounds and derive energy and carbon, they are best suited biocatalytic machinery for bioconversion of steroid into useful compounds. ChOx can detect cholesterol and transform it into cholest-4-en-3-one metabolite (Randolph et al. 1988; Bru et al. 1989). ChOx first provide oxidation of 3β-hydroxyl group of steroid and then degradation of side chain starts. Later the main core structure of steroid is metabolized. In strict sense, ChOx can oxidize any 3β-hydroxysteroids to the corresponding ketones (Aparicio and Martin 2008). A number of pharmaceutical steroids and 3β-hydroxysteroids have been transformed into the synthetic steroid hormones by microbial ChOx (Guo et al. 2003). For example, ChOx from *Rhodococcus* sp. are exclusively recognized for biotransformation ability of steroids into hydroxylated steroid metabolites (Lashkarian et al. 2010).

12.7 Future Prospective

ChOx have potential application in different industries. However, its application and demand as a therapeutic agent are increasing day by day. To fulfill this increasing demand is seemingly not possible with conventional enzyme production methods; hence involvement of modern tools and fermentation techniques is necessary. Researchers have also suggested to use recombinant ChOx with improved stabilities and catalytic ability amenable for industrial use. Recently, Yamada et al. (2019) showed that the need of complex medium containing whole-yeast cells essential for the production of ChOx by *S. lavendulae* strain, which is a natural high producer of ChOx, can be skipped by the expression of ChOx gene heterologously in *Streptomyces lividans* and *Streptomyces albus*. By using suitable vector, overexpression of ChOx gene in *S. lavendulae* strain also increased comparative yields of ChOx (Yamada et al. 2019).

Overproduction of ChOx in *Chromobacterium* sp. DS1 by optimization of various parameters (host strain, culture media, induction time, isopropyl ß-D-1-thiogalactopyranoside concentration, as well as post-induction incubation time and temperature) was studied by Fazaeli et al. (2019). They showed that optimization increased the yield of recombinant ChOx significantly from 92 U/L to 2115 U/L.

The whole-cell biotransformation of cholesterol compounds by using microbial ChOx is getting popularized in pharmaceutical industries but application of immobilized cells also faces challenges of structural- and thermo-stability (Stepankova et al. 2013). In this context self-assembled organic–inorganic hybrid nanoflowers is a new approach for enzyme immobilization. For example, in a study Hao et al. (2019) prepared *Brevibacterium* ChOx-Cu hybrid nanoflowers which showed enhanced structural and thermostability, tolerance to biphasic mixture, and catalytic efficiency of bioconversion of cholesterol.

Currently immobilizations of enzymes on nanomatrices have attracted massive interest particularly in cutting down the cost of enzyme production (Binod et al. 2013). Immobilized ChOx will provide more space for the enzymatic activity, so it could also be useful in the development of high quality of biosensors for clinical use. In the near future, ChOx-containing nanozymes could also be used in the fabrication of different types of diagnostic kits useful in the detection of lethal diseases for structural and catalytic properties.

12.8 Conclusion

The enzyme ChOx has a great potential in several industries. ChOx utilization in measuring the value of cholesterol in clinical samples, bioconversion of cholesterol into useful compounds, preparation of food stuffs, and insecticidal activity against cotton weevil have been well explored and exploited commercially. Although research showed that native or wild-type microbial strains hold great promise of ChOx production, newer approaches such as use of recombinant systems would be needed for settling the industrial demand of ChOx. Nanotechnology-based approach could be useful to strengthen the ChOx-based biosensors in clinical use.

Acknowledgments We are thankful to Naveen Kumar Arora and Jitendra Mishra for providing editorial contribution. We are also gratified to Jitendra Mishra for preparing color illustration. The authors thank Dr. Qazi Mohd. Sajid Jamal (Department of Heath Informatics, College of Public Health Informatics, Qassim University, Saudi Arabia) for providing the assistance in computational modelling data.

References

Allain CC, Poon LS, Chan CS, Richmond W, Fu PC (1974) Enzymatic determination of total serum cholesterol. Clin Chem 20:470–475

Aparicio JF, Martin JF (2008) Microbial cholesterol oxidases: bioconversion enzymes or signal proteins. Mol BioSyst 4:804–809

Arya SK, Datta M, Malhotra BD (2008) Recent advances in cholesterol biosensor. Biosens Bioelectron 23:1083–1100

Atrat PG, Wagner B, Wagner M, Schumann G (1992) Localization of the cholesterol oxidase in *Rhodococcus erythropolis* IMET 7185 studied by immunoelectron microscopy. J Steroid Biochem Mol Bio 42:193–200

Binod P, Sindhu R, Pandey A (2013) The alcohol fermentation step: The most common ethanologenic microorganisms among yeasts, bacteria and filamentous fungi. In: Faraco V (ed) Lignocellulose conversion: Enzymatic and microbial tools for bioethanol production. Springer, Berlin, pp 131–150

Bongiovanni C, Ferri T, Poscia A, Varalli M, Santucci R, Desideri A (2001) An electrochemical multienzymatic biosensor for determination of cholesterol. Bioelectrochemistry 54:17–22

Bru R, Sánchez FA, García CF (1989) Characterization of cholesterol oxidase activity in AOT-isooctane reverse micelles and its dependence on micelle size. Biotechnol Lett 11:237–242

Brzostek A, Dziadek B, Rumijowska GA, Pawelczyk J, Dziadek J (2007) Cholesterol oxidase is required for virulence of *Mycobacterium tuberculosis*. FEMS Microbiol Lett 275:106–112

Caldinelli L, Iametti S, Barbiroli A, Bonomi F, Fessas D, Molla G, Pilone MS, Pollegioni L (2005) Dissecting the structural determinants of the stability of cholesterol oxidase containing covalently bound flavin. J Biol Chem 280:22572–22581

Cheetham PSJ, Dunnill P, Lilly MD (1982) The characterization and interconversion of three forms of cholesterol oxidase extracted from Nocardia rhodochrous. Biochem J 201:512–521

Chen L, Lyubimov AY, Brammer L, Vrielink A, Sampson NS (2007) The binding and release of oxygen and hydrogen peroxide are directed through a hydrophobic tunnel in cholesterol oxidase. Biochemistry 47:5368–5377

Coulombe R, Yue KQ, Ghisla S, Vrielink A (2001) Oxygen access to the active site of cholesterol oxidase through a narrow channel is gated by an Arg-Glu pair. J Biol Chem 276:30435–30441

Croteau N, Vrielink A (1996) Crystallization and preliminary X-ray analysis of cholesterol oxidase from *Brevibacterium sterolicum* containing covalently bound FAD. J Struct Biol 116:317–319

Dewan SS (2011) Global markets for enzymes in industrial applications (BIO030G). BCC Research

Dewan SS (2012) Global Markets and Technologies for Biofuel Enzymes (EGY099A). BCC Research

Doukyu N (2009) Characteristics, biotechnological applications of microbial cholesterol oxidases. Appl Microbiol Biotechnol 83:825–837

Doukyu N, Aono R (2001) Cloning, sequence analysis and expression of a gene encoding and organic solvent and detergent-tolerant cholesterol oxidase of *Burkholderia cepacia* strain ST-200. Appl Microbiol Biotechnol 57:146–152

Doukyu N, Nihei S (2015) Cholesterol oxidase with high catalytic activity from Pseudomonas aeruginosa: screening, molecular genetic analysis, expression and characterization. J Biosci Bioeng 120(1):24–30

Doukyu N, Shibata K, Ogino H, Sagermann M (2008) Purification and characterization of *Chromobacterium* sp. DS-1 cholesterol oxidase with thermal, organic solvent and detergent tolerance. Appl Microbiol Biotechnol 80:59–70

Edwards DP, O'Conner JL, Bransome ED, Braselton WE (1976) Human placental 3b-hydroxysteroid dehydrogenase: D5-Isomerase. J Biol Chem 251:1632–1638

El-Naggar NE, Deraz SF, Soliman HM, El-Deeb NM, El-Shweihy NM (2017) Purification, characterization and amino acid content of cholesterol oxidase produced by Streptomyces aegyptia NEAE 102. BMC Microbiol 17(1):76

Ernst ND, Cleeman JI (2002) National cholesterol education program keeps a priority on lifestyle modification to decrease cardiovascular disease risk. Curr Opin Lipidol 13:69–73

Espenshade PJ, Hughes AL (2007) Regulation of sterol synthesis in eukaryotes. Annu Rev Genet 41:401–427

Eventoff W, Rossmann MG (1975) The evolution of dehydrogenases and kinases. CRC Crit Rev Biochem 3:111–140

Fazaeli A, Golestani A, Lakzaei M, Rasi Varaei SS, Aminian M (2019) Expression optimization, purification, and functional characterization of cholesterol oxidase from Chromobacterium sp. DS1. PLoS One 14:e0212217

Fukuyama M, Miyake Y (1979) Purification and some properties of cholesterol oxidase from *Schizophyllum commune* with covalently bound flavin. J Biochem 85:1183–1193

Ghoshroy KB, Zhu W, Sampson NS (1997) Investigation of membrane disruption in the reaction catalyzed by cholesterol oxidase. Biochemistry 36(20):6133–6140. https://doi.org/10.1021/bi962190p

Giorgi V, Menéndez P, García-Carnelli C (2019) Microbial transformation of cholesterol: reactions and practical aspects—an update. World J Microbiol Biotechnol 35(9):131

Greenplate JT, Duck NB, Pershing JC, Purcell JP (1995a) Cholesterol oxidase: an oostatic and larvicidal agent active against the cotton boll weevil *Anthonomus grandis*. Entomol Exp Appl 74:253–258

Greenplate JT, Duck NB, Pershing JC, Purcell JP (1995b) Cholesterol oxidase: an oostatic and larvicidal agent active against the cotton boll weevil, *Anthonomus grandis*. Entomol Exp Appl 74:253–258

Guo FQ, Okamoto M, Crawford NM (2003) Identification of a plant nitric oxide synthase gene involved in hormonal signaling. Science 302:100–103

Hao M, Fan G, Zhang Y, Xin Y, Zhang L (2019) Preparation and characterization of copper-Brevibacterium cholesterol oxidase hybrid nanoflowers. Int J Biol Macromol 126:539–548

Hol WG, Duijnen VPT, Berendsen HJ (1978) The alpha-helix dipole and the properties of proteins. Nature 273:443–446

Illanes A (2008) Enzyme production. In: Illanes A (ed) Enzyme biocatalysis: principles and applications. Springer Netherlands, Dordrecht, pp 57–106

Isobe K, Mori N, Wakao N (2003a) The second cholesterol oxidase produced by gamma-proteobacterium Y-134. J Biosci Bioeng 96(3):257–261

Isobe K, Shoji K, Nakanishi Y, Yokoe M, Wakao N (2003b) Purification and some properties of cholesterol oxidase stable in detergents from gamma- *Proteobacterium* Y-134. J Biosci Bioeng 95:257–263

Kim K, Rhee C, Park H (2002) Degradation of cholesterol by *Bacillus subtilis* SFF34 isolated from Korean traditional fermented flatfish. Lett Appl Microbiol 35:468–472

Kreit J, Lefebvre G, Germain P (1994) Membrane-bound cholesterol oxidase from *Rhodococcus* sp. cells: production and extraction. J Biotechnol 33:271–282

Lartillot S, Kedziora P (1990) Production, purification and some properties of cholesterol oxidase from a *Streptomyces* sp. Prep Biochem 20:51–62

Lashkarian H, Raheb J, Shahzamani K, Shahbani H, Shamsara M (2010) Extracellular cholesterol oxidase from *Rhodococcus* sp.: isolation and molecular characterization. Iran Biomed J 14(1–2):49–57

Lee SY, Rhee HI, Tae WC, Shin JC, Park BK (1989) Purification and characterization of cholesterol oxidase from *Pseudomonas* sp. and taxonomic study of the strain. Appl Microbiol Biotechnol 31:542–546

Li S, Yang X, Yang S, Zhu M, Wang X (2012) Technology prospecting on enzymes: application, marketing and engineering. Comput Struct Biotechnol J 2:e201209017. https://doi.org/10.5936/csbj.201209017

Lin Y, Fu J, Song X (2012) Purification and characterization of an extracellular cholesterol oxidase from a *Bordetella* sp. Process Biochem 45:1563–1569

Liu WH, Meng MH, Chen KS (1988) Purification and some properties of cholesterol oxidases produced by an inducible and a constitutive mutant of *Arthrobacter simplex* (microbiology and fermentation industry). Agric Biol Chem 52:413–418

MacLachlan J, Wotherspoon AT, Ansell RO, Brooks CJ (2000) Cholesterol oxidase: sources, physical properties and analytical applications. J Steroid Biochem Mol Biol 72:169–195

Matsui I, Nakashima K, Taniguchi T et al. (1982) Preparation of cholesterol oxidase. Patent No JP5757122792

Mendes MV, Recio E, Anton N, Guerra SM, Santos AJ, Martin JF, Aparicio JF (2007) Cholesterol oxidases act as signaling proteins for the biosynthesis of the polyene macrolide pimaricin. Chem Biol 14:279–290

Navas J, Gonzalez-Zorn B, Ladron N, Garrido P, Vazquez-Boland JA (2001) Identification and mutagenesis by allelic exchange of choE, encoding a cholesterol oxidase from the intracellular pathogen *Rhodococcus equi*. J Bacteriol 183:4796–4805

Norouzi P, Faridbod F, Nasli Esfahani E, Larijani B, Ganjali M (2010) Cholesterol biosensor based on MWCNTs-MnO2 nanoparticles using FFT continuous cyclic voltammetry. Int J Electrochem Sci 5:1008–1017

Ohlsson I, Nordstrom B, Branden CI (1974) Structural and functional similarities within the coenzyme binding domains of dehydrogenases. J Mol Biol 89:339–354

Ohta T, Fujishiro K, Yamaguchi K, Tamura Y, Aisaka K, Uwajima T, Hasegawa M (1991) Sequence of gene choB encoding cholesterol oxidase of *Brevibacterium sterolicum*: comparison with choA of *Streptomyces* sp. SA-COO. Gene 103:93–96

Pei Y, Dupont C, Sydor T, Haas A, Prescott JF (2006) Cholesterol oxidase (ChoE) is not important in the virulence of *Rhodococcus equi*. Vet Microbiol 118:240–246

Pollegioni L, Piubelli L, Molla G (2009) Cholesterol oxidase: biotechnological applications. FEBS J 276:6857–6870

Prakash D, Nawani N, Prakash M, Bodas M, Mandal A, Khetmalas M, Kapadnis B (2013) Actinomycetes: a repertory of green catalysts with a potential. Biomed Res Int 2013:264020, 1-8

Praveen V, Srivastava A, Tripathi C (2011) Purification and characterization of the enzyme cholesterol oxidase from a new isolate of *Streptomyces* sp. Appl Biochem Biotechnol 165:1414–1426

Purcell JP, Greenplate JT, Jennings MG, Ryerse JS, Pershing JC, Sims SR, Prinsen MJ, Corbin D, Tran RM, Sammons RD (1993a) Cholesterol oxidase: a potent insecticidal protein active against boll weevil larvae. Biochem Biophys Res Commun 196:1406–1413

Purcell JP, Greenplate JT, Jennings MG, Ryerse JS, Pershing JC, Sims SR, Prinsen MJ, Corbin DR, Tran M, Sammons RD, Stonard RJ (1993b) Cholesterol oxidase: a potent insecticidal protein active against boll weevil larvae. Biochem Biophys Res Commun 196(3):1406–1413

Purcell JP, Greenplate JT, Jennings MG, Ryerse JS, Pershing JC, Sims SR et al (1993c) Cholesterol oxidase: a potent insecticidal protein active against boll weevil larvae. Biochem Biophys Res Commun 196(3):1406–1413

Qiaocui S, Tuzhi P, Yunu Z, Yang CF (2005) An electrochemical biosensor with cholesterol oxidase sol-gel film on a nanoplatinum/carbon nanotube electrode. Electroanalysis 17:857–861

Randolph T, Clark D, Blanch H, Prausnitz J (1988) Enzymatic oxidation of cholesterol aggregates in supercritical carbon dioxide. Science 239:387–390

Reiss R, Faccio G, Thöny-Meyer L, Richter M (2014) Cloning, expression and biochemical characterization of the cholesterol oxidase CgChoA from *Chryseobacterium gleum*. BMC Biotechnol 14:46

Richmond W (1973) Preparation and properties of a cholesterol oxidase from *Nocardia* sp. and its application to the enzymatic assay of total cholesterol in serum. Clin Chem 19:1350–1356

Sahu S, Shera SS, Banik RM (2019) Optimization of process parameters for cholesterol oxidase production by *Streptomyces olivaceus* MTCC 6820. Open Biotechnol J 13(1)

Salva TJ, Liserre AM, Moretto AL, Zullo MA, Ventrucci G, Menezes TJ (1999) Some enzymatic properties of cholesterol oxidase produced by *Brevibacterium* sp. Rev Microbiol 30(4):315–323

Sampson NS, Vrielink A (2003) Cholesterol oxidases: a study of nature's approach to protein design. Acc Chem Res 36:713–722

Schatz A, Savard K, Pintner IJ (1949) The ability of soil microorganisms to decompose steroids. J Bacteriol 58:117–125

Shi QC, Zhi TP (2005) A novel cholesterol oxidase biosensor based on pt-nanoparticle/carbon nanotube modified electrode. Chin Chem Lett 16(8):1081–1084

Shirokane YNK, Mizusawa K (1977) Purification and some properties of an extracellular 3-hydroxysteroid oxidase produced by *Corynebacterium cholesterolicum*. J Ferment Tech 55:337–345

Smith AG, Brooks CJW (1976) Cholesterol oxidases: properties and applications. J Steroid Biochem 7:705–713

Smith M, Zahnley J, Pfeifer D, Goff D (1993) Growth and cholesterol oxidation by *Mycobacterium* species in Tween 80 medium. Appl Environ Microbiol 59:1425–1429

Sojo M, Bru R, Lopez-Molina D, Garcia-Carmona F, Arguelles JC (1997) Cell linked and extracellular cholesterol oxidase activities from *Rhodococcus erythropolis*. Isolation and physiological characterization. Appl Microbiol Biotechnol 47:583–589

Srivastava A, Singh V, Shafiul H, Pandey S, Mishra M et al (2018a) Response surface methodology-genetic algorithm based medium optimization, purification, and characterization of cholesterol oxidase from *Streptomyces rimosus*. Sci Rep 8:10913

Srivastava A, Singh V, Tripathi CKM (2018b) Scale up and optimization of cholesterol oxidase production from Streptomyces rimosus MTCC10792 in a 3-L bioreactor. Environ Sustain 1(1)

Stepankova V, Bidmanova S, Koudelakova T, Prokop Z, Chaloupkova R, Damborsky J (2013) Strategies for stabilization of enzymes in organic solvents. ACS Catal 3(12):2823–2836

Turfitt GE (1944) The microbiological degradation of steroids: 2. Oxidation of cholesterol by *Proactinomyces* sp. Biochem J 38:492–496

Varma R, Nene S (2003) Biosynthesis of cholesterol oxidase by *Streptomyces lavendulae* NCIM 2421. Enzym Microb Technol 33:286–291

Vidal JC, Espuelas J, Castillo JR (2004) Amperometric cholesterol biosensor based on in situ reconstituted cholesterol oxidase on an immobilized monolayer of flavin adenine dinucleotide cofactor. Anal Biochem 333:88–98

Vrielink A, Ghisla S (2009) Cholesterol oxidase: biochemical properties. FEBS J 276:6826–6843

Watanabe K, Aihara H, Nakamura R, Suzuki KI, Momagata K (1986) Isolation and identification of cholesterol degradation *Rhodococcus* strains food of animal origin and their cholesterol oxidase activities. J Gen Appl Microbiol 32:137–147

Wilmanska D, Sedlaczek L (1988) The kinetics of biosynthesis and some properties of an extracellular cholesterol oxidase produced by *Arthrobacter* sp. IM 79. Acta Microbiol Pol 37:45–51

Yamada K, Koroleva A, Laughlin M, Oksanen N, Akhgari A, Safronova V, Yakovleva E, Kolodyaznaya V, Buldakova T, Metsä-Ketelä M (2019) Characterization and overproduction of cell-associated cholesterol oxidase ChoD from *Streptomyces lavendulae* YAKB-15. Sci Rep 9(1):11850

Yazdi MT, Malekzadeh F, Zarrini G, Faramarzi MA, Kamranpour N, Khaleghparast S (2001) Production of cholesterol oxidase by a newly isolated *Rhodococcus* sp. World J Microbiol Biotechnol 17:731–737

Ye D, Lei J, Li W, Ge F, Wu K, Xu W, Yong B (2008) Purification and characterization of extracellular cholesterol oxidase from *Enterobacter* sp. World J Microbiol Biotechnol 24:2227–2233

Yıldırımoğlu F, Arslan F, Çete S, Yaşar A (2009) Preparation of a polypyrrole polyvinyl sulphonate composite film biosensor for determination of cholesterol based on entrapment of cholesterol oxidase. Sensors (Basel) 9:6435–6445

Yue QK, Kass IJ, Sampson NS, Vrielink A (1999) Crystal structure determination of cholesterol oxidase from *Streptomyces* and structural characterization of key active site mutants. Biochemistry 38:4277–4286

Therapeutic Applications of Microbial Enzymes in the Management of Kidney Stone Diseases

13

Shruti Gupta and Shamsher Singh Kanwar

Abstract

Nephrolithiasis is a terrible pathological condition marked by the presence and formation of kidney stones. It affects around 3–20% of the community in the world. Several environmental, physiological, and nutritional conditions influence this disease. Not only the food sources but also the body's own metabolism add up oxalate content in the human body. The increased intake of oxalate leads to hyperoxaluria, which often results in the formation of calcium oxalate stones, commonly known as kidney stones. The incidences of kidney stone are very common, and the current therapeutic measure of its cure is not much effective. Therefore, new therapeutic approaches are needed. In the last few years, the use of gut microbiome with oxalate-degrading activity has emerged as an excellent therapeutic approach to treat kidney stones. As the genes responsible for oxalate-degrading enzymes are not found in humans use of bacterial enzymes with the ability to degrade oxalate in intestinal digestion has a significant therapeutic impact. This chapter summarizes the roles of microbial enzymes produced by gut microflora involved in the solubilization of the dietary oxalates, and their potential applications in kidney stone diseases.

13.1 Introduction

Kidney stone or urolithiasis is a condition primarily attributed to the deposition of an enhanced level of calcium oxalate in the form of crystals due to supersaturation (of calcium oxalate) during removal of water from urine (Peck et al. 2016). Although oxalic acid is a general component present in human diets, it is also endogenously

S. Gupta · S. S. Kanwar (✉)
Department of Biotechnology, Himachal Pradesh University, Shimla, India

N. K. Arora et al. (eds.), *Microbial Enzymes: Roles and Applications in Industries*, Microorganisms for Sustainability 11,
https://doi.org/10.1007/978-981-15-1710-5_13

produced during amino acid metabolism. Moreover, oxalic acid is absorbed in the stomach, small bowel, and colon from the dietary sources (Nazzal et al. 2016). Binding with different cations such as sodium, potassium, magnesium, and calcium results in the formation of different oxalate salts but mostly calcium oxalate (Mogna et al. 2014). Ingestion and exposure of a high amount of oxalate lead to building up of oxalate crystals in the kidneys, which might be lethal and cause hypocalcemia, azotemia, and hemorrhage in the visceral organs (Aslani et al. 2011). Oxalates of calcium and phosphate are the main constituents of kidney stones (Bungash et al. 2011). Apart from the formation of stones in the kidney, oxalate crystals can destruct epithelium in the oral cavity and gastrointestinal tract, causing inflammation, diarrhea, and gastric hemorrhage which indirectly becomes a cause of death (Ellis et al. 2015).

As humans lack the enzyme for directly metabolizing oxalate, alternate pathways are used to regulate this potentially toxic compound (Mogna et al. 2014). Current remedial strategies which are used for kidney stones are inefficient and have been proven to be unsuccessful in preventing the recurrence of the disease (Sutherland et al. 1985). However, therapeutic measures such as allopurinol, thiazide, potassium alkali, and tiopronin along with dietary modifications and intake of adequate fluids have been used for a long time to limit urolithiasis (Trinchieri 2013). Hence the evolution of new therapeutic strategies aiming to prevent recurrent stone formation has become the need of the hour. Since a decade, attempts have been made to use plants and oxalate-degrading microbial enzymes to solubilize oxalate kidney stones, and some success has been achieved (Peck et al. 2016). The roles of gut enzymes produced through microflora in the solubilization of the dietary oxalates are a new frontier area for treating kidney stone disease. This chapter provides a brief insight into current research and the roles of gut microbial enzymes for the treatment of kidney diseases (Fig. 13.1).

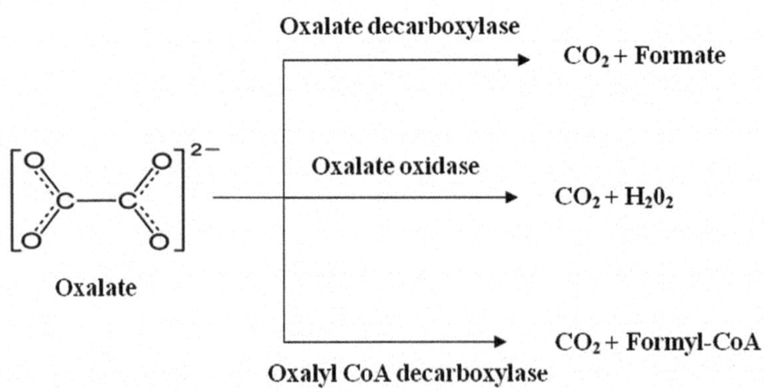

Fig. 13.1 Enzymatic degradation of oxalate

13.2 Role of the Gut Microbiome in Oxalate Degradation

Formation of oxalate stones in humans may be prevented by two symbiotically existing bacterial genera, *Oxalobacter* and *Lactobacillus,* in the gut. Both bacterial genera have been found to act on some biochemical pathways by the intervention of their oxalate-degrading enzymes (Sadaf et al. 2017). It has also been hypothesized that the *Oxalobacter formigenes*, a Gram-negative, obligate anaerobe found in the gastrointestinal tract and in humans, performs a significant role in mediating mammalian oxalate homeostasis (Svedruzic et al. 2005). The bacterium *O. formigenes* colonizes the gut in nearly 70–80% of the healthy population and utilizes oxalate as the sole material for energy and carbon source. Formyl-CoA transferase and oxalyl-CoA decarboxylase are the two enzymes from *O. formigenes,* which catalyze oxalate for biosynthesis (Hoppe et al. 2005). By transferring the coenzyme-A moiety to lactic acid and oxalic acid that is connected with calcium oxalate and calcium phosphate, degradation reaction occurs which results in elevation of oxalate and lactate level (Salminen et al. 2010).

Further, oxalate is broken down into CO_2 and formate, which is further metabolized and excreted via the feces (Hoppe et al. 2005). It has been found that in standard colonization conditions *O. formigenes* can degrade more than 1 g of oxalate per day. However, attempts to culture this bacterium out of fecal specimens have given low colony counts, i.e.. up to 10^6 CFU per gram of wet sample (Allison and Cook 1981). As investigated by Peck et al. (2016) in most of the cases gut of children between the age of 1 and 6 years is more naturally colonized by *O. formigenes,* while 20–25% of the colonization is lost during early adulthood and adolescence in healthy populations (Peck et al. 2016). In addition to *O. formigenes*, other oxalate-degrading bacterial genera are *Lactobacillus*, *Enterococcus*, *Eubacteria*, and *Bifidobacterium*. Amongst them, *Enterococcus faecalis* uses oxalate as a sole carbon and energy source in a nutrient-deficit environment; otherwise it can also consume other substrates for growth (Miller and Dearing 2013). In some circumstances, along with other microflora, natural colonization of *O. formigenes* in the gut is affected. However, continuous use of antibiotics, e.g., in patients with cystic fibrosis, or therapeutic use in diseases such as Crohn's disease also exacerbates kidney stone formation (Kumar et al. 2004; Hatch 2014).

13.3 Probiotic Therapies for the Treatment of Kidney Stones

Use of probiotics as a therapeutic and preventive measure in kidney stone and hyperoxaluria has gained much attention. It has been found that in the form of probiotics, aerotolerant *Lactobacillus* and obligatory anaerobe *Bifidobacterium* present in the intestine show oxalate-degrading activity, which is considered useful for the prevention of stone formation (Abratt and Reid 2010). Studies confirmed that through treatment with *Bifidobacterium lactis* DSM 10140, *Bifidobacterium longum* MB 282, and *Bifidobacterium adolescentis* MB 238 strains, the degradation of oxalate could be achieved up to 61%, 35.2%, and 57%, respectively (Turroni et al. 2007; Abratt and

Reid 2010). Both *Lactobacillus* and *Bifidobacterium* break down oxalate only in the presence of glucose and lactose; however, they do not use oxalate as a sole source of carbon and hence they are also known as "generalist oxalobacters" (Sadaf et al. 2017). Studies reveal that *Lactobacillus acidophilus* NCFM contains genes that code for the oxalate CoA decarboxylase (Oxc) and oxalate CoA transferase (frc) enzymes and constitute the functional oxalate-degrading formyl-CoA. A number of natural sources such as milk, yogurt, pickles, tomato, cucumber, spinach, and dieffenbachia plant are found to contain natural population of *Lactobacillus* and *Oxalobacter* probiotics used in the prevention of kidney stones (Gomathi et al. 2014). In a study, Lieske et al. (2010) reported that application of mixed cultures of *Bifidobacterium infantis, L. acidophilus, Streptococcus thermophilus*, and *Lactobacillus brevis* sold under the brand name of "Oxadrop" with a low-oxalate diet did not produce any effect on the inhibition of kidney stone formation but when given with a normal diet it reduced oxalate excretion. The probiotic capability of *O. formigenes* in the prevention of kidney stone formation has also been reported. However, studies have demonstrated that only an unabated inoculation of *O. formigenes* with an oxalate-rich diet reduced the concentration of urinary oxalate and restoring back to low-oxalate diet resulted in low oxalate degradation with apparent loss of *O. formigenes* colonization (Miller and Dearing 2013).

13.4 Oxalate Degradation by Microbial Enzymes

Absence, deficiency, or complete lack of oxalate degradation enzymes evokes the formation of calcium oxalate. Hence, utilization of oxalate-degrading enzymes in the prevention and treatment of calcium oxalate stones has suddenly increased (Cai et al. 2018). Three major types of microbial enzymes (Table 13.1) reported for oxalate degradation are (1) oxalate decarboxylase (ODC, oxalate carboxylyase, EC 4.1.1.2), (2) oxalate oxidase (OXO, oxalate: oxygen oxidoreductase, EC 1.2.3.4), and (3) oxalyl-CoA decarboxylase (oxalyl-CoA carboxylyase, EC 4.1.1.8) (Mäkelä et al. 2010).

13.4.1 Oxalate Decarboxylase

Oxalate decarboxylase (EC 4.1.1.2) was first discovered in basidiomycetes fungi, *Collybia (Flammulina) velutipes* and *Coriolus hirsutus* (Twahir et al. 2015). Apart from fungal sources, in some cases animal tissue (liver of guinea pigs) has also been described to exhibit oxalate decarboxylase activity (Murthy et al. 1981). Later, bacteria, plants, and fungi were characterized as established sources of oxalate decarboxylase (Svedruzic et al. 2005). Basically, in the presence of dioxygen, which acts as a co-catalyst, the enzyme produces formate and carbon dioxide by the heterolytic cleavage of unreactive carbon–carbon bond in oxalic acid. A little bit of oxalate oxidase activity leading to the formation of carbon dioxide and hydrogen peroxide in the place of formate has also been reported (Twahir et al. 2015).

Table 13.1 Sources and mechanism of action of enzymes of oxalate degradation

Enzyme	Source	Mechanism of action	References
Oxalate decarboxylase	**Bacteria**	Cleaves the oxalate carbon–carbon bond heterolytically to formate and CO_2 through a radical based catalytic cycle that involves electron transfer from the coordinated Mn^{2+} ion to the bound dioxygen	Yu-Hu et al. (2008); Mäkelä et al. (2010); Alberta et al. (2017)
	Agrobacterium tumefaciens *Bacillus subtilis* *Thermotoga maritima* and *Pandorea* sp.		
	Fungi		
	Trametes hirsuta (Coriolus hirsutus) *Flammulina (Collybia) velutipes* *Agaricus bisporus* *Postia placenta* *Pleurotus ostreatus* and *Aspergillus* sp.		
Oxalate oxidase	**Plant materials**	Oxalic oxidase at first gets oxidized by O_2 which upon catalysis cleaves oxalic acid into two CO_2 molecules along with generation of H_2O_2	Svedruzic et al. (2005); Hu et al. (2015)
	Barley seedlings, stems, and roots *Amaranthus* leaves Beet stems and leaves Sorghum leaves Maize, oats, rice, and rye Banana, azalea		
	Fungi		
	White-rot fungi basidiomycetes		
Oxalyl-CoA decarboxylase	**Bacteria**	Converts activated oxalyl-CoA to formyl-CoA and CO_2 employing thiamin pyrophosphate as a cofactor	Svedruzic et al. (2005); Mäkelä et al. (2010)
	Pseudomonas oxalaticus *Bacillus oxalophilus* *O. formigenes* *Bifidobacterium lactis* *Lactobacillus acidophilus* and *Thiobacillus novellus*		

Fungal and bacterial oxalate decarboxylases belong to a functionally varied superfamily of proteins known as the cupins and contain a range of conserved residues forming β-barrels which support the binding of different metal cofactors (Yu-Hu et al. 2008). Cupin proteins share primary and tertiary structure with two conserved histidine-containing Mn^{2+}-binding motifs separated by an inter-motif region, which varies in length (Mäkelä et al. 2010). Functional oxalate decarboxylase consists of two trimers of the bicupin subunits, therefore, probably making it a hexameric enzyme (Anand et al. 2002). The fungal oxalate decarboxylases are secretory enzymes while bacterial ones are involved in the energy metabolism and are probably confined to cytosol (Yu-Hu et al. 2008). The most thoroughly studied oxalate decarboxylase belongs to *Bacillus subtilis* (Anand et al. 2002). The expression of oxalate decarboxylase gene *oxdC* in *B. subtilis* in response to low pH is regulated by sigma factor, YvrI, and its co-regulators, YvrHa and YvrL, which function as an anti-sigma factor (Just et al. 2007; MacLellan et al. 2008; MacLellan et al. 2009). It is unexpectedly present in vesicles on the cell wall (Antelmann et al. 2007). *B. subtilis* oxalate decarboxylase consists of a pentapeptide loop (amino acid residues 161–165) that makes up the lid structure which is involved in determining the reaction specificity and enzyme's catalytic efficiency (Burrell et al. 2007; Svedruzic et al. 2007).

Moreover, oxalate decarboxylase activity may convert into oxalate oxidase activity by forming H_2O_2 due to a mutation in the amino acids of the lid region (Burrell et al. 2007). Earlier it was proposed that the activity of *B. subtilis* oxalate decarboxylase to convert oxalate into formate and CO_2 is conserved in its N-terminal domain (Just et al. 2004; Burrell et al. 2007; Svedruzic et al. 2007) but later evidence showed that both N- and C-terminal domains may catalyze the decarboxylation reaction (Tabares et al. 2009). The structural and spectroscopic studies revealed that site 1 acts as the catalytic site, in the presence of two manganese-binding sites in *B. subtilis*. The data also suggests that site 1 contains formate bound to it in one crystal structure, that the lid carries a suitable proton donor Glu162 that can cause isolation of site 1 in solution, and that site 2 shows marked inaccessibility to solvents in both known structures (Just et al. 2007). A mutation leading to the replacement of the Glu162 results in no oxalate decarboxylase activity and significant oxalate oxidase activity (Just et al. 2004).

Although the activity of oxalate decarboxylase has been observed in the cell wall or released in the culture media or bound to the extracellular polysaccharide matrices, fungal oxalate decarboxylase is known to show intracellular enzyme activity which is predominantly confined close to the plasma membrane or in vesicles (Sato et al. 2007). In several ascomycetous and basidiomycetous species, the enzyme's translated genes contain N-terminal secretion leader peptides that aid in the release of oxalate decarboxylase of fungal origin (Sato et al. 2007; Mäkelä 2009; Mäkelä et al. 2009).

The relevance of oxalate decarboxylase in biotechnology has been discovered way back in the 1960s when the enzyme was analyzed in a brewing process for the removal of oxalic acid (Haas and Fleischman 1961). Later, the enzyme was applied in clinical samples as a diagnostic tool for knowing the oxalate levels in clinical samples. Plants expressing oxalate decarboxylase were also used in the control of plant pathogens (Kesarwani et al. 2000; Dias et al. 2006; Jin et al. 2007; Walz et al. 2008). However, the therapeutic use of this enzyme in kidney stone removal and prevention of hyperoxaluria is more widely accepted (Grujic et al. 2009; Jeong et al. 2009; Kolandaswamy et al. 2009; Cowley et al. 2010; Mäkelä et al. 2010).

In the last few years, use of food-grade probiotics products with oxalate decarboxylase activity has emerged as an effective therapeutic option for lowering the concentration of dietary oxalates (Fig. 13.2). The impact of probiotics with oxalate decarboxylase enzymes has been evaluated on human gut and result showed that probiotic properties make them a potentially safe option for prophylaxis of calcium oxalate stone disease. Transgenic plants expressing fungal oxalate decarboxylase may lower the nutritional stress of oxalate content in herbivores (Dias et al. 2006). Breakdown of intestinal oxalate and oxalic acid using oxalate decarboxylase is a prominent solution to oxalate degradation in humans (Cowley et al. 2010). Studies also confirmed that recombinant *B. subtilis* oxalate decarboxylase expressed in *Escherichia coli* given orally to rat was able to decrease oxalate concentration in urine (Jeong et al. 2009), while in other experiments on mice, the treatment with OxDc-CLEC®, a crystalline, cross-linked formulation containing recombinant *B. subtilis* oxalate decarboxylase, showed substantial decrease in symptoms of hyperoxaluria, urolithiasis, and nephrocalcinosis (i.e., increased level of calcium in the kidneys) as well (Grujic et al. 2009).

13.4.2 Oxalate Oxidase

Oxalate oxidase (EC 1.2.3.4) was initially discovered in a mold, and after that it has been reported from various plant sources such as barley seedlings and roots, beet stems, and sorghum leaves (Koyama 1988). Along with the formation of hydrogen

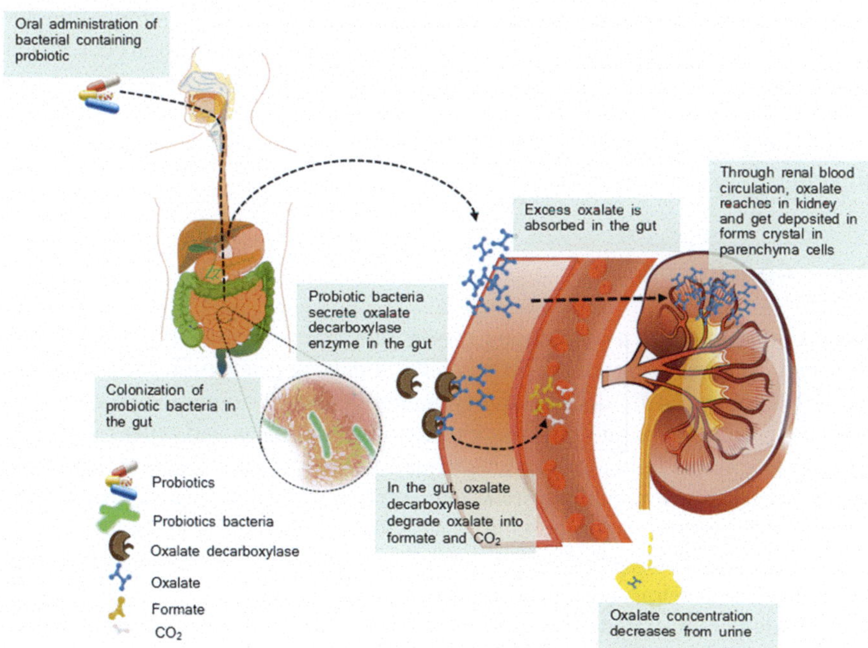

Fig. 13.2 Use of food-grade probiotics with oxalate decarboxylase enzyme activity in kidney stone removal

peroxide in an oxygen-dependent manner, oxalate oxidase catalyzes the oxidation of oxalic acid into carbon dioxide (Whittaker and Whittaker 2002). It has been assumed that H_2O_2 production by oxalate oxidase is applied as a defense mechanism against pathogenic infections (Svedruzic et al. 2005). Intracellular oxalate oxidase activity has been studied in white-rot basidiomycetous fungi *Ceriporiopsis subvermispora* and *Abortiporus biennis* even though the enzyme is principally native to plants (Aguilar et al. 1999; Grąz et al. 2009). Indeed, the activity of both these oxalate-degrading enzymes, i.e., oxalate oxidase and oxalate decarboxylase, was first reported in the fungal species *Ceriporiopsis subvermispora* (Aguilar et al. 1999; Watanabe et al. 2005). Oxalate oxidase present in the cell wall of plants has a role in cell morphogenesis, and it also promotes plant's defense mechanisms against diseases and other environmental stresses. Oxalate oxidase found in higher plants, fungi, and bacteria is now part of preventive therapy of hyperoxaluria, urolithiasis, and medical diagnosis of oxalate content in urine, whereas the food and papermaking industries also use this enzyme for various applications (Hu et al. 2015).

13.4.3 Oxalyl-CoA Decarboxylase

Oxalyl-CoA decarboxylase (EC 4.1.1.8), a thiamin-dependent oxalate-degrading enzyme, performs the catalysis of oxalyl-CoA to formyl-CoA and CO_2 (Svedruzic et al. 2005). The enzyme was discovered around 50 years ago and is mainly found in bacterial species including *B. lactis, Oxalobacter formigenes*, *L. acidophilus,* and *Thiobacillus novellus* (Federici et al. 2004; Turroni et al. 2007; Mäkelä et al. 2010). In *O. formigenes*, oxalyl-CoA decarboxylase is involved in oxalate-dependent ATP synthesis. Along with the degradation of oxalate by oxalyl-CoA decarboxylase, a proton-motive force that drives ATP synthesis is generated in *O. formigenes* due to antiporting of oxalate and formate (Mäkelä et al. 2010).

13.5 Conclusion

Oxalic acid is found in a vast range of foods and often consumed by the humans. It is a well-established fact that whether dietary intake or production during metabolism, oxalic acid can be detrimental to human health. Assimilation of oxalate is highly toxic to humans and ultimately causes hyperoxaluria and other related ailments. Owing to the limitations and inadequate success of current therapeutic drugs used in the treatment of kidney stone, the need for novel and better prophylactic measures have become an important issue. Although the use of probiotic bacteria has attracted significant attention, the use of crude enzyme with oxalate-degrading potential showed astonishing results. Oxalate decarboxylase and oxalate oxidase have already demonstrated great capabilities to dissolve calcium oxalate crystals in in vitro investigations. However, putting these enzymes to work in clinical practice still requires great investigation and research.

Acknowledgments We are thankful to Naveen Kumar Arora and Jitendra Mishra for providing editorial contribution. We are also gratified to Jitendra Mishra for preparing color illustration.

References

Abratt VR, Reid SJ (2010) Oxalate degradation bacteria of the human gut as probiotic in the management of kidney stones diseases. Adv Appl Microbiol 72:63–87

Aguilar C, Urzúa U, Koenig C, Vicuña R (1999) Oxalate oxidase from *Ceriporiopsis subvermispora*: biochemical and cytochemical studies. Arch Biochem Biophys 366:275–282

Alberta A, Tiwaria V, Paula E, Ganesana D, Ayyavub M, Kujura R, Ponnusamyc S, Shanmugamd K, Sasoe L, Selvama GS (2017) Expression of heterologous oxalate decarboxylase in HEK293 cells confers protection against oxalate induced oxidative stress as a therapeutic approach for calcium oxalate stone disease. J Enzyme Inhib Med Chem 32:426 433

Allison MJ, Cook HM (1981) Oxalate degradation by microbes of the large bowel of herbivores: the effect of dietary oxalate. Science 212:675–676

Anand R, Dorrestein P, Kinsland C, Begley T, Ealick S (2002) Structure of oxalate decarboxylase from *Bacillus subtilis* at 1. 75Å resolution. Biochemistry 41:7659–7669

Antelmann H, Towe S, Albrecht D, Hecker M (2007) The phosphorus source phytate changes the composition of the cell wall proteome in *Bacillus subtilis*. J Proteome Res 6:897–903

Aslani MR, Movassaghi AR, Najarnezhad V, Pirouz HJ, Bami MH (2011) Acute oxalate intoxication associated to ingestion of eshnan (*Seidlitzia rosmarinus*) in sheep. Trop Anim Health Prod 43:1065–1068

Bungash K, Shigri F, Jamal A, Anwar K (2011) Spectrum of renal stones composition; chemical analysis of kidney stones. Int J Pathol 9:63–66

Burrell MR, Just VJ, Bowater L, Fairhurst SA, Requena L, Lawson DM, Bornemann S (2007) Oxalate decarboxylase and oxalate oxidase activities can be interchanged with a specificity switch of up to 282000 by mutating an active site lid. Biochemistry 46:12327–12336

Cai XH, Lin RH, Wu J, He JB, Wu YC, Wang XY (2018) Adsorption of ethylenediaminetetraacetic dianhydride modified oxalate decarboxylase on calcium oxalate. Biotech Histochem 93:220. https://doi.org/10.1080/10520295.2017.1420820

Cowley AB, Poage DW, Dean RR, Meschter CL, Ghoddusi M, Li Q-S, Sidhu H (2010) 14-day repeat-dose oral toxicity evaluation of oxazyme in rats and dogs. Int J Toxicol 29:20–31

Dias BBA, Cunha WG, Morais LS, Vianna GR, Rech EL, de Capdeville G, Aragão FJL (2006) Expression of an oxalate decarboxylase gene from *Flammulina* sp. in transgenic lettuce (*Lactuca sativa*) plants and resistance to *Sclerotinia sclerotiorum*. Plant Pathol 55:187–193

Ellis ML, Shaw KJ, Jackson SB, Daniel SL, Knight J (2015) Analysis of commercial kidney stone probiotic supplements. Urology 85:517–521

Federici F, Vitali B, Gotti R, Pasca MR, Gobbi S, Peck AB, Brigidi P (2004) Characterization and heterologous expression of the oxalyl coenzyme A decarboxylase gene from *Bifidobacterium lactis*. Appl Environ Microbiol 70:5066–5073

Gomathi S, Sasikumar P, Anbazhagan K, Sasikumar S, Kavitha M, Selvi MS, Selvam GS (2014) Screening of indigenous oxalate degrading lactic acid bacteria from human faeces and South Indian fermented foods: assessment of probiotic potential. Sci World J:648059

Grąz M, Jarosz-Wilkołazka A, Pawlikowska-Pawlęga B (2009) *Abortiporus biennis* tolerance to insoluble metal oxides: oxalate secretion, oxalate oxidase activity, and mycelia morphology. Biometals 22:401–410

Grujic D, Salido EC, Shenoy BC, Langman CB, McGrath ME, Patel RJ, Rashid A, Mandapati S, Jung CW, Margolin AL (2009) Hyperoxaluria is reduced and nephrocalcinosis prevented with an oxalate-degrading enzyme in mice with hyperoxaluria. Am J Nephrol 29:86–93

Haas GJ, Fleischman AI (1961) The rapid enzymatic determination of oxalate in wort and beer. Agric Food Chem 9:451–452

Hatch M (2014) Intestinal adaptations in chronic kidney disease and the influence of gastric bypass surgery. Exp Physiol 99:1163–1167

Hoppe B, Unruh G, Hesse NLA, Sidhu H (2005) Oxalate degrading bacteria: new treatment option for patients with primary and secondary hyperoxaluria? Urol Res 33:372–375

Hu Y, Xiang M, Jin C, Chen Y (2015) Characteristics and heterologous expressions of oxalate degrading enzymes "oxalate oxidases" and their applications on immobilization, oxalate detection, and medical usage potential. J Biotechnol Res 6:63–75

Jeong BC, Han DH, Seo SI, Jeon SS, Lee HM, Choi HY, Park YH, Kim HH (2009) YvrK gene recombinant *E. coli* reduce the concentration of urine oxalate in transient hyperoxaluria rat model. J Urol 181:660

Jin Z-X, Wang C, Chen W, Chen X, Li X (2007) Induction of oxalate decarboxylase by oxalate in a newly isolated *Pandoraea* sp. OXJ-11 and its ability to protect against *Sclerotinia sclerotiorum* infection. Can J Microbiol 53:1316–1322

Just VJ, Stevenson CEM, Bowater L, Tanner A, Lawson DM, Bornemann S (2004) A closed conformation of *Bacillus subtilis* oxalate decarboxylase OxdC provides evidence for the true identity of the active site. J Biol Chem 279:19867–19874

Just VJ, Burrell MR, Bowater L, McRobbie I, Stevenson CEM, Lawson DM, Bornemann S (2007) The identity of the active site of oxalate decarboxylase and the importance of the stability of active-site lid conformations. Biochem J 407:397–406

Kesarwani M, Azam M, Natarajan K, Mehta A, Datta A (2000) Oxalate decarboxylase from *Collybia velutipes*. Molecular cloning and its overexpression to confer resistance to fungal infection in transgenic tobacco and tomato. J Biol Chem 275:7230–7238

Kolandaswamy A, George L, Sadasivam S (2009) Heterologous expression of oxalate decarboxylase in *Lactobacillus plantarum* NC8. Curr Microbiol 58:117–121

Koyama H (1988) Purification and characterization of oxalate oxidase from *Pseudomonas* sp. OX-53. Agric Biol Chem 52:743–748

Kumar R, Ghoshal UC, Singh G, Mittal RD (2004) Infrequency of colonization with *Oxalobacter formigenes* in inflammatory bowel disease: possible role in renal stone formation. J Gastroenterol Hepatol 19:1403–1409

Lieske JC, Tremaine WJ, Simone C, O'Connor HM, Li X, Bergstralh EJ, Goldfarb SD (2010) Diet, but not oral probiotics, effectively reduces urinary oxalate excretion and calcium oxalate supersaturation. Kidney Int 78:1178–1185

MacLellan SR, Wecke T, Helmann JD (2008) A previously unidentified factor and two accessory proteins regulate oxalate decarboxylase expression in *Bacillus subtilis*. Mol Microbiol 69:954–967

MacLellan SR, Helmann JD, Antelmann H (2009) The yvri alternative factor is essential for acid stress induction of oxalate decarboxylase in *Bacillus subtilis*. J Bacteriol 191:931–939

Mäkelä MR (2009) The white-rot fungi *Phlebia radiata* and *Dichomitus squalens* in wood-based cultures: expression of laccases, lignin peroxidases, and oxalate decarboxylase. Ph.D. thesis, University of Helsinki, Helsinki

Mäkelä MR, Hildén K, Hatakka A, Lundell TK (2009) Oxalate decarboxylase of the white-rot fungus *Dichomitus squalens* demonstrates a novel enzyme primary structure and noninduced expression on wood and in liquid cultures. Microbiology 155:2726–2738

Mäkelä MR, Hildén K, Lundell TK (2010) Oxalate decarboxylase: biotechnological update and prevalence of the enzyme in filamentous fungi. Appl Microbiol Biotechnol 87:801–814

Miller AW, Dearing D (2013) The metabolic and ecological interactions of oxalate-degrading Bacteria in the mammalian gut. Pathogens 2:636–652

Mogna L, Pane M, Nicola S, Raiteri E (2014) Screening of different probiotic strains for their *in vitro* ability to metabolise oxalates. J Clin Gastroenterol 48:S91–S95

Murthy MSR, Talwar HS, Nath R, Thind SK (1981) Oxalate decarboxylase from guinea pig liver. IRCS Med Sci 9:683–684

Nazzal L, Puri S, Goldfarb DS (2016) Enteric hyperoxaluria: an important cause of end-stage kidney disease. Nephrol Dial Transplant 31:375–382

Peck AB, Canales BK, Nguyen CQ (2016) Oxalate-degrading microorganisms or oxalate-degrading enzymes: which is the future therapy for enzymatic dissolution of calcium-oxalate uroliths in recurrent stone disease. Urolithiasis 44:45–50

Sadaf H, Raza SI, Hassan SW (2017) Role of gut microbiota against calcium oxalate. Microb Pathog 109:287–291

Salminen S, Nybom S, Meriluoto J, Maria CC, Satu V, Nezami H (2010) Interaction of probiotics and pathogens benefits to human health? Curr Opin Biotechnol 21:157–167

Sato S, Liu F, Koc H, Tien M (2007) Expression analysis of extracellular proteins from *Phanerochaete chrysosporium* grown on different liquid and solid substrates. Microbiology 153:3023–3033

Sutherland JW, Parks JH, Coe FL (1985) Recurrence after a single renal stone in a community practice. Miner Electrolyte Metab 11:267–269

Svedruzic D, Jonssona S, Toyotaa CG, Reinhardtb LA, Ricagnoc S, Lindqvistb Y, Richardsa NGJ (2005) The enzymes of oxalate metabolism: unexpected structures and mechanisms. Arch Biochem Biophys 433:176–192

Svedruzic D, Liu Y, Reinhardt LA, Wroclawska E, Cleland WW, Richards NGJ (2007) Investigating the roles of putative active site residues in the oxalate decarboxylase from Bacillus subtilis. Arch Biochem Biophys 464:36–47

Tabares LC, Gätjens J, Hureau C, Burrell MR, Bowater L, Pecoraro VL, Bornemann S, Un S (2009) pH-dependent structures of the manganese binding sites in oxalate decarboxylase as revealed by high-field electron paramagnetic resonance. J Phys Chem B 113:9016–9025

Trinchieri A (2013) Diet and renal stone formation. Minerva Med 104:41–54

Turroni S, Vitali B, Bendazzoli C, Candela M, Gotti R, Federici F, Pirovano F, Brigidi P (2007) Oxalate consumption by lactobacilli: evaluation of oxalyl-CoA decarboxylase and formyl-CoA transferase activity in *Lactobacillus acidophilus*. J Appl Microbiol 103:1600–1609

Twahir U, Molina L, Ozarowski A, Angerhofer A (2015) Immobilization of *Bacillus subtilis* oxalate decarboxylase on a Zn-IMAC resin. Biochem Biophys Rep 4:98–103

Walz A, Zingen-Sell I, Theisen S, Kortekamp A (2008) Reactive oxygen intermediates and oxalic acid in the pathogenesis of the necrotrophic fungus *Sclerotinia sclerotiorum*. Eur J Plant Pathol 120:317–330

Watanabe T, Hattori T, Tengku S, Shimada M (2005) Purification and characterization of NAD-dependent formate dehydrogenase from the white-rot fungus *Ceriporiopsis subvermispora* and a possible role of the enzyme in oxalate metabolism. Enzym Microb Technol 37:68–75

Whittaker MM, Whittaker JW (2002) Characterization of recombinant barley oxalate oxidase expressed by *Pichia pastoris*. J Biol Inorg Chem 7:136–145

Yu-Hu S, Liu RJ, Wang HQ (2008) Oxalate decarboxylase from *Agrobacterium tumefaciens* C58 is translocated by a twin arginine translocation system. J Microbiol Biotechnol 18:1245–1251

The manufacturer's authorised representative in the EU is Springer
Nature Customer Service Centre GmbH, Europaplatz 3, 69115 Heidelberg,
Germany. If you have any concerns regarding our products, please
contact ProductSafety@springernature.com

Printed and bound by CPI Group (UK) Ltd, Croydon, CR0 4YY
30/09/2026
02225248-0007